S0-CKV-455

RECIPROCITY IN ELASTODYNAMICS

The reciprocity theorem has been used for over 100 years to establish interesting and useful relations between different loading states of a body. In a well-known application, reciprocity relations have been used to formulate problems in a way suitable for numerical calculation by the boundary element method.

This book presents a discussion of current and novel uses of reciprocity relations for the determination of elastodynamic fields. The author, who is internationally distinguished for his contributions to theoretical and applied mechanics, presents a novel method to solve for wave fields by the reciprocity of the actual field with a so-called virtual wave, shedding new light on the use of reciprocity relations for dynamic fields in an elastic body. Professor Achenbach describes the use of reciprocity in acoustics, in one-dimensional elastodynamics, and in two- and three-dimensional elastodynamics. Integral representations and integral equations for application of the boundary element method are discussed in some detail, with applications to scattering by inhomogeneities and related inverse problems. Chapters are also devoted to reciprocity for coupled acousto-elastic systems, such as elastic bodies submerged in a fluid, and to reciprocity for piezoelectric systems.

The material presented in the book is relevant to several fields in engineering and applied physics. Examples are ultrasonics for medical imaging and non-destructive evaluation, acoustic microscopy, seismology, exploratory geophysics, structural acoustics, the response of structures to high-rate loads and the determination of material properties by ultrasonic techniques.

OTHER TITLES IN THIS SERIES

All the titles listed below can be obtained from good booksellers or from Cambridge University Press. For a complete series listing visit

http://publishing.cambridge.org/stm/mathematics/cmma

Topographic Effects in Stratified Flows
PETER BAINES

Inverse Methods in Physical Oceanography
ANDREW F. BENNETT

Wave Interactions and Fluid Flows
A.D.D. CRAIK

Theory and Computation in Hydrodynamic Stability
W. CRIMINALE, T.L. JACKSON AND R.D. JOSLIN

Theory of Solidification
STEPHEN H. DAVIS

Dynamic Fracture Mechanics
L.B. FREUND

Finite Plastic Deformation of Crystalline Solids
K.S. HAVNER

Turbulence, Coherent Structures, Dynamical Systems and Symmetry
PHILIP HOLMES, J.L. LUMLEY AND GAL BERKOOZ

Acoustics of Fluid-Structure Interactions
M.S. HOWE

The Dynamics of Fluidized Particles
ROY JACKSON

Benard Cells and Taylor Vortices
E.L. KOSCHMIEDER

Ocean Acoustic Tomography
WALTER MUNK, PETER WORCESTER, AND CARL WUNSCH

Turbulent Combustion
NORBERT PETERS

Colloidal Dispersions
W.B. RUSSELL, D.A. SAVILLE, AND W.R. SCHOWALTER

Vortex Dynamics
P.G. SAFFMAN

The Structure of Turbulent Shear Flow
A.A. TOWNSEND

Buoyancy Effects in Fluids
J.S. TURNER

RECIPROCITY IN ELASTODYNAMICS

J. D. ACHENBACH
McCormick School of Engineering and Applied Science
Northwestern University
Evanston, USA

BOWLING GREEN STATE
UNIVERSITY LIBRARIES

PUBLISHED BY THE PRESS SYNDICATE OF THE UNIVERSITY OF CAMBRIDGE
The Pitt Building, Trumpington Street, Cambridge, United Kingdom

CAMBRIDGE UNIVERSITY PRESS
The Edinburgh Building, Cambridge, CB2 2RU, UK
40 West 20th Street, New York, NY 10011–4211, USA
477 Williamstown Road, Port Melbourne, VIC 3207, Australia
Ruiz de Alarcón 13, 28014 Madrid, Spain
Dock House, The Waterfront, Cape Town 8001, South Africa

http://www.cambridge.org

© Cambridge University Press 2003

This book is in copyright. Subject to statutory exception
and to the provisions of relevant collective licensing agreements,
no reproduction of any part may take place without
the written permission of Cambridge University Press.

First published 2003

Printed in the United Kingdom at the University Press, Cambridge

Typeface Times 10/13 pt. *System* LATEX 2ε [TB]

A catalog record for this book is available from the British Library

Library of Congress Cataloging in Publication data
Achenbach, J. D.
Reciprocity in elastodynamics / J. D. Achenbach.
p. cm. – (Cambridge monographs on mechanics)
Includes bibliographical references and index.
ISBN 0 521 81734 X
1. Elasticity. 2. Elastic solids. 3. Reciprocity theorems. I. Title. II. Series.
QA931.A34 2003 531′.382 – dc21 2003055420

ISBN 0 521 81734 X hardback

Contents

Preface

Reciprocity relations are among the most interesting and intriguing relations in classical physics. At first acquaintance these relations promise to be a goldmine of useful information. It takes some ingenuity, however, to unearth the nuggets that are not immediately obvious from the formulation. In the theory of elasticity of solid materials the relevant reciprocity theorem emanated from the work of Maxwell, Helmholtz, Lamb, Betti and Rayleigh, towards the end of the nineteenth century, and several applications have appeared in the technical literature since that time. This writer has always believed, however, that more information than is generally assumed can be wrested from reciprocity considerations. I have wondered in particular whether reciprocity considerations could be used to actually determine by analytical means the elastodynamic fields for the high-rate loading of structural configurations. I have explored this question for a number of problems and obtained the actual fields generated by loading from a reciprocity relation in conjunction with an auxiliary solution, a free wave called the "virtual" wave. These recent results comprise an important part of the book.

To my knowledge, the topic of reciprocity in elastodynamics has not been discussed in a comprehensive manner in the technical literature. It is hoped that this book will fill that void. Various forms of the reciprocity theorem are presented, with an emphasis on those for time-harmonic fields, together with numerous applications, general and specific, old and new.

The book should be of interest to research workers in such fields as the ultrasonics of solids, particularly the detection of defects and the determination of elastic constants, seismology, exploratory geophysics, the dynamic response of structures and structural acoustics.

Parts of the book were read by graduate students in a course on wave propagation in elastic solids. My colleague John Harris also read a number of chapters, as did David Feit and A. T. de Hoop. Their comments are gratefully acknowledged.

A special word of thanks goes to Linda Kearfott who typed and retyped the manuscript as the material was arranged, rearranged and revised.

Over the years my work in the area of this book and in related areas has been consistently supported by the Office of Naval Research (ONR). The monitor of my research project in recent years has been Dr Y. N. Rajapakse. The ONR funding is gratefully acknowledged.

The constant support of my wife, Marcia, made it possible to complete the book. I can never thank her enough for all she has given me.

1
Introduction

1.1 Reciprocity

Reciprocity is a good thing. Something is given and something else, equally or more valuable, is returned. So it is in reciprocity for states of deformation of elastic bodies. What is received in return is the main benefit from the reciprocal relationship. From a known solution to one loading case, some important aspect of, or the complete solution to, another loading case is returned. The return is, however, not always a complete solution, but sometimes an equation for computing such a solution.

For dynamic systems the concept of reciprocity goes back to the nineteenth century. A pertinent reciprocity theorem was first formulated by von Helmholtz (1860). Lord Rayleigh (1873, 1877), subsequently derived a quite general reciprocity relation for the time-harmonic motion of a linear dynamic system with a finite or infinite number of degrees of freedom. Rayleigh's formulation included the effects of dissipation. In a later work Lamb (1888) attributed the following general reciprocity theorem to von Helmholtz (1886):

> Consider any natural motion of a conservative system between two configurations A and A' through which it passes at times t and t' respectively, and let $t' - t = \tau$. Let $q_1, q_2 \ldots$, be the coordinates of the system, and $p_1, p_2, \ldots$ the component momenta, at time t, and let the values of the same quantities at time t' be distinguished by accents. As the system is passing through the configuration A, let a small impulse δp_r of any type be given to it; and let the consequent alteration in any coordinate q_s after the time τ be denoted by $\delta q_s'$. Next consider the *reversed* motion of the system, in virtue of which it would, if undisturbed, pass from the configuration A' to the configuration A in the time τ. Let a small impulse $\delta p_s'$ be applied as it is passing through the configuration A', and let the consequent change in the coordinate q_r, after a time τ, be δq_r. The reciprocity theorem asserts that
>
> $$\delta q_r : \delta p_s' = \delta q_s' : \delta p_r. \qquad (1.1.1)$$

If the coordinates q_r, q_s be of the same kind (e.g., both lines or both angles), the statement of the theorem may be simplified by supposing $\delta p'_s = \delta p_r$, in which case

$$\delta q_r = \delta q'_s. \tag{1.1.2}$$

In words, the change produced in the time τ by a small initial impulse of any type in the coordinate of any other (or of the same) type, in the *direct* motion, is equal to the change produced in the same time by a small initial impulse of the second type in the coordinate of the first type, in the *reversed* motion.

Lamb (1888) asserted that the reciprocity theorems of von Helmholtz (1886), formulated in his paper on the theory of least action and in earlier papers on acoustics and optics, and of Lord Rayleigh in acoustics, were particular cases of a general result derived in his 1888 paper.

Reciprocity theorems in elasticity theory provide a relation between displacements, traction components and body forces for two different loading states of a single body or two bodies of the same geometry. As discussed by Love (1892), for the elastostatic case the principal theorem is due to Betti (1872). A more general theorem, which includes the elastodynamic case, was given by Raleigh (1873). Statements of elastodynamic reciprocity theorems using contemporary notation can be found in, among others, the books by Achenbach (1973), Achenbach, Gautesen and McMaken (1982) and de Hoop (1995).

In the present text we concern ourselves with applications of reciprocity to time-harmonic elastodynamic states. The reciprocity theorem can be used to obtain a number of interesting relations between two such states. A well-known result is the relation between the magnitudes and directions of the forces and displacements at the points of application of two forces applied to an elastic body of finite dimensions or infinite extent. An important application of the reciprocity theorem produces boundary integral equations. If, for bodies with boundaries subjected to loads or displacements, the solution to a concentrated load in an unbounded solid of the same material is known then an equation (a boundary integral equation) for fields on the boundary, complementary to the applied fields, can be obtained, as shown in the sequel.

Of course, it would be desirable to obtain more than just the relations between elastic states or boundary integral equations from reciprocity relations. The purpose of the present work is therefore to give also direct applications to the computation of elastodynamic displacement fields. As will be discussed, it is possible to obtain a complete solution for certain configurations and concentrated loading cases by the use of elastodynamic reciprocity. The selected auxiliary solution for these cases we call a "virtual" wave. This is a wave motion that satisfies appropriate conditions on the boundaries and is a solution of the elastodynamic equations. It is shown that combining the desired solution as state A with a virtual wave as state B provides explicit expressions for state A. By using a virtual wave for other cases, particularly problems involving scattering by obstacles in waveguides, simple general expressions can be obtained for reflection and transmission coefficients and for scattering coefficients.

A formal definition of an elastodynamic reciprocity theorem may be given as follows: "A reciprocity theorem relates, in a specific manner, two admissible elastodynamic states that can occur in the same time-invariant linearly elastic body. Each of the two states can be associated with its own set of time-invariant material parameters and its own set of loading conditions. The domain to which the reciprocity theorem applies may be bounded or unbounded."

Elastodynamics covers both vibrations, which are standing waves, and propagating waves. The applications of reciprocity considerations discussed in this book are, however, primarily to propagating waves.

The propagation of mechanical waves is a common occurrence in nature. The simplest example is sound in an acoustic medium, which is a special case of elastodynamics. Not all sound is audible to the human ear. Sound with frequencies too low (infrasound), or too high (ultrasound, above 20 000 Hz), cannot normally be heard. Pierce (1981) briefly discussed the history of acoustics. Observation of the propagation of water waves, produced for example by a pebble dropped in a pond, led the Greeks and Romans to speculate that sound was a wave phenomenon. Aristotle (384–22 BC) stated that air motion is generated by a source "thrusting forward in like manner the adjoining air, so that sound travels unaltered in quality as far as the disturbance of the air manages to reach." The first mathematical theory of sound propagation was formulated by Isaac Newton (1642–1717), whose ideas included a mechanical interpretation of sound as being "pressure pulses transmitted through neighboring fluid particles." Newton also made a first determination of the speed of sound. A theory of sound propagation resting on firmer mathematical and physical concepts was developed during the eighteenth century by Euler, Lagrange and d'Alembert.

The best-known example of wave motion in a solid is probably provided by the ground motion due to an earthquake, generated by the sliding of tectonic plates over a fault surface, often deep in the interior of the earth. Waves in solids occur, however, in a multitude of other cases of dynamic excitation.

Generally speaking, the high-rate application of a load to a solid body gives rise to wave motion. Waves so induced are used in geophysical prospecting and in quantitative non destructive evaluation. The reflection and scattering of externally excited incident waves can be used for imaging or other techniques of data processing, with as the ultimate goal the detection of inhomogeneities, including defects, or the determination of material properties.

Engineering applications of wave propagation in solids are concerned with the performance of structures under high rates of loading. Other applications are related to high-speed machinery, ultrasonics and piezoelectric phenomena, as well as to such civil engineering practices as pile driving. By now the study of wave propagation effects has become well established in the field of the mechanics of solids.

The study of wave propagation in elastic solids has a long and distinguished history. The early work on elastic waves received its impetus from the view, which was prevalent until the end of the nineteenth century, that light could be regarded as the propagation of a disturbance in an elastic aether. This view was espoused by such great mathematicians as Cauchy and Poisson and to a large extent motivated them to develop what is now generally known as the theory of elasticity. The early investigations on the propagation of waves in elastic solids carried out by Poisson, Ostrogradsky, Cauchy, Green, Lamé, Stokes, Clebsch and Christoffel are discussed in the historical introduction in Love's treatise on the mathematical theory of elasticity (1892).

1.2 Static reciprocity for an elastic body subjected to concentrated loads

In its simplest form an elastostatic reciprocity theorem can be established by considering an elastic body of arbitrary shape. The body is supported in such a way that displacement as a rigid body is impossible. Two states of loading are defined. In the first state the load is P_1, applied at point 1, and in the second state the load is P_2, applied at point 2 (Fig. 1.1). The corresponding displacements at the points of application of the loads and in the direction of load application are δ_{11}, δ_{21}, δ_{22} and δ_{12}. Here δ_{11} is the displacement at point 1 in the direction of P_1 due to the application of P_1, while δ_{21} is the displacement at point 2 in the direction of P_2 due to the application of P_1 at point 1. Analogous definitions hold for δ_{22} and δ_{12}.
The reciprocity relation states that

$$P_1\delta_{12} = P_2\delta_{21}. \tag{1.2.1}$$

For $P_1 = P_2$ we obtain what is known as Maxwell's theorem (1864),

$$\delta_{12} = \delta_{21}. \tag{1.2.2}$$

In words, for forces of equal magnitude P_1 and P_2, the displacement produced by the force P_1 at the point of application of the force P_2 and in the direction

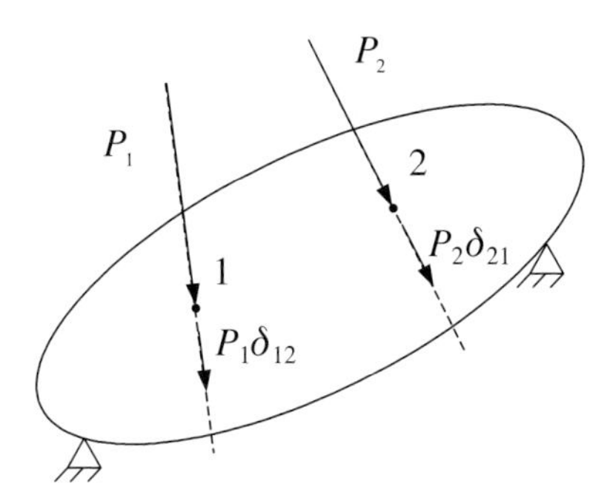

Figure 1.1 Body of volume V and boundary S subjected to concentrated loads P_1 and P_2.

of that force is equal to the displacement produced by the force P_2 at the point of application of the force P_1 and in the direction of P_1.

The usual proof of Eq. (1.2.1) is based on the result that the energy stored in an elastic body after application of the two forces is independent of their sequence of application.

If P_1 is applied first and then P_2, the stored energy is

$$U = \tfrac{1}{2}P_1\delta_{11} + P_1\delta_{12} + \tfrac{1}{2}P_2\delta_{22}. \tag{1.2.3}$$

However, when P_2 is applied first we have

$$U = \tfrac{1}{2}P_2\delta_{22} + P_2\delta_{21} + \tfrac{1}{2}P_1\delta_{11}. \tag{1.2.4}$$

Since the stored energy must be independent of the order of application of the two forces, equating the two expressions for U immediately yields Eq. (1.2.1).

1.3 Static reciprocity for distributed body forces and surface loads

The elastostatic result given by Eq. (1.2.1) can be extended to distributions of body forces, surface tractions and displacements. Let us define the two elastostatic states by superscripts A and B:

$$f_i^A(\boldsymbol{x}), \quad t_i^A(\boldsymbol{x}) \quad \text{and} \quad u_i^A(\boldsymbol{x}) \tag{1.3.1}$$

and

$$f_i^B(\boldsymbol{x}), \quad t_i^B(\boldsymbol{x}) \quad \text{and} \quad u_i^B(\boldsymbol{x}), \tag{1.3.2}$$

where f_i, t_i and u_i denote the components of body forces, surface tractions and displacements, respectively. The following relationship is derived in detail in Section 2.5:

$$\int_V \left(f_i^A u_i^B - f_i^B u_i^A\right) dV + \int_S \left(t_i^A u_i^B - t_i^B u_i^A\right) dS = 0, \tag{1.3.3}$$

where S is the boundary of the domain of volume V; see Fig. 1.1. Equation (1.3.3) expresses Betti's reciprocity theorem, one of the most elegant and useful theorems in the linear theory of elasticity.

1.4 The wave equation in one dimension

Every book on wave propagation contains a discussion of the wave equation in one dimension. The equation governs vibrations and wave propagation in many one-dimensional physical systems. Examples are a column of gas, a thin elastic rod and a taut string. The equation is

$$\frac{\partial^2 u}{\partial x^2} = \frac{1}{c^2}\frac{\partial^2 u}{\partial t^2}, \tag{1.4.1}$$

where c is a velocity (m/s).

A general solution to Eq. (1.4.1) can be obtained by introducing the new variables

$$\alpha = t - x/c, \qquad \beta = t + x/c.$$

Equation (1.4.1) then becomes

$$\frac{\partial^2 u}{\partial \alpha \partial \beta} = 0.$$

The general solution of this equation is

$$u = f(\alpha) + g(\beta)$$

or

$$u(x,t) = f(t - x/c) + g(t + x/c). \tag{1.4.2}$$

This general solution is due to d'Alembert (1747). For the first term the argument remains unchanged if increments in time, Δt, and position, Δx, are related by $\Delta x = c\Delta t$. This implies that, for a position that propagates in the x-direction with velocity c, the argument (or phase) of $f(t - x/c)$, and therefore the magnitude of $f(t - x/c)$, remains unchanged. The function $f(t - x/c)$ is said to represent a wave propagating in the x-direction with phase velocity c.

The shape of the wave does not change as it propagates through the medium, and hence this wave is called non-dispersive. Similarly $g(t + x/c)$ represents a non-dispersive wave propagating in the negative x-direction.

A special case of $u(x, t) = f(t - x/c)$ is a harmonic wave of the form

$$u(x, t) = U \cos[k(ct - x)]. \tag{1.4.3}$$

For fixed t (a snapshot), $u(x, t)$ is a periodic function of x with wavelength $\lambda = 2\pi/k$. Also, U is the amplitude and k is the wavenumber, $k = 2\pi/\lambda$. For fixed x, $u(x, t)$ is a harmonic function of time with circular frequency ω, where

$$\omega = kc. \tag{1.4.4}$$

Equation (1.4.3) may also be written as

$$u(x, t) = U \cos(kx - \omega t). \tag{1.4.5}$$

We define the period as $T = 2\pi/\omega$, and the frequency as $f = 1/T = \omega/(2\pi)$. The frequency, f, is expressed in Hertz (1 Hz = 1 cycle/sec).

The particle velocity follows from (1.4.5) as

$$\dot{u}(x, t) = Ukc \sin[k(x - ct)].$$

Hence

$$(\dot{u}/c)_{\max} = Uk = 2\pi U/\lambda$$

In a linear theory we must have $U/\lambda \ll 1$. Hence

$$\dot{u} \ll c.$$

It is convenient to use complex notation:

$$u(x, t) = u(x)e^{-i\omega t}, \qquad i = \sqrt{-1}, \tag{1.4.6}$$

where

$$u(x) = Ue^{ikx}.$$

It is understood that the physical quantity is either the real or the imaginary part of Eq. (1.4.6).

1.5 Use of a virtual wave in reciprocity considerations

Let us illustrate the use of a virtual wave by a simple one-dimensional example. For this example we consider wave motion on an infinitely long taut string. The governing equation for a string subjected to a distributed load $q(x,t)$ is

$$\frac{\partial^2 w}{\partial t^2} = c^2 \frac{\partial^2 w}{\partial x^2} + \frac{1}{\rho} q(x,t), \qquad \text{where} \qquad c^2 = \frac{T}{\rho} \tag{1.5.1}$$

and $w(x,t)$ is the deflection, ρ the mass density per unit length and T the tension in the string. For the time-harmonic case with time factor $\exp(-i\omega t)$, where $w(x,t)$ is defined analogously to Eq. (1.4.6), Eq. (1.5.1) becomes

$$c^2 \frac{d^2 w}{dx^2} + \omega^2 w + \frac{1}{\rho} q(x) = 0, \tag{1.5.2}$$

where $w \equiv w(x)$ and $q(x)$ depend on x only.

Now let us consider the two states $w^A(x)$ and $w^B(x)$, corresponding to $q^A(x)$ and $q^B(x)$, respectively. When Eq. (1.5.2) is written for states A and B, the equation for state A is multiplied by $w^B(x)$ and the equation for state B is multiplied by $w^A(x)$, and one of the equations is subsequently subtracted from the other, we obtain the local reciprocity relation

$$\frac{d^2 w^A}{dx^2} w^B + \frac{1}{\rho c^2} q^A w^B - \frac{d^2 w^B}{dx^2} w^A - \frac{1}{\rho c^2} q^B w^A = 0. \tag{1.5.3}$$

Integration of Eq. (1.5.3) over x from $x = a$ to $x = b$, using integration by parts, yields the global reciprocity relation

$$\left[\frac{dw^A}{dx} w^B - \frac{dw^B}{dx} w^A \right]_{x=a}^{x=b} + \frac{1}{\rho c^2} \int_a^b (q^A w^B - q^B w^A) dx = 0. \tag{1.5.4}$$

Now for state A we select the solution to wave motion generated by a concentrated load applied at $x = 0$, i.e.,

$$q^A(x) = P\delta(x), \tag{1.5.5}$$

where $a < x = 0 < b$ and $\delta(x)$ is the Dirac delta function. Because of the symmetry with respect to $x = 0$, and because we know the general form of the solution from the homogeneous form of Eq. (1.5.2) we can write

$$x > 0: \qquad w^A(x) = Re^{ikx}, \tag{1.5.6}$$

$$x < 0: \qquad w^A(x) = Re^{-ikx}, \tag{1.5.7}$$

where it has been taken into account that the time factor is $\exp(-i\omega t)$, R is a radiation constant and

$$k = \frac{\omega}{c}. \tag{1.5.8}$$

For state B we select the following virtual wave:

$$q^B \equiv 0 \qquad \text{and} \qquad w^B = e^{ikx}. \tag{1.5.9}$$

It is noted that, for $x > 0$, $w^A(x)$ and $w^B(x)$ propagate in the same direction, while for $x < 0$ they form a system of counter-propagating waves. Substitution of Eqs. (1.5.5) and (1.5.6)–(1.5.9) into Eq. (1.5.4) yields

$$ikRe^{2ikb} - ikRe^{2ikb} + 2ikR = -\frac{P}{\rho c^2}. \tag{1.5.10}$$

Superposition of the two waves at $x = b$ yields wave forms of the type exp($2ikb$). However, the two terms at $x = b$ cancel. At $x = a$ the superimposed wave forms eliminate each other, yielding constants that add to $2ikR$. This particular kind of interaction between propagating and counter-propagating waves forms the basis for the combined use of reciprocity considerations and a virtual wave. The result shows that we only need to take into account the terms from the side where the actual wave and the virtual wave are counter-propagating.

Equation (1.5.10) has the solution

$$R = \frac{i}{2}\frac{1}{k}\frac{P}{T}. \tag{1.5.11}$$

Thus, we have solved the problem of wave radiation from a concentrated force at $x = 0$ by combining reciprocity considerations with the use of a virtual wave. The problem can be solved in various other ways, one being the use of the exponential Fourier transform with respect to x.

Equation (1.5.4) can also be used conveniently to determine the radiation from a distributed load. Let the distributed load be defined by

$$q^A(x) = f(x) \qquad \text{for} \qquad 0 \le x \le l. \tag{1.5.12}$$

Equation (1.5.10) then becomes

$$2ikR = -\frac{1}{\rho c^2}\int_0^l f(x)e^{ikx}dx. \tag{1.5.13}$$

The expression for $w^A(x)$, Eq. (1.5.6), will be valid for $x > l$ and $x < 0$.

1.6 Synopsis

To save the reader the trouble of looking elsewhere for relevant information on elastodynamic theory, and to establish the notation that is employed throughout the book, Chapter 2 presents a brief discussion of elastodynamic theory. The chapter also includes a summary of equations governing the linear theory of viscoelasticity as well as a statement of the governing equations for an acoustic medium.

Chapter 3 discusses wave motion in an unbounded, homogeneous, isotropic, linearly elastic solid. Plane waves are considered first. The presence of a surface gives rise to reflected waves. The reflection of plane waves incident at an arbitrary angle on a plane surface free of surface tractions is discussed. Expressions for wave motion generated by a point load and a line load are presented, together with their far-field approximations.

Chapters 4, 5 and 6 are concerned with reciprocity in acoustics, in one-dimensional elastodynamics and in three-dimensional elastodynamics, respectively. Each chapter states the pertinent reciprocity relations in the time domain, the frequency domain and the Laplace transform domain. Reciprocity considerations are most easily applied in the frequency domain and the applications are, therefore, primarily concerned with time-harmonic solutions. Simple applications to dynamic problems of the use of reciprocity considerations in conjunction with a virtual wave are presented in these chapters. The purpose is to give the reader a sense of the applicability of reciprocity as a tool for obtaining solutions for acoustic and elastodynamic problems. Most of the examples are very simple. In the literature their solutions are usually obtained by the use of Fourier transform techniques. Some of the cases in Chapter 6, such as the anti-plane elastodynamic field generated by a line load, can be solved in an equally simple or simpler way by a number of other methods. An interesting case is provided by the determination of surface waves on a half-space when an anti-plane line load is applied in its interior and the surface of the half-plane is constrained by an impedance condition. Other cases deal with acoustic wave motion with polar symmetry, reciprocity for elastic waves reflected from a free surface, reciprocity for fields generated by point forces in bounded and unbounded bodies and formulations of scattering and related inverse problems. The chapter ends with a brief summary of examples from the technical literature.

Chapter 7 contains new material for wave motion guided by a carrier wave on a preferred plane. It is shown that the carrier wave satisfies a simple two-dimensional reduced wave equation. Motion away from the plane of the carrier wave is represented by the same expressions, irrespective of the form of the carrier wave. As examples we discuss Rayleigh surface waves and Lamb waves. A few forms of the carrier wave are discussed.

In Chapter 8 we combine this formulation with reciprocity considerations and the use of a virtual wave to determine the surface wave motion of an elastic half-space generated by a sub-surface line load or a point load of arbitrary direction. For the case of the point load, the virtual wave motion that is employed consists of the sum of a converging and a diverging wave. For the special, axially symmetric, case of surface waves generated by a point load applied normally to the free surface the results are compared with those obtained by the use of the Hankel transform technique.

Chapter 9 is concerned with reciprocity in the case of an elastic layer. The plane of the carrier wave is the mid-plane of the layer. As shown earlier, in Chapter 7, the thickness shear motion and thickness stretch motion through the thickness of the layer are governed by the well-known Rayleigh–Lamb frequency equations, whatever form the carrier wave takes. Reciprocity considerations together with a virtual wave are used to obtain orthogonality conditions for the wave modes in the layer, for the two-dimensional case of plane strain as well as for Lamb waves in polar coordinates.

The dynamic responses of the layer to a line load and to a point load of arbitrary direction applied at an arbitrary location are considered in Chapter 10. A solution in terms of wave modes is obtained by the use of reciprocity and virtual wave modes. For the normal-load case a comparison with the results of plate theory is presented.

In Chapter 11 we move on to new territory. Here the reciprocity relation is used together with the basic elastodynamic solution to derive integral representations and integral equations for elastodynamic fields. Boundary integral equations over the surface of a body are of particular interest since they form the basis for the use of the boundary element method. Both the two- and three-dimensional cases are considered. Examples of boundary integral equations are given for scattering by a cavity and by a crack. The chapter also includes the derivation of interesting reciprocity relations for the scattering of plane elastic waves by obstacles of arbitrary shape in unbounded elastic bodies. In the final sections the Kirchhoff approximation for scattering by a crack and reciprocity for a diffraction problem are discussed.

Further applications of reciprocity to the scattering of an incident wave by a defect in a layer or a half-space are given in Chapter 12. These problems are of interest for the detection of cracks by ultrasonic techniques. Other interesting cases discussed in this chapter deal with a thin film on an inhomogeneous substrate, with a disbond between film and substrate, and with the use of a scattering matrix for a body of arbitrary shape containing a defect. The transducer response to the scattering of an incident wave by a crack is discussed in some detail.

Reciprocity for coupled systems is discussed in Chapters 13 and 14. We distinguish the coupling of wave phenomena as either configurational or physical. For the configurational case, coupling comes about because bodies of different constitutive behaviors are in contact, as is the case for acousto-elastic systems; this is described in Chapter 13. For physical coupling, however, the wave interaction takes place in the same body when different physical phenomena are coupled by the constitutive equations. That is the case for electromagnetoelastic coupling, specifically piezoelectricity, which is discussed in Chapter 14. Both chapters contain several relevant examples.

2
Some elastodynamic theory

2.1 Introduction

Problems of the motion and deformation of solids are rendered amenable to mathematical analysis by introducing the concept of a continuous medium. In this idealization it is assumed that properties averaged over a very small element, for example the mean mass density and the mean displacement and stress, are continuous functions of position and time. Although it might seem that the microscopic structure of real materials is not consistent with the concept of a continuum, the idealization produces very useful results, simply because the lengths characterizing the microscopic structure of most materials are generally much smaller than any lengths arising in the deformation of the medium. Even if in certain special cases the microstructure gives rise to significant phenomena, these can be taken into account within the framework of the continuum theory by appropriate generalizations.

Continuum mechanics is a classical subject that has been discussed in great generality in numerous treatises. The theory of continuous media is built upon the basic concepts of stress, motion and deformation, upon the laws of conservation of mass, linear momentum, moment of momentum (angular momentum) and energy and on the constitutive relations. The constitutive relations characterize the mechanical and thermal response of a material while the basic conservation laws abstract the common features of mechanical phenomena irrespective of the constitutive relations.

The governing equations used in this book are for homogeneous, isotropic, linearly elastic solids. These equations are valid if it may be assumed that the strains are small and that the stress components are proportional to the strains.

The theory of linear elasticity, one of the triumphs of solid mechanics, was developed more than a century ago. A renowned work entitled *Mathematical Theory of Elasticity* by A. E. H. Love was published in 1892. The last edition of this book appeared in 1927, and it has since been reprinted many times, e.g. Love (1944). In recent years several books have appeared specifically dealing with wave propagation in linearly elastic solids. We mention the books by Achenbach (1973), Pao and Mow (1973), Graff (1975), Eringen and Suhubi (1975) and Miklowitz (1978). These books have summaries of the relevant elastodynamic theory.

In addition to the basic equations for linear elastodynamics, this chapter also presents a summary of the corresponding equations for a linearly viscoelastic solid and for an acoustic medium.

2.2 Linear elastodynamics

In this section we briefly review the concepts of stress, motion and deformation, conservation laws and constitutive relations. Mechanical phenomena are governed by conservation laws of mass, linear momentum, moment of momentum and energy. Constitutive relations between the stress components and the deformation variables (the strains) model the particular material response.

Notation

Both indicial notation and vector notation are used. In a Cartesian coordinate system with coordinates denoted by x_j, the vector $\boldsymbol{u}(\boldsymbol{x}, t)$ is represented by

$$u = u_1 i_1 + u_2 i_2 + u_3 i_3 \tag{2.2.1}$$

where the i_j $(j = 1, 2, 3)$ are a set of orthonormal base vectors. Since summations of the type (2.2.1) occur frequently, it is convenient to introduce the *summation convention*, whereby a repeated italic subscript implies a summation. Equation (2.2.1) may then be rewritten as

$$\boldsymbol{u} = u_j \boldsymbol{i}_j.$$

The following notation is used for the field variables:

position vector	$\boldsymbol{x}$	(coordinates x_i),
displacement vector	$\boldsymbol{u}$	(components u_i),
small-strain tensor	$\boldsymbol{\varepsilon}$	(components ε_{ij}),
stress tensor	$\boldsymbol{\tau}$	(components τ_{ij}).

It may generally be assumed that the functions $u_i(x_1, x_2, x_3, t)$ are differentiable. A shorthand notation for the nine partial derivatives $\partial u_i(x_1, x_2, x_3, t)/\partial x_j$ is

$$\partial u_i/\partial x_j = u_{i,j} \tag{2.2.2}$$

It can be shown that the $u_{i,j}$ are the components of a second-rank tensor.

A time derivative is often indicated by a dot over the quantity, i.e., $\partial u_i/\partial t = \dot{u}_i$.

Deformation

The field defining the displacement at position $\boldsymbol{x}$ at time t is denoted by $\boldsymbol{u}(\boldsymbol{x}, t)$. As a direct implication of the notion of a continuum, the deformation of the medium can be expressed in terms of the gradients of the displacement vector. Within the restrictions of the linearized theory the deformation is described in a very simple manner by the small-strain tensor $\boldsymbol{\varepsilon}$, with components

$$\varepsilon_{ij} = \tfrac{1}{2}(u_{i,j} + u_{j,i}). \tag{2.2.3}$$

It is evident that $\varepsilon_{ij} = \varepsilon_{ji}$, i.e., $\boldsymbol{\varepsilon}$ is a symmetric tensor of rank two. It is also useful to introduce the rotation tensor $\boldsymbol{\omega}$, whose components are defined as

$$\omega_{ij} = \tfrac{1}{2}(u_{i,j} - u_{j,i}). \tag{2.2.4}$$

We note that $\boldsymbol{\omega}$ is an antisymmetric tensor, $\omega_{ij} = -\omega_{ji}$.

Linear momentum and the stress tensor

A basic postulate in the theory of continuous media is that the mechanical action of the material points situated on one side of an arbitrary surface inside a body on those on the other side can be completely accounted for by prescribing a suitable surface traction on this surface. Thus if a surface element has a unit outward normal $\boldsymbol{n}$ we introduce the surface traction $\boldsymbol{t}$, defining a force per unit area. The surface tractions generally depend on the orientation of $\boldsymbol{n}$ as well as on the location $\boldsymbol{x}$.

Suppose that we remove from a body a closed region $V + S$, where S is the boundary and V is the volume of the region. The surface S is subjected to a distribution of surface tractions $\boldsymbol{t}(\boldsymbol{x}, t)$. At each location, the body may be subjected to a body force per unit volume $\boldsymbol{f}(\boldsymbol{x}, t)$. According to the principle of balance of linear momentum, the instantaneous rate of change of the linear momentum of the body is equal to the resultant external force acting on the body at the particular instant of time. In the linear theory this leads to the equation

$$\int_S \boldsymbol{t}\, dS + \int_V \boldsymbol{f}\, dV = \int_V \rho \ddot{\boldsymbol{u}}\, dV, \tag{2.2.5}$$

where $\ddot{\boldsymbol{u}} = \partial^2 \boldsymbol{u}(\boldsymbol{x}, t)/\partial t^2$. By considering Eq. (2.2.5) for a tetrahedron, shown in Fig. 2.1, the limit of Eq. (2.2.5) as the volume of the tetrahedron shrinks to

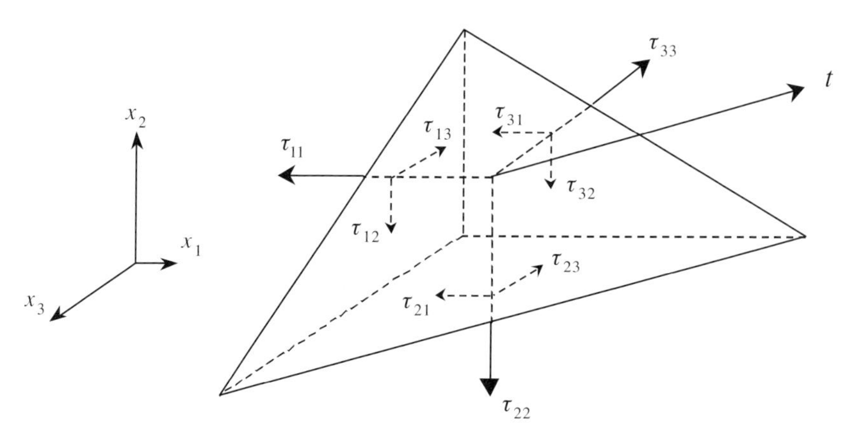

Figure 2.1 Stress components $\tau_{k\ell}$ and stress vector $\boldsymbol{t}$ on a tetrahedron.

zero leads to the definition of the stress tensor $\boldsymbol{\tau}$ with components $\tau_{k\ell}$, where

$$t_\ell = \tau_{k\ell} n_k. \tag{2.2.6}$$

Equation (2.2.6) is the Cauchy stress formula. Physically τ_{kl} is the component in the x_ℓ-direction of the traction on the surface with unit normal $\boldsymbol{i}_k$, as shown in Fig. 2.1.

By substitution of $t_\ell = \tau_{k\ell} n_k$, Eq. (2.2.5) is rewritten in indicial notation as

$$\int_S \tau_{k\ell} n_k \, dS + \int_V f_\ell \, dV = \int_V \rho \ddot{u}_\ell \, dV. \tag{2.2.7}$$

The surface integral can be transformed into a volume integral by Gauss's theorem, and so

$$\int_V (\tau_{k\ell,k} + f_\ell - \rho \ddot{u}_\ell) \, dV = 0.$$

Since V can be an arbitrary part of the body it follows that wherever the integrand is continuous we have

$$\tau_{k\ell,k} + f_\ell = \rho \ddot{u}_\ell. \tag{2.2.8}$$

This is Cauchy's first law of motion.

Balance of moment of momentum

A development similar to the one for linear momentum can be followed for moment of momentum. For the linearized theory the balance of moment of momentum yields the result

$$\tau_{\ell m} = \tau_{m\ell}, \tag{2.2.9}$$

i.e., the stress tensor is symmetric.

Stress–strain relations

In general form, the linear relation between the components of the stress tensor and the components of the strain tensor (Hooke's law) is

$$\tau_{ij} = C_{ijk\ell}\varepsilon_{k\ell}, \tag{2.2.10}$$

where

$$C_{ijk\ell} = C_{jik\ell} = C_{k\ell ij} = C_{ij\ell k}.$$

It follows that 21 of the 81 components of the tensor $C_{ijk\ell}$ are independent. The solid is homogeneous if the coefficients $C_{ijk\ell}$ do not depend on $\boldsymbol{x}$. It is isotropic when there are no preferred directions. It can be shown that elastic isotropy implies that the constants $C_{ijk\ell}$ may be expressed as

$$C_{ijk\ell} = \lambda\delta_{ij}\delta_{k\ell} + \mu(\delta_{ik}\delta_{j\ell} + \delta_{i\ell}\delta_{jk})$$

where δ_{ij} is the Kronecker delta, whose components are

$$\delta_{ij} = \begin{cases} 1 & \text{if} \quad i = j, \\ 0 & \text{if} \quad i \neq j. \end{cases}$$

Hooke's law then assumes the well-known form

$$\tau_{ij} = \lambda\varepsilon_{kk}\delta_{ij} + 2\mu\varepsilon_{ij}. \tag{2.2.11}$$

Equation (2.2.11) contains two elastic constants λ and μ, which are known as Lamé's elastic constants, μ being the shear modulus.

Stress and strain deviators

The stress tensor can be written as the sum of two tensors, one of which represents a spherical or hydrostatic stress in which each normal stress component is $\frac{1}{3}\tau_{kk}$ and all shear stresses vanish. The complementary tensor is called the *stress deviator* and is denoted by s_{ij}. Thus, the components of the stress deviator are defined by

$$s_{ij} = \tau_{ij} - \tfrac{1}{3}\tau_{kk}\delta_{ij}. \tag{2.2.12}$$

Table 2.1 *Relationships between the isotropic elastic constants; E is Young's modulus, B is the bulk modulus, υ is Poisson's ratio and λ, μ are the Lamé constants, μ being the shear modulus*

	E, υ	E, μ	λ, μ
λ	$\dfrac{E\upsilon}{(1+\upsilon)(1-2\upsilon)}$	$\dfrac{\mu(E-2\mu)}{3\mu-E}$	λ
μ	$\dfrac{E}{2(1+\upsilon)}$	μ	μ
E	E	E	$\dfrac{\mu(3\lambda+2\mu)}{\lambda+\mu}$
B	$\dfrac{E}{3(1-2\upsilon)}$	$\dfrac{\mu E}{3(3\mu-E)}$	$\lambda+\dfrac{2}{3}\mu$
υ	υ	$\dfrac{E-2\mu}{2\mu}$	$\dfrac{\lambda}{2(\lambda+\mu)}$

In the same manner we can define the *strain deviator* e_{ij} by

$$e_{ij} = \varepsilon_{ij} - \tfrac{1}{3}\varepsilon_{kk}\delta_{ij}. \tag{2.2.13}$$

From Eq. (2.2.11) it can now quite easily be shown that the following simple relation exists between s_{ij} and e_{ij}:

$$s_{ij} = 2\mu e_{ij}, \tag{2.2.14}$$

where μ is the shear modulus. In addition we have

$$\tfrac{1}{3}\tau_{kk} = B\varepsilon_{kk} \qquad \text{where} \qquad B = \lambda + \tfrac{2}{3}\mu \tag{2.2.15}$$

is the bulk modulus. Equations (2.2.14) and (2.2.15) are completely equivalent to Hooke's law (2.2.11), and thus these equations may also be considered as the constitutive equations for a homogeneous, isotropic, linearly elastic solid.

Other elastic constants that often appear in linear elasticity are Young's modulus E and Poisson's ratio υ. A number of useful relationships among the isotropic elastic constants are summarized in table 2.1.

Problem statement for elastodynamics

We consider a body B occupying a regular region in space, which may be bounded or unbounded, with interior V and boundary S. The system of equations governing the motion of a homogeneous, isotropic, linearly elastic solid consists of the stress equations of motion, Hooke's law and the strain–displacement relations:

$$\tau_{ij,j} + \rho f_i = \rho \ddot{u}_i, \tag{2.2.16}$$

$$\tau_{ij} = \lambda \varepsilon_{kk} \delta_{ij} + 2\mu \varepsilon_{ij} \tag{2.2.17}$$

and

$$\varepsilon_{ij} = \tfrac{1}{2}(u_{i,j} + u_{j,i}), \tag{2.2.18}$$

respectively. If the strain–displacement relations are substituted into Hooke's law and the expressions for the stresses are subsequently substituted in the stress equations of motion, the displacement equations of motion are obtained as

$$\mu u_{i,jj} + (\lambda + \mu) u_{j,ji} + f_i = \rho \ddot{u}_i. \tag{2.2.19}$$

Equations (2.2.19) represent a system of coupled partial differential equations for the displacement components. The system of equations can of course be uncoupled by eliminating two of the three displacement components, but this results in partial differential equations of higher order. A far more convenient approach is to express the components of the displacement vector in derivatives of potentials. It is possible to select potentials that satisfy uncoupled wave equations, as shown in Section 3.8.

In vector notation the system of displacement equations of motion (2.2.19) becomes

$$\mu \nabla^2 \boldsymbol{u} + (\lambda + \mu) \nabla \nabla \cdot \boldsymbol{u} + \boldsymbol{f} = \rho \ddot{\boldsymbol{u}}. \tag{2.2.20}$$

On the surface S of the undeformed body, boundary conditions must be prescribed. The following boundary conditions are most common:

Displacement boundary conditions the three components u_i are prescribed on the boundary.

Traction boundary conditions the three traction components t_i are prescribed on the boundary at a position with unit normal $\boldsymbol{n}$. Through Cauchy's formula

$$t_i = \tau_{ji} n_j, \tag{2.2.21}$$

this case actually corresponds to conditions on three components of the stress tensor.

Displacement boundary conditions on part S_1 of the boundary and *traction boundary conditions* on the remaining part $S - S_1$.

Other conditions on the boundary of the body are possible.

To complete the problem statement we define initial conditions; in V we have at time $t = 0$

$$u_i(\boldsymbol{x}, 0) = u_i^0(\boldsymbol{x}),$$
$$\dot{u}_i(\boldsymbol{x}, 0^+) = v_i^0(\boldsymbol{x}).$$

In the next two sections we consider the special cases of one- and two-dimensional elastodynamics.

2.3 One-dimensional problems

If the body forces and the components of the stress tensor depend on only one spatial variable, say x_1, the stress equations of motion reduce to

$$\tau_{i1,1} + f_i = \rho \ddot{u}_i. \tag{2.3.1}$$

Three cases can be considered.

Longitudinal strain Of all the displacement components only the longitudinal displacement $u_1(x_1, t)$ does not vanish. The only non-zero strain component is $\varepsilon_{11} = \partial u_1/\partial x_1$. By employing (2.2.11) the components of the stress tensor are obtained as

$$\tau_{11} = (\lambda + 2\mu)u_{1,1}, \qquad \tau_{22} = \tau_{33} = \lambda u_{1,1},$$

and the equation of motion (2.3.1) becomes

$$(\lambda + 2\mu)u_{1,11} + f_1 = \rho \ddot{u}_1. \tag{2.3.2}$$

Wave propagation in one-dimensional stress is considered next.

Longitudinal stress Now the longitudinal stress τ_{11}, which is a function of x_1 and t only, is the one non-vanishing stress component. Equating the transverse normal stresses τ_{22} and τ_{33} to zero, we obtain the following relations:

$$\varepsilon_{22} = \varepsilon_{33} = -\frac{\lambda}{2(\lambda + \mu)}\varepsilon_{11} = -\upsilon\varepsilon_{11},$$

where υ is Poisson's ratio. Subsequent substitution of these results in the expression for τ_{11} yields

$$\tau_{11} = E\varepsilon_{11}, \tag{2.3.3}$$

where

$$E = \frac{\mu(3\lambda + 2\mu)}{\lambda + \mu}.$$

The constant E is known as Young's modulus of elasticity. The equation of motion follows by substitution of (2.3.3) into (2.3.1). It should be noted that whereas longitudinal strain is an exact case of deformation in a body of infinite transverse dimensions, longitudinal stress is an approximation, frequently used for waves in thin rods.

Shear In this case the displacement is in a plane normal to the x_1-axis,

$$\mathbf{u} = u_2(x_1, t)\mathbf{i}_2 + u_3(x_1, t)\mathbf{i}_3.$$

The stresses are

$$\tau_{21} = \mu u_{2,1}, \qquad \tau_{31} = \mu u_{3,1}.$$

Clearly, the equations of motion reduce to uncoupled wave equations for u_2 and for u_3, respectively. Deformation in shear corresponds to a transverse wave.

2.4 Two-dimensional problems

In two-dimensional problems the body forces and the components of the stress tensor are independent of one of the coordinates, say x_3. The stress equations of motion can be derived from (2.2.8) by setting the derivatives with respect to x_3 equal to zero. We find that the system of equations splits up into two uncoupled systems. They are

$$\tau_{3\beta,\beta} + f_3 = \rho\ddot{u}_3 \tag{2.4.1}$$

and

$$\tau_{\alpha\beta,\beta} + f_\alpha = \rho\ddot{u}_\alpha. \tag{2.4.2}$$

In Eqs. (2.4.1) and (2.4.2), and throughout this section, the Greek indices can assume the values 1 and 2 only.

Anti-plane shear

A deformation described by a displacement distribution $u_3(x_1, x_2, t)$ is called an anti-plane shear deformation. The corresponding stress components follow from Hooke's law:

$$\tau_{3\beta} = \mu u_{3,\beta}. \tag{2.4.3}$$

Eliminating $\tau_{3\beta}$ from Eqs. (2.4.1) and (2.4.3), we find that $u_3(x_1, x_2, t)$ is governed by the scalar wave equation

$$\mu u_{3,\beta\beta} + f_3 = \rho \ddot{u}_3. \tag{2.4.4}$$

Pure shear motions governed by (2.4.4) are usually called horizontally polarized shear motions.

In-plane motions

It follows from Eq. (2.4.2) that the in-plane displacements u_α depend on x_1, x_2 and t only. With regard to the dependence of u_3 on the spatial coordinates and time, two separate cases are encompassed by Eq. (2.4.2).

Plane strain In a plane-strain deformation *all* field variables are independent of x_3 and the displacement in the x_3-direction vanishes identically. Hooke's law then yields the following relations:

$$\tau_{\alpha\beta} = \lambda u_{\gamma,\gamma}\delta_{\alpha\beta} + \mu(u_{\alpha,\beta} + u_{\beta,\alpha}) \tag{2.4.5}$$

$$\tau_{33} = \lambda u_{\gamma\gamma}, \tag{2.4.6}$$

where, as noted earlier, the Greek indices can assume the values 1 and 2 only.

Elimination of $\tau_{\alpha\beta}$ from (2.4.2) and (2.4.5) leads to

$$\mu u_{\alpha,\beta\beta} + (\lambda + \mu)u_{\beta,\beta\alpha} + f_\alpha = \rho \ddot{u}_\alpha. \tag{2.4.7}$$

Of course, (2.4.7) can also be derived directly from (2.2.19) by setting $u_3 \equiv 0$ and $\partial/\partial x_3 \equiv 0$.

Plane stress A two-dimensional stress field is called a plane stress field if τ_{33}, τ_{23} and τ_{13} are identically zero. From Hooke's law it then follows that ε_{33} is related to $\varepsilon_{11} + \varepsilon_{22}$ by

$$\varepsilon_{33} = -\frac{\lambda}{\lambda + 2\mu} u_{\gamma,\gamma}. \tag{2.4.8}$$

Substitution of (2.4.8) into the expression for $\tau_{\alpha\beta}$ yields

$$\tau_{\alpha\beta} = \frac{2\mu\lambda}{\lambda + 2\mu} u_{\gamma,\gamma}\delta_{\alpha\beta} + \mu(u_{\alpha,\beta} + u_{\beta,\alpha}) \tag{2.4.9}$$

Substituting (2.4.9) into (2.4.2), we obtain the displacement equations of motion. As far as the governing equations are concerned, the difference between plane strain and plane stress is merely a matter of different constant coefficients.

It should be noted that (2.4.8) implies a linear dependence of u_3 on the coordinate x_3. The case of plane stress is often used for an approximate description of the in-plane motions of a thin sheet.

The results of this section show that wave motions in two dimensions are the superposition of horizontally polarized motions and in-plane motions. These motions are governed by uncoupled equations.

2.5 Linearly viscoelastic solid

In this section we summarize the governing equations for a class of materials for which loads and deformations are linearly related but for which the deformation depends not only on the present magnitude of the loads but also on the history of the loading process. These materials are called linearly viscoelastic. The theory of linear viscoelasticity was discussed in detail by Christensen (1972).

Let us consider a thought experiment in which an infinitesimal element of a material is instantaneously brought into a state of homogeneous longitudinal stress defined by $\tau_x \neq 0$, $\tau_y = \tau_z \equiv 0$. In a perfectly elastic element the stress τ_x instantaneously gives rise to strains, in particular to a homogeneous extensional strain of magnitude $\varepsilon_x = \tau_x/E$, where E is the extensional or Young's modulus. In an element of a viscoelastic material the instantaneous response will be followed, however, by an additional strain that increases with time. This phenomenon, which is called creep, is characteristic of viscoelastic materials. The extensional strain response to a homogeneous longitudinal stress of unit magnitude is called the creep function $J_E(t)$. The creep function increases monotonically with time. Conversely, if the element is instantaneously brought into a state of extensional strain ε_x, combined with $\tau_y = \tau_z \equiv 0$, the instantaneous stress response is followed by a decrease of the stress level. This phenomenon is called relaxation. The longitudinal stress response to a strain ε_x of unit magnitude is termed the relaxation function $G_E(t)$ and decreases monotonically with time. By virtue of the linearity of the process the longitudinal stress due to an extensional strain of arbitrary time dependence may then be expressed as a superposition integral over $G_E(t)$ and $\varepsilon_x(t)$:

$$\tau_x(t) = \varepsilon_x(0) G_E(t - t_0) + \int_{t_0+}^{t} G_E(t - s) \frac{d\varepsilon_x}{ds} \, ds, \tag{2.5.1}$$

where it is assumed that the process starts at time $t = t_0$. Equation (2.5.1) can also be written in the form

$$\tau_x(t) = G_E(0) \varepsilon_x(t) + \int_{t_0+}^{t} G_E'(t - s) \varepsilon_x(s) \, ds, \tag{2.5.2}$$

where a prime denotes the derivative with respect to the argument. Equations (2.5.1) and (2.5.2) show one way of representing linear viscoelastic constitutive behavior for the case of one-dimensional stress. A more compact way of writing (2.5.1) is

$$\tau_x(t) = \int_{t_0}^{t} G_E(t-s)\,d\varepsilon_x. \tag{2.5.3}$$

Constitutive equations in three dimensions

In an isotropic elastic solid the mechanical behavior can be completely described by two elastic constants. As shown in Section 2.2 a convenient choice consists of the shear modulus μ and the bulk modulus B. The advantage of using these constants is that they have definite physical interpretations and that they can be measured. In terms of μ and B the elastic stress–strain relations are

$$s_{ij} = 2\mu e_{ij}, \qquad \tfrac{1}{3}\tau_{kk} = B\varepsilon_{kk}, \tag{2.5.4}$$

where the stress deviator s_{ij} and the strain deviator e_{ij} are defined by Eqs. (2.2.12) and (2.2.13), respectively. In analogy to Eq. (2.5.3) the viscoelastic relations corresponding to Eq. (2.5.4) may be written in the forms

$$s_{ij}(\mathbf{x},t) = 2\int_0^t G_S(t-s)\,de_{ij}(\mathbf{x},s) \tag{2.5.5}$$

$$\tau_{kk}(\mathbf{x},t) = 3\int_0^t G_B(t-s)\,d\varepsilon_{kk}(\mathbf{x},s), \tag{2.5.6}$$

where $G_S(t)$ and $G_B(t)$ are the shear and bulk relaxation functions, respectively. The corresponding relation between τ_{ij} and ε_{ij} is

$$\tau_{ij}(\mathbf{x},t) = \delta_{ij}\int_0^t [G_B(t-s) - \tfrac{2}{3}G_S(t-s)]\,d\varepsilon_{kk}(\mathbf{x},s) + 2\int_0^t G_S(t-s)\,d\varepsilon_{ij}(\mathbf{x},s). \tag{2.5.7}$$

Complex modulus

Suppose that the strain history is specified as a harmonic function of time, so that the strain deviator may be written as

$$e_{ij}(\mathbf{x},t) = e_{ij}(\mathbf{x};\omega)e^{-i\omega t}.$$

Assuming that a steady state has been reached, the stress deviator is then also time harmonic with frequency ω and may be written in the form

$$s_{ij}(\mathbf{x},t) = 2G_S(\omega)e_{ij}(\mathbf{x};\omega)e^{-i\omega t}, \tag{2.5.8}$$

where the complex modulus $G_S(\omega)$ is a complex function of the frequency. We write

$$G_S = G'_S(\omega) + iG''_S(\omega), \tag{2.5.9}$$

or

$$G_S = |G_S| e^{i\varphi_S}(\omega), \tag{2.5.10}$$

where

$$|G_S| = \left\{ \left[G'_S(\omega) \right]^2 + \left[G''_S(\omega) \right]^2 \right\}^{1/2} \tag{2.5.11}$$

and

$$\varphi_S(\omega) = \tan^{-1} \frac{G''_S(\omega)}{G'_S(\omega)}. \tag{2.5.12}$$

It follows that the stress and the strain are out of phase, which implies that energy is dissipated during steady-state oscillations. It can be shown that the real and imaginary parts of the complex modulus may be expressed in terms of the relaxation function; see Christensen (1982).

For steady-state time-harmonic oscillations Eq. (2.5.7) becomes

$$\tau_{ij}(\boldsymbol{x};\omega) = \delta_{ij}[G_B(\omega) - \tfrac{2}{3}G_S(\omega)]\varepsilon_{kk}(\boldsymbol{x};\omega) + 2G_S(\omega)\varepsilon_{ij}(\boldsymbol{x};\omega), \tag{2.5.13}$$

where

$$\tau_{ij}(\boldsymbol{x},t) = \tau_{ij}(\boldsymbol{x};\omega)e^{-i\omega t}, \qquad \varepsilon_{ij}(\boldsymbol{x},t) = \varepsilon_{ij}(\boldsymbol{x};\omega)e^{-i\omega t}. \tag{2.5.14}$$

2.6 Acoustic medium

The homogeneous, isotropic, linearly elastic solid discussed in the preceding sections is characterized by its resistance to shear and volume deformations. An acoustic medium (or an ideal fluid) cannot offer resistance to shear deformation and thus cannot sustain shear stresses. Hence the state of stress in an acoustic medium is purely hydrostatic, and is described by

$$\tau_{ij} = -p\delta_{ij}, \tag{2.6.1}$$

where p is the pressure. Note that the pressure is positive when the traction vector on a surface points into the material. The equation of state for an acoustic medium defines the pressure as a function of the mass density, $p = p(\rho)$.

A wave motion in an acoustic medium is a dynamic perturbation of a static equilibrium state defined by p_0 and ρ_0. The perturbations are denoted by $p'(\boldsymbol{x},t)$, $\rho'(\boldsymbol{x},t)$ and $v(\boldsymbol{x},t)$, where $v(\boldsymbol{x},t)$ is the particle velocity. The derivation of the linearized equations that relate $p'(\boldsymbol{x},t)$, $\rho'(\boldsymbol{x},t)$ and $v(\boldsymbol{x},t)$ can be found in books on acoustics; see for example Pierce (1981) and Dowling and Ffowcs Williams (1983). They can also be obtained from the equations of linearized elastodynamics by setting $\mu = 0$. For $\mu = 0$, Eq. (2.2.11) yields

$$\tau_{11} = \tau_{22} = \tau_{33} = \lambda \varepsilon_{kk} = -p', \tag{2.6.2}$$

$$\tau_{12} = \tau_{13} = \tau_{23} = 0. \tag{2.6.3}$$

In liquids λ, (2.2.11), is quite large, but it has only moderate values for gases.

For an acoustic medium Cauchy's first law of motion, Eq. (2.2.8), yields:

$$-p'_{,i} + f_i = \rho_0 \dot{v}_i. \tag{2.6.4}$$

Here $f_i(\boldsymbol{x}, t)$ defines the density of the volume forces (in $\mathrm{N/m^3}$). Equation (2.6.2) can be rewritten as a relation between v and $\dot{p}'$:

$$v_{i,i} = -\kappa \dot{p}', \tag{2.6.5}$$

where

$$\kappa = 1/\lambda. \tag{2.6.6}$$

Equation (2.6.5) relates the rate of change of the volume of an element to the pressure perturbation rate. For the case where there is external injection of fluid, an inhomogeneous term should be included on the right-hand side of Eq. (2.6.5), to give

$$v_{i,i} + \kappa \dot{p}' = q, \tag{2.6.7}$$

where $q(\boldsymbol{x}, t)$ defines the volume-source density-of-injection rate (in $\mathrm{s^{-1}}$).

The constant κ, which is called the compressibility of the acoustic medium, can be related to the usual equation of state $p = p(\rho)$ by noting that for a small pressure perturbation we have

$$p'(\boldsymbol{x}, t) = \rho' \frac{dp}{d\rho}(\rho_0) \tag{2.6.8}$$

or

$$p'(\boldsymbol{x}, t) = c^2 \rho', \tag{2.6.9}$$

where

$$c^2 = \frac{dp}{d\rho}(\rho_0). \tag{2.6.10}$$

Then, using the conservation of mass as expressed by

$$\frac{\partial \rho'}{\partial t} + \rho_0 \nabla \cdot \boldsymbol{v} = 0, \tag{2.6.11}$$

and eliminating $\partial\rho'/\partial t$ from the time derivative of Eq. (2.6.8) by the use of (2.6.11), we find

$$\dot{p}'(\boldsymbol{x}, t) = -\rho_0 \frac{dp}{d\rho}(\rho_0)\nabla \cdot \boldsymbol{v} = -\rho_0 c^2 \nabla \cdot \boldsymbol{v}, \tag{2.6.12}$$

where c^2 is defined by Eq. (2.6.10). Comparison of (2.6.5) and (2.6.12) yields

$$\frac{1}{\kappa} = \rho_0 \frac{dp}{d\rho}(\rho_0) = \rho_0 c^2. \tag{2.6.13}$$

If there is only a distribution of volume forces, i.e., $f_i \neq 0$ but $q(\boldsymbol{x}, t) \equiv 0$, a simple wave equation for the pressure perturbation $p'(\boldsymbol{x}, t)$ can be obtained from Eqs. (2.6.4) and (2.6.5). From Eq. (2.6.4) we have

$$\rho_0 \dot{v}_{i,i} = f_{i,i} - p'_{,ii}. \tag{2.6.14}$$

For $q \equiv 0$ it follows from Eq. (2.6.7) that

$$\dot{v}_{i,i} + \kappa \ddot{p}' = 0. \tag{2.6.15}$$

Elimination of the particle velocity from Eqs. (2.6.15) and (2.6.14) yields

$$\frac{1}{c^2}\frac{\partial^2 p'}{\partial t^2} - \nabla^2 p' = F(\boldsymbol{x}, t), \tag{2.6.16}$$

where ∇^2 is the three-dimensional Laplacian, and

$$F(\boldsymbol{x}, t) = -f_{i,i} = -\operatorname{div} \boldsymbol{f}, \tag{2.6.17}$$

and $c^2 = 1/(\rho_0 \kappa)$, as given by Eq. (2.6.13). It is noted that the pressure perturbation satisfies the classical wave equation. The inhomogeneous term $F(\boldsymbol{x}, t)$ represents a distribution of dipoles.

If there are no volume forces, i.e., $f(\boldsymbol{x}, t) \equiv 0$, differentiation of Eq. (2.6.4) yields

$$p'_{,ii} + \rho_0 \dot{v}_{i,i} = 0. \tag{2.6.18}$$

From (2.6.7) it follows that

$$\dot{v}_{i,i} = -\kappa \ddot{p}' + \dot{q}. \tag{2.6.19}$$

Substitution in (2.6.18) yields

$$\frac{1}{c^2}\frac{\partial^2 p'}{\partial t^2} - \nabla^2 p' = \rho_0 \dot{q}. \tag{2.6.20}$$

The inhomogeneous term $\rho_0 \dot{q}$ represents a distribution of monopoles.

Similarly, again for $f(\boldsymbol{x}, t) \equiv 0$, differentiation of Eq. (2.6.4) with respect to t, taking the gradient of Eq. (2.6.7), and combining the results yields the wave equation for $\boldsymbol{v}$ as

$$\nabla^2 \boldsymbol{v} - \frac{1}{c^2}\frac{\partial^2 \boldsymbol{v}}{\partial t^2} = \nabla q. \tag{2.6.21}$$

Another set of equations for the acoustic wave field can be obtained by employing a velocity potential $\varphi(\boldsymbol{x}, t)$. Let us take the curl of the homogeneous version of Eq. (2.6.4). Since $\nabla \wedge \nabla p' \equiv 0$,

$$\nabla \wedge \left(\nabla p' + \rho_0 \frac{\partial \boldsymbol{v}}{\partial t}\right) = \rho_0 \frac{\partial}{\partial t} \nabla \wedge \boldsymbol{v} = 0, \tag{2.6.22}$$

This equation implies that $\boldsymbol{v}(\boldsymbol{x}, t)$ is irrotational and may thus be written as the gradient of a potential $\varphi(\boldsymbol{x}, t)$, so that

$$v_i = \varphi_{,i}\,. \tag{2.6.23}$$

A second relation is now obtained by substituting $v_i = \varphi_{,i}$ into the homogeneous form of Eq. (2.6.4). We obtain

$$\left(p' + \rho_0 \frac{\partial \varphi}{\partial t}\right)_{,i} = 0, \tag{2.6.24}$$

and thus we conclude that

$$p' = -\rho_0 \frac{\partial \varphi}{\partial t}. \tag{2.6.25}$$

Substitution of the above expressions for v_i and p' into Eq. (2.6.7) yields

$$\nabla^2 \varphi - \frac{1}{c^2}\frac{\partial^2 \varphi}{\partial t^2} = q(\boldsymbol{x}, t). \tag{2.6.26}$$

2.7 Governing equations for linear elasticity in cylindrical coordinates

For future reference we list here the governing equations of the theory of elasticity in polar coordinates r, θ and z.

The Laplacian is

$$\nabla^2 = \frac{\partial^2}{\partial r^2} + \frac{1}{r}\frac{\partial}{\partial r} + \frac{1}{r^2}\frac{\partial^2}{\partial \theta^2} + \frac{\partial^2}{\partial z^2}. \tag{2.7.1}$$

In cylindrical coordinates the strain–displacement relations are given by

$$\varepsilon_r = \frac{\partial u}{\partial r}, \qquad \varepsilon_\theta = \frac{u}{r} + \frac{1}{r}\frac{\partial v}{\partial \theta}, \qquad \varepsilon_z = \frac{\partial w}{\partial z}, \tag{2.7.2a, b, c}$$

$$2\varepsilon_{r\theta} = 2\varepsilon_{\theta r} = \frac{\partial v}{\partial r} - \frac{v}{r} + \frac{1}{r}\frac{\partial u}{\partial \theta}, \tag{2.7.3}$$

$$2\varepsilon_{\theta z} = 2\varepsilon_{z\theta} = \frac{1}{r}\frac{\partial w}{\partial \theta} + \frac{\partial v}{\partial z}, \tag{2.7.4}$$

$$2\varepsilon_{zr} = 2\varepsilon_{rz} = \frac{\partial u}{\partial z} + \frac{\partial w}{\partial r}. \tag{2.7.5}$$

The stress–strain relations are

$$\tau_r = \lambda\left(\frac{\partial u}{\partial r} + \frac{u}{r} + \frac{1}{r}\frac{\partial v}{\partial \theta} + \frac{\partial w}{\partial z}\right) + 2\mu\frac{\partial u}{\partial r}, \tag{2.7.6}$$

$$\tau_\theta = \lambda\left(\frac{\partial u}{\partial r} + \frac{u}{r} + \frac{1}{r}\frac{\partial v}{\partial \theta} + \frac{\partial w}{\partial z}\right) + 2\mu\left(\frac{u}{r} + \frac{1}{r}\frac{\partial v}{\partial \theta}\right), \tag{2.7.7}$$

$$\tau_z = \lambda\left(\frac{\partial u}{\partial r} + \frac{u}{r} + \frac{1}{r}\frac{\partial v}{\partial \theta} + \frac{\partial w}{\partial z}\right) + 2\mu\frac{\partial w}{\partial z}, \tag{2.7.8}$$

$$\tau_{r\theta} = \mu\left(\frac{\partial v}{\partial r} - \frac{v}{r} + \frac{1}{r}\frac{\partial u}{\partial \theta}\right), \tag{2.7.9}$$

$$\tau_{\theta z} = \mu\left(\frac{1}{r}\frac{\partial w}{\partial \theta} + \frac{\partial v}{\partial z}\right), \tag{2.7.10}$$

$$\tau_{zr} = \mu\left(\frac{\partial u}{\partial z} + \frac{\partial w}{\partial r}\right). \tag{2.7.11}$$

3

Wave motion in an unbounded elastic solid

3.1 Introduction

In this chapter the equations governing linear, isotropic and homogeneous elasticity are used to describe the propagation of mechanical disturbances in elastic solids. Some well-known wave-propagation results are summarized as a preliminary to their use in subsequent chapters.

There are essential differences between waves in elastic solids and acoustic waves in fluids and gases. Some of these differences are exhibited by plane waves. For example, *two* kinds of plane wave (longitudinal and transverse waves) can propagate in a homogeneous, isotropic, linearly elastic solid. These waves may propagate independently, i.e., uncoupled, in an unbounded solid. Generally, longitudinal and transverse waves are however, coupled by conditions on boundaries. Most boundary conditions or internal source mechanisms generate both kinds of wave simultaneously.

Plane waves in an unbounded domain are discussed in Section 3.2. The flux of energy in plane time-harmonic waves is considered in Section 3.3. The presence of a surface gives rise to reflected waves. The details of the reflection of plane waves incident at an arbitrary angle on a free plane surface are discussed in Section 3.4. As is well known, the incidence of a plane wave, say a longitudinal wave, gives rise to the reflection of both a longitudinal and a transverse wave. This wave-splitting effect happens, of course, also for an incident transverse wave. The reflection coefficients and other relevant results are listed in Section 3.4. Energy partition due to wave splitting is discussed in Section 3.5.

The most important elastic wave field is the basic singular solution. This is the elastodynamic field induced by the application of a point force in an unbounded elastic solid. A point force generates both longitudinal and transverse spherical waves, which radiate from its point of application. Explicit solutions for the displacement and stress fields are presented in Sections 3.8 and 3.9. By superposition considerations the field radiated by a distribution of point forces can be obtained in the form of an integral.

In Section 3.11 superposition is used to write the field radiated by a line load in the form of line integrals. For a spatially uniform time-harmonic line load, which defines the case of plane strain, the line integrals can be recognized as Hankel functions. Far-field expressions for the displacements and the stress components generated by point forces and line loads are presented in Sections 3.10 and 3.11.

3.2 Plane waves

In a Cartesian coordinate system, the relation

$$\mathbf{x} \cdot \mathbf{p} = x_p, \tag{3.2.1}$$

where $\mathbf{p}$ is a unit vector, defines the projection of the position vector $\mathbf{x}$ onto the direction of $\mathbf{p}$. For a specified value of x_p, Eq. (3.2.1) represents a plane perpendicular to the direction of $\mathbf{p}$. Analogously to the one-dimensional case, Eq. (1.4.2), the function

$$\mathbf{u}(\mathbf{x}, t) = f(\mathbf{x} \cdot \mathbf{p} - ct)\mathbf{d} = f(x_p - ct)\mathbf{d} \tag{3.2.2}$$

therefore represents a wave propagating with phase velocity c in the direction of the unit vector $\mathbf{p}$. In Eq. (3.2.2), $\mathbf{d}$ is a unit vector that defines the direction of the displacement. Since the phase of the wave, $x_p - ct$, is spatially uniform in the plane defined by Eq. (3.2.1), we say that Eq. (3.2.2) represents a plane wave propagating in three-dimensional space in the direction of the propagation vector $\mathbf{p}$.

The question of interest is whether for a specified propagation vector $\mathbf{p}$ the unit vector $\mathbf{d}$ and the phase velocity c are arbitrary or must satisfy certain relations. To answer this question, the expression for a plane wave, Eq. (3.2.2), is substituted into the homogeneous form of the displacement equation of motion given by Eq. (2.2.20). By employing the relations

$$\begin{aligned}
\nabla \cdot \mathbf{u} &= (\mathbf{p} \cdot \mathbf{d}) f'(\mathbf{x} \cdot \mathbf{p} - ct), \\
\nabla\nabla \cdot \mathbf{u} &= (\mathbf{p} \cdot \mathbf{d}) f''(\mathbf{x} \cdot \mathbf{p} - ct)\, \mathbf{p}, \\
\nabla^2 \mathbf{u} &= f''(\mathbf{x} \cdot \mathbf{p} - ct)\, \mathbf{d}, \\
\ddot{\mathbf{u}} &= c^2 f''(\mathbf{x} \cdot \mathbf{p} - ct)\, \mathbf{d},
\end{aligned}$$

where the prime denotes differentiation with respect to the argument, we obtain

$$[(\lambda + \mu)(\mathbf{p} \cdot \mathbf{d})\mathbf{p} + \mu \mathbf{d} - \rho c^2 \mathbf{d}] f''(\mathbf{x} \cdot \mathbf{p} - ct) = 0.$$

Hence

$$(\mu - \rho c^2)\boldsymbol{d} + (\lambda + \mu)(\boldsymbol{p} \cdot \boldsymbol{d})\boldsymbol{p} = 0. \tag{3.2.3}$$

Since $\boldsymbol{p}$ and $\boldsymbol{d}$ are different unit vectors, Eq. (3.2.3) can be satisfied in two ways only. One way is to set

$$\boldsymbol{d} = \pm \boldsymbol{p}.$$

Equation (3.2.3) then yields

$$c^2 = c_L^2 = (\lambda + 2\mu)/\rho. \tag{3.2.4}$$

In this case the motion is parallel to the direction of propagation, and the wave is therefore called a *longitudinal* or L wave. If $\boldsymbol{d} \neq \pm \boldsymbol{p}$, both terms in (3.2.3) have to vanish independently, yielding

$$\boldsymbol{p} \cdot \boldsymbol{d} = 0, \qquad \text{and} \qquad c^2 = c_T^2 = \mu/\rho. \tag{3.2.5}$$

Now the motion is normal to the direction of propagation, and the wave is called a *transverse* or T wave. The displacement can have any direction in a plane normal to the direction of propagation, but usually we choose the x_1x_2-plane to contain the vector $\boldsymbol{p}$ and consider transverse motions in the x_1x_2-plane or normal to the x_1x_2-plane. If the motion in the x_1x_2-plane is transverse then the wave called a "vertically" polarized transverse or TV wave, while a wave motion normal to the x_1x_2-plane is called a "horizontally" polarized transverse or TH wave.

From Eqs. (3.2.4) and (3.2.5) it follows that

$$\frac{c_L}{c_T} = \kappa = \left[\frac{\lambda + 2\mu}{\mu}\right]^{1/2} = \left[\frac{2(1-\upsilon)}{1-2\upsilon}\right]^{1/2}, \tag{3.2.6}$$

where υ is Poisson's ratio. Since $0 \leq \upsilon \leq 0.5$, it follows that $c_L > c_T$. For metals, the phase velocities of both longitudinal and transverse waves are generally very large. Thus we find for structural steel $c_L = 590\,000$ cm/sec and $c_T = 320\,000$ cm/sec. For a few materials, representative values of ρ, c_L, c_T and κ are listed in Table 3.1 .

By substituting the components of (3.2.2) into Hooke's law, Eq. (2.2.11), the components of the stress tensor are obtained as

$$\tau_{lm} = [\lambda \delta_{lm}(d_j p_j) + \mu(d_l p_m + d_m p_l)] f'(x_n p_n - ct), \tag{3.2.7}$$

where the prime denotes a derivative of the function $f(\ )$ with respect to the argument $x_n p_n - ct$.

Table 3.1 *Approximate values of* ρ, c_L, c_T *and* κ

Material	ρ (kg/m^3)	c_L (m/s)	c_T (m/s)	κ
air	1.2	340		
water	1000	1480		
steel	7800	5900	3200	1.845
copper	8900	4600	2300	2
aluminum	2700	6300	3100	2.03
glass	2500	5800	3400	1.707
rubber	930	1040	27	38.5

A plane harmonic displacement wave propagating with phase velocity c and propagation vector $\boldsymbol{p}$ is represented by

$$\boldsymbol{u} = A\boldsymbol{d} \exp[ik(\boldsymbol{x} \cdot \boldsymbol{p} - ct)], \tag{3.2.8}$$

where the amplitude A is independent of x and t, and $k = \omega/c$ is the wavenumber. Here ω is the circular frequency. Equation (3.2.8) is clearly a special case of (3.2.2).

For further discussion a convenient representation of a plane harmonic displacement wave is given by

$$\boldsymbol{u} = A\boldsymbol{d}e^{i\eta} \tag{3.2.9}$$

where

$$\eta = k(\boldsymbol{x} \cdot \boldsymbol{p} - ct). \tag{3.2.10}$$

By the use of Eq. (3.2.7) the components of the stress tensor then take the form

$$\tau_{lm} = [\lambda\delta_{lm}(d_j p_j) + \mu(d_l p_m + d_m p_l)]ikAe^{i\eta}. \tag{3.2.11}$$

3.3 Flux of energy in plane time-harmonic waves

For a surface element of unit area, the instantaneous rate of work of the surface traction is the scalar product of the surface traction vector $\boldsymbol{t}$ and the particle velocity $\dot{\boldsymbol{u}}$. This scalar product is called the power per unit area and is denoted by P:

$$P = \boldsymbol{t} \cdot \dot{\boldsymbol{u}}. \tag{3.3.1}$$

The power per unit area defines the rate at which energy is transmitted per unit area of the surface. Hence it represents the energy flux across the surface element. If the outer normal at the surface element is $\boldsymbol{n}$, we have $t_l = \tau_{lm} n_m$, and thus

$$P = \tau_{lm} n_m \dot{u}_l, \tag{3.3.2}$$

where use of the summation convention is implied.

Let us examine a longitudinal time-harmonic wave with propagation vector $\boldsymbol{p}$, and let

$$d_1 = p_1 = \sin\theta, \qquad d_2 = p_2 = \cos\theta, \qquad d_3 = p_3 = 0.$$

For a surface element normal to the direction of propagation we have

$$n_1 = -p_1 = -\sin\theta, \qquad n_2 = -p_2 = -\cos\theta, \qquad n_3 = 0.$$

The components of the stress tensor appearing in (3.3.2) are then τ_{11}, τ_{12}, τ_{21} and τ_{22}, which are obtained from (3.2.11) as

$$\begin{aligned} \tau_{11} &= (\lambda + 2\mu \sin^2\theta) ikAe^{i\eta}, \\ \tau_{12} = \tau_{21} &= (2\mu \sin\theta\cos\theta) ikAe^{i\eta}, \\ \tau_{22} &= (\lambda + 2\mu\cos^2\theta) ikAe^{i\eta}, \end{aligned}$$

where η follows from Eq. (3.2.10) as

$$\eta = k(x_1 \sin\theta + x_2\cos\theta - c_L t).$$

Substituting these results into (3.3.2) and assuming that A is real-valued, we find in the longitudinal case

$$P_L = (\lambda + 2\mu) c_L k^2 A^2 \mathcal{R}(ie^{i\eta}) \mathcal{R}(ie^{i\eta}).$$

The time average of the power over a period T is then obtained as

$$\langle P_L \rangle = \tfrac{1}{2}(\lambda + 2\mu) c_L k^2 A^2 = \tfrac{1}{2}(\lambda + 2\mu)\frac{\omega^2}{c_L} A^2. \tag{3.3.3}$$

Similarly, for a plane transverse wave the time-averaged power is

$$\langle P_T \rangle = \tfrac{1}{2}\mu \frac{\omega^2}{c_T} A^2. \tag{3.3.4}$$

Using a single formula, we can write

$$\langle P_\gamma \rangle = \tfrac{1}{2}\rho\omega^2 c_\gamma (A_\gamma)^2, \tag{3.3.5}$$

where $\gamma = L$ or T. Comparing Eqs. (3.3.3) and (3.3.4) we note that for a given frequency and amplitude the time-averaged rate of energy transmission is larger for an L wave than for a T wave.

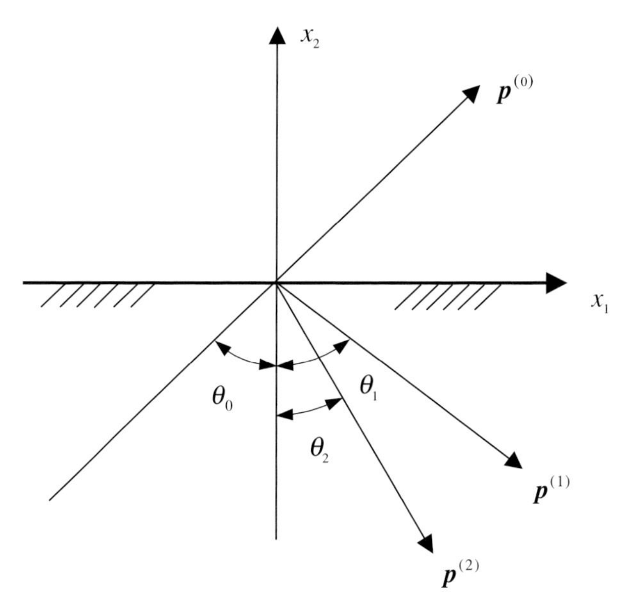

Figure 3.1 Reflection of an L wave at a free surface.

3.4 Reflection of time-harmonic plane waves at a free surface

Incident longitudinal wave

Let us investigate the reflection of plane longitudinal waves whose displacement vectors and propagation vectors are situated in the x_1x_2-plane. Employing the notation introduced in Section 3.2, the incident as well as the reflected waves are denoted as

$$\mathbf{u}^{(n)} = A_n \mathbf{d}^{(n)} \exp(i\eta_n), \tag{3.4.1}$$

where different values of the index n serve to label the various types of wave that occur when a longitudinal wave is reflected, and where η_n follows from Eq. (3.2.10). For reflection at the plane $x_2 = 0$ (Fig. 3.1) the relevant stresses are $\tau_{2j}^{(n)}$, where $j = 1, 2$. These components are readily computed by the use of Hooke's law, Eq. (3.2.11), as

$$\tau_{22}^{(n)} = ik_n\left[(\lambda + 2\mu)\, d_2^{(n)} p_2^{(n)} + \lambda d_1^{(n)} p_1^{(n)}\right] A_n \exp(i\eta_n), \tag{3.4.2}$$

$$\tau_{21}^{(n)} = ik_n\mu\left[d_2^{(n)} p_1^{(n)} + d_1^{(n)} p_2^{(n)}\right] A_n \exp(i\eta_n). \tag{3.4.3}$$

The displacements and the stresses at $x_2 = 0$ are obtained by replacing in (3.4.1)–(3.4.3) the variable η_n by $\bar{\eta}_n$, where

$$\bar{\eta}_n = k_n\left(x_1 p_1^{(n)} - c_n t\right). \tag{3.4.4}$$

The index n is assigned the value $n = 0$ for the incident L wave. It follows from Fig. 3.1 that we have

$$d_1^{(0)} = \sin\theta_0, \qquad d_2^{(0)} = \cos\theta_0,$$
$$p_1^{(0)} = \sin\theta_0, \qquad p_2^{(0)} = \cos\theta_0,$$
$$c_0 = c_L.$$

In the plane $x_2 = 0$ the displacements and the stresses of the incident wave then become

$$u_1^{(0)} = A_0 \sin\theta_0 \exp(i\bar{\eta}_0), \tag{3.4.5}$$
$$u_2^{(0)} = A_0 \cos\theta_0 \exp(i\bar{\eta}_0), \tag{3.4.6}$$
$$\tau_{22}^{(0)} = ik_0(\lambda + 2\mu\cos^2\theta_0)A_0 \exp(i\bar{\eta}_0), \tag{3.4.7}$$
$$\tau_{21}^{(0)} = 2ik_0\mu \sin\theta_0 \cos\theta_0 A_0 \exp(i\bar{\eta}_0), \tag{3.4.8}$$

where

$$\bar{\eta}_0 = k_0(x_1 \sin\theta_0 - c_L t).$$

The reflected L wave is labeled by $n = 1$. Referring to Fig. 3.1 we can write

$$d_1^{(1)} = \sin\theta_1, \qquad d_2^{(1)} = -\cos\theta_1,$$
$$p_1^{(1)} = \sin\theta_1, \qquad p_2^{(1)} = -\cos\theta_1,$$
$$c_1 = c_L \qquad .$$

It is anticipated that an incident L wave will also give rise to a reflected transverse wave with displacement polarized in the x_1x_2-plane. That type of transverse wave was earlier introduced as a TV wave. The reflected TV wave is labeled $n = 2$, and we have, see again Fig. 3.1:

$$d_1^{(2)} = \cos\theta_2, \qquad d_2^{(2)} = \sin\theta_2$$
$$p_1^{(2)} = \sin\theta_2, \qquad p_2^{(2)} = -\cos\theta_2,$$
$$c_2 = c_T.$$

The relations governing the reflection depend on the boundary conditions at the reflecting plane $x_2 = 0$. The following conditions may be considered.

(i) The plane $x_2 = 0$ is free of tractions: $\tau_{22} = \tau_{21} = 0$ (free boundary).
(ii) The displacements vanish at $x_2 = 0$: $u_1 = u_2 = 0$ (clamped boundary).

(iii) The normal displacement and the tangential stress vanish: $u_2 = 0$, $\tau_{21} = 0$ (smooth boundary).
(iv) The tangential displacement and the normal stress vanish: $u_1 = 0$, $\tau_{22} = 0$.

The two cases of "mixed" boundary conditions defined by (c) and (d) are unfortunately physically somewhat unrealistic. Of more practical significance are the reflections from a free or a clamped surface. For a free or a clamped surface an incident longitudinal wave will generate not only a reflected L wave but also a reflected TV wave.

For a free surface the sum of the three tractions must vanish at $x_2 = 0$, and we obtain

$$\tau_{22} = \tau_{22}^{(0)} + \tau_{22}^{(1)} + \tau_{22}^{(2)} \equiv 0,$$

that is,

$$ik_0(\lambda + 2\mu\cos^2\theta_0)A_0 \exp(i\bar{\eta}_0) + ik_1(\lambda + 2\mu\cos^2\theta_1)A_1 \exp(i\bar{\eta}_1) - 2ik_2\mu \sin\theta_2 \cos\theta_2\, A_2 \exp(i\bar{\eta}_2) = 0 \qquad (3.4.9)$$

and

$$\tau_{21} = \tau_{21}^{(0)} + \tau_{21}^{(1)} + \tau_{21}^{(2)} \equiv 0,$$

that is,

$$2ik_0\mu \sin\theta_0 \cos\theta_0\, A_0 \exp(i\bar{\eta}_0) - 2ik_1\mu \sin\theta_1 \cos\theta_1\, A_1 \exp(i\bar{\eta}_1) + ik_2\mu(\sin^2\theta_2 - \cos^2\theta_2)A_2 \exp(i\bar{\eta}_2) = 0. \qquad (3.4.10)$$

Equations (3.4.9) and (3.4.10) must be valid for all values of x_1 and t, and the exponentials must therefore be the same in both equations, which implies that

$$\bar{\eta}_0 = \bar{\eta}_1 = \bar{\eta}_2.$$

Inspection of $\bar{\eta}_n$ from (3.4.4) then leads to the following conclusions:

$$k_0 \sin\theta_0 = k_1 \sin\theta_1 = k_2 \sin\theta_2 = k = \text{ apparent wavenumber},$$
$$k_0 c_L = k_1 c_L = k_2 c_T = \omega = \text{ circular frequency}.$$

These results provide, in turn, the simpler relations

$$k_1 = k_0,$$
$$\frac{k_2}{k_0} = \frac{c_L}{c_T} = \kappa,$$

$$\theta_1 = \theta_0,$$
$$\sin\theta_2 = \frac{1}{\kappa}\sin\theta_0.$$

The last relation is known as Snell's law. The material constant κ was defined by Eq. (3.2.6) as

$$\kappa = \left[\frac{2(1-\upsilon)}{1-2\upsilon}\right]^{1/2}.$$

Since $\kappa > 1$ and $\theta_2 \le \pi/2$, it is apparent that $\theta_2 < \theta_1$. The wavenumber $k = k_0 \sin\theta_0$ refers to the wave propagating along the surface $x_2 = 0$. The phase velocity along the surface $x_2 = 0$ is then obtained as

$$c = \frac{\omega}{k} = \frac{c_L}{\sin\theta_0}.$$

With the aid of (3.4.9), (3.4.10) the algebraic equations for A_1/A_0 and A_2/A_0 can now be simplified to

$$(\lambda + 2\mu\cos^2\theta_0)(A_1/A_0) - \kappa\mu\sin 2\theta_2(A_2/A_0) = -(\lambda + 2\mu\cos^2\theta_0),$$
$$-\mu\sin 2\theta_0\,(A_1/A_0) - \kappa\mu\cos 2\theta_2\,(A_2/A_0) = -\mu\sin 2\theta_0.$$

The solutions to this set of equations are

$$R_L^L = \frac{A_1}{A_0} = \frac{\sin 2\theta_0 \sin 2\theta_2 - \kappa^2\cos^2 2\theta_2}{\sin 2\theta_0 \sin 2\theta_2 + \kappa^2\cos^2 2\theta_2}, \tag{3.4.11}$$

$$R_T^L = \frac{A_2}{A_0} = \frac{2\kappa \sin 2\theta_0 \cos 2\theta_2}{\sin 2\theta_0 \sin 2\theta_2 + \kappa^2\cos^2 2\theta_2}. \tag{3.4.12}$$

Here R_L^L and R_T^L are the reflection coefficients; the superscript defines the incident wave and the subscript the reflected wave.

It is a little more descriptive to replace θ_0 by θ_L and θ_2 by θ_T. Equations (3.4.11) and (3.4.12) may then be rewritten as functions of the angle of incidence:

$$R_L^L(\theta_L) = \frac{\sin 2\theta_L \sin 2\theta_T - \kappa^2\cos^2 2\theta_T}{R(\theta_L, \theta_T)}, \tag{3.4.13}$$

$$R_T^L(\theta_L) = \frac{2\kappa \sin 2\theta_L \cos 2\theta_T}{R(\theta_L, \theta_T)}, \tag{3.4.14}$$

where

$$R(\theta_L, \theta_T) = \sin 2\theta_L \sin 2\theta_T + \kappa^2\cos^2 2\theta_T \tag{3.4.15}$$

and

$$\sin\theta_T = \frac{1}{\kappa}\sin\theta_L. \tag{3.4.16}$$

Incident transverse wave

Next we consider the reflection of an incident transverse wave. As before, we assign $n = 0$ to the incident wave, so that

$$p_1^{(0)} = \sin\theta_0, \qquad p_2^{(0)} = \cos\theta_0,$$
$$d_1^{(0)} = -\cos\theta_0, \qquad d_2^{(0)} = \sin\theta_0,$$
$$c_0 = c_T.$$

In the plane $x_2 = 0$, the displacements and the stresses of the incident wave then are of the following forms.

$$u_1^{(0)} = -A_0\cos\theta_0\exp(i\bar{\eta}_0),$$
$$u_2^{(0)} = A_0\sin\theta_0\exp(i\bar{\eta}_0),$$
$$\tau_{22}^{(0)} = 2ik_0\mu\sin\theta_0\cos\theta_0\,A_0\exp(i\bar{\eta}_0),$$
$$\tau_{21}^{(0)} = ik_0\mu(\sin^2\theta_0 - \cos^2\theta_0)A_0\exp(i\bar{\eta}_0),$$

where

$$\bar{\eta} = k_0(x_1\sin\theta_0 - c_T t).$$

When the reflecting surface, $x_2 = 0$, is free of surface tractions, the conditions at $x_2 = 0$ can be satisfied only if both a reflected L wave and a reflected TV wave are generated. The reflection coefficients are found to be

$$R_L^T = -\frac{\kappa\sin 4\theta_0}{\sin 2\theta_0\sin 2\theta_1 + \kappa^2\cos^2 2\theta_0},$$

$$R_T^T = \frac{\sin 2\theta_0\sin 2\theta_1 - \kappa^2\cos^2 2\theta_0}{\sin 2\theta_0\sin 2\theta_1 + \kappa^2\cos^2 2\theta_0},$$

where θ_1 is the angle of the reflected longitudinal wave. Analogously to Eq. (3.4.13) and (3.4.14) the reflection coefficients may be written in terms of the angle of incidence θ_T:

$$R_L^T(\theta_T) = -\frac{\kappa\sin 4\theta_T}{R(\theta_L,\theta_T)}, \tag{3.4.17}$$

$$R_T^T(\theta_T) = \frac{\sin 2\theta_T\sin 2\theta_L - \kappa^2\cos^2 2\theta_T}{R(\theta_L,\theta_T)}, \tag{3.4.18}$$

where for a given θ_T the angle θ_L follows from Eq. (3.4.16).

A number of interesting observations can be made regarding the nature of the reflected waves on the basis of the expressions for the reflection coefficients, see Achenbach (1973) and Ewing, Jardetzky and Press (1957).

3.5 Energy partition

For a plane wave of type α the time-averaged rate of energy transmission per unit area is given by Eq. (3.3.5). Let us now consider a beam of plane waves of type α incident on the plane $x_2 = 0$. The cross-sectional area of the beam is ΔS_α. The corresponding beams of reflected waves of type γ ($\gamma =$ L, T) have cross-sectional areas ΔS_γ. The common intersection of the incident and reflected beams with the free surface is ΔS. Clearly,

$$\Delta S_\alpha = \Delta S \cos\theta_\alpha \qquad \text{and} \qquad \Delta S_\gamma = \Delta S \cos\theta_\gamma. \tag{3.5.1}$$

Since the surface area ΔS is free of tractions, and since no energy is dissipated, the time-averaged rate of energy transmission across ΔS_α must equal the sum of the time-averaged rate of energy transmissions across ΔS_γ ($\gamma =$ L, T). Hence

$$\tfrac{1}{2}\rho\omega^2 c_\alpha \Delta S_\alpha = \tfrac{1}{2}\rho\omega^2 c_L \left(R_L^\alpha\right)^2 \Delta S_L + \tfrac{1}{2}\rho\omega^2 c_T \left(R_T^\alpha\right)^2 \Delta S_T. \tag{3.5.2}$$

By the use of Eq. (3.5.1) we find

$$\left(R_L^\alpha\right)^2 \frac{c_L}{c_\alpha}\frac{\cos\theta_L}{\cos\theta_\alpha} + \left(R_T^\alpha\right)^2 \frac{c_T}{c_\alpha}\frac{\cos\theta_T}{\cos\theta_\alpha} = 1. \tag{3.5.3}$$

It may be checked that the previously derived reflection coefficients (3.4.13), (3.4.14) and (3.4.17), (3.4.18) satisfy (3.5.3). Equation (3.5.3) shows a partitioning of the incident energy flow over the two reflected waves.

3.6 Solving the wave equation for polar symmetry

In Section 2.6 it was shown that the generation and propagation of acoustic wave motion is governed by the classical wave equation, with or without inhomogeneous terms. The classical wave equation is one of the most important equations in classical physics, and in this section we examine some of its basic solutions for the case of polar symmetry. This case has much in common with wave propagation in one dimension.

For polar symmetry the equation for the acoustic potential, when wave motion is generated by a distribution of fluid injection sources of volume density $q(R,t)$, follows from Eq. (2.6.26) as

$$\frac{1}{R^2}\frac{\partial}{\partial R}\left(R^2\frac{\partial\varphi}{\partial R}\right) - \frac{1}{c^2}\frac{\partial^2\varphi}{\partial t^2} = q(R,t). \tag{3.6.1}$$

In Eq. (3.6.1), R is the radial coordinate, i.e., the distance from the origin.

Let us first consider the homogeneous form of Eq. (3.6.1). The general solution to that equation can be obtained by introducing the representation

$$\varphi(R,t) = \frac{1}{R}\Phi(R,t). \tag{3.6.2}$$

Substitution of this expression for $\varphi(R,t)$ in the homogeneous wave equation produces a one-dimensional wave equation for $\Phi(R,t)$:

$$\frac{\partial^2 \Phi}{\partial R^2} - \frac{1}{c^2}\frac{\partial^2 \Phi}{\partial t^2} = 0. \tag{3.6.3}$$

The general solution to this equation was derived in Section 1.4. It follows that $\varphi(R,t)$ is of the form

$$\varphi(R,t) = \frac{1}{R}F\left(t - \frac{R}{c}\right) + \frac{1}{R}G\left(t + \frac{R}{c}\right). \tag{3.6.4}$$

The first term in Eq. (3.6.4) represents a spherical wave propagating away from the origin with an amplitude decay governed by $1/R$. The second term represents a wave propagating towards the origin with an increasing amplitude governed by $1/R$.

Next we consider the case where the source distribution on the right-hand side of Eq. (3.6.1) is of the form

$$q(\boldsymbol{x},t) = q(t)\delta(\boldsymbol{x}). \tag{3.6.5}$$

By using an analogous representation of the delta function in terms of R, so that

$$\delta(\boldsymbol{x}) = \frac{\delta(R)}{4\pi R^2}, \tag{3.6.6}$$

we can write Eq. (3.6.1) as

$$\frac{1}{R^2}\frac{\partial}{\partial R}\left(R^2\frac{\partial \varphi}{\partial R}\right) - \frac{1}{c^2}\frac{\partial^2 \varphi}{\partial t^2} = q(t)\frac{\delta(R)}{4\pi R^2}. \tag{3.6.7}$$

The well-known solution to this equation is

$$\varphi(R,t) = -\frac{1}{4\pi R}q\left(t - \frac{R}{c}\right). \tag{3.6.8}$$

By virtue of Eq. (2.6.23) the corresponding particle velocity is

$$v_R(R,t) = \frac{1}{4\pi R^2}q\left(t - \frac{R}{c}\right) + \frac{1}{4\pi R}\frac{1}{c}q'\left(t - \frac{R}{c}\right), \tag{3.6.9}$$

where the prime indicates a derivative with respect to the argument.

The function $\varphi(R, t)$ given by Eq. (3.6.8) defines the fundamental solution of the wave equation. It is of interest to evaluate the integral

$$\lim_{\varepsilon \to 0} \iint v_R(R, t)\, dR \tag{3.6.10}$$

over the spherical surface of radius ε. By the use of Eq. (3.6.9) it is found that this limit, which represents the rate at which material enters the domain, is equal to $q(t)$, as it should be from the definition of $q(t)$.

For the case that the point source is time harmonic, i.e.,

$$q(\boldsymbol{x}, t) = Q e^{-i\omega t} \delta(\boldsymbol{x}), \tag{3.6.11}$$

we have

$$\varphi(R, t) = -\frac{Q}{4\pi R} e^{-i\omega(t - R/c)}. \tag{3.6.12}$$

The solutions derived in this section are basic singular solutions, i.e., Green's functions. By superposition they can be used to derive solutions for arbitrary source distributions.

3.7 The Helmholtz decomposition

The representation of a vector $\boldsymbol{U}$ by

$$\boldsymbol{U} = \nabla P + \nabla \wedge \boldsymbol{Q}, \tag{3.7.1}$$

where

$$\nabla \cdot \boldsymbol{Q} = 0$$

is called the Helmholtz decomposition of the vector $\boldsymbol{U}$. The construction of the scalar P and the vector $\boldsymbol{Q}$ in terms of the vector $\boldsymbol{U}$ is discussed in this section.

Let the vector $\boldsymbol{U}(\boldsymbol{x})$ be piecewise differentiable in a finite open region V of space. With each point of space we now associate the vector

$$\boldsymbol{W}(\boldsymbol{x}) = -\frac{1}{4\pi} \iiint_V \frac{\boldsymbol{U}(\boldsymbol{\xi})}{|\boldsymbol{x} - \boldsymbol{\xi}|}\, dV_\xi, \tag{3.7.2}$$

where

$$dV_\xi = d\xi_1\, d\xi_2\, d\xi_3$$

and

$$|\boldsymbol{x} - \xi| = \left[(x_1 - \xi_1)^2 + (x_2 - \xi_2)^2 + (x_3 - \xi_3)^2\right]^{1/2}.$$

It is well known that at interior points where $\boldsymbol{U}$ is continuous, $\boldsymbol{W}(\boldsymbol{x})$ then satisfies the vector equation

$$\nabla^2 \boldsymbol{W} = \boldsymbol{U}(\boldsymbol{x}), \tag{3.7.3}$$

while we have

$$\nabla^2 \boldsymbol{W} = 0$$

at points outside the region V.

Now we employ the vector identity

$$\nabla^2 \boldsymbol{W} = \nabla\nabla \cdot \boldsymbol{W} - \nabla \wedge \nabla \wedge \boldsymbol{W}. \tag{3.7.4}$$

Let $\boldsymbol{W}(\boldsymbol{x})$ be as defined by Eq. (3.7.2). By using (3.7.3), Eq. (3.7.4) can then be rewritten as

$$\boldsymbol{U} = \nabla[\nabla \cdot \boldsymbol{W}] + \nabla \wedge [-\nabla \wedge \boldsymbol{W}]. \tag{3.7.5}$$

If we set

$$P = \nabla \cdot \boldsymbol{W}, \tag{3.7.6}$$

$$\boldsymbol{Q} = -\nabla \wedge \boldsymbol{W}, \tag{3.7.7}$$

then Eq. (3.7.5) becomes

$$\boldsymbol{U} = \nabla P + \nabla \wedge \boldsymbol{Q}. \tag{3.7.8}$$

It can be shown that P and $\boldsymbol{Q}$ are everywhere definite and continuous and are differentiable at interior points where $\boldsymbol{U}$ is continuous.

To prove the Helmholtz decomposition we have thus provided a recipe for the construction of P and $\boldsymbol{Q}$. Given the vector field $\boldsymbol{U}(\boldsymbol{x})$, the vector $W(\boldsymbol{x})$ can be constructed from (3.7.2), whereupon $P(\boldsymbol{x})$ and $\boldsymbol{Q}(\boldsymbol{x})$ are obtained from (3.7.6) and (3.7.7), respectively. Since the divergence of a curl vanishes we observe that

$$\nabla \cdot \boldsymbol{Q} = 0. \tag{3.7.9}$$

The Helmholtz decomposition is also valid for an infinite domain provided that $|\boldsymbol{U}| = O(R^{-2})$, i.e., provided that $|\boldsymbol{U}|$ decreases to zero at large distances r from the origin at least as rapidly as a constant times R^{-2}.

3.8 Displacement potentials

Next we consider displacement potentials. The waves discussed in the previous sections are free waves that can be investigated without specific reference to a source mechanism. In actuality wave motions are generated by excitation at the external boundaries of a body or by forces applied in the interior. When forces are applied to the boundary, the displacement equations of motion form a complicated system of homogeneous partial differential equations with boundary conditions. This system of governing equations can be simplified by the use of displacement potentials.

First, we rewrite the governing equations in vector notation as

$$\mu\nabla^2\boldsymbol{u} + (\lambda + \mu)\nabla\nabla\cdot\boldsymbol{u} = \rho\ddot{\boldsymbol{u}}. \tag{3.8.1}$$

Let us now consider the displacement vector in the form

$$\boldsymbol{u} = \nabla\psi^L + \nabla\wedge\boldsymbol{\psi}^T \tag{3.8.2}$$

where ψ^L is a scalar potential and $\boldsymbol{\psi}^T$ is a vector potential. Substitution of (3.8.2) into (3.8.1) yields

$$\begin{aligned}\mu\nabla^2[\nabla\psi^L + \nabla\wedge\boldsymbol{\psi}^T] &+ (\lambda+\mu)\nabla\nabla\cdot[\nabla\psi^L + \nabla\wedge\boldsymbol{\psi}^T]\\ &= \rho\frac{\partial^2}{\partial t^2}[\nabla\psi^L + \nabla\wedge\boldsymbol{\psi}^T].\end{aligned}$$

Since $\nabla\cdot\nabla\psi^L = \nabla^2\psi^L$ and $\nabla\cdot\nabla\wedge\boldsymbol{\psi}^T = 0$, we obtain upon rearranging terms

$$\nabla[(\lambda+2\mu)\nabla^2\psi^L - \rho\ddot{\psi}^L] + \nabla\wedge[\mu\nabla^2\boldsymbol{\psi}^T - \rho\ddot{\boldsymbol{\psi}}^T] = 0.$$

Clearly, the displacement representation (3.8.2) satisfies the equation of motion (3.8.1) if

$$\nabla^2\psi^L = \frac{1}{c_L^2}\ddot{\psi}^L \tag{3.8.3}$$

and

$$\nabla^2\boldsymbol{\psi}^T = \frac{1}{c_T^2}\ddot{\boldsymbol{\psi}}^T, \tag{3.8.4}$$

where c_L and c_T are the velocities of longitudinal and transverse waves, given by (3.2.4) and (3.2.5), respectively. Equations (3.8.3) and (3.8.4) are uncoupled wave equations.

Although the scalar potential ψ^L and the components of the vector potential $\boldsymbol{\psi}^T$ are generally coupled through the boundary conditions, which still causes substantial mathematical complications, the use of the displacement decomposition simplifies the analysis. To determine the solution to a boundary-initial value problem one may simply select appropriate particular solutions of Eqs. (3.8.3) and (3.8.4). If these solutions can subsequently be chosen such that the boundary conditions and the initial conditions for the elastic body are satisfied, then the problem has been solved. The solution so obtained is unique by virtue of the uniqueness theorem of elastodynamic theory; see Achenbach (1973).

It should be noted that Eq. (3.8.2) relates the three components of the displacement vector to four other functions: the scalar potential and the three components of the vector potential. This indicates that ψ^L and the components of $\boldsymbol{\psi}^T$ should be subjected to an additional constraint. Generally the components of $\boldsymbol{\psi}^T$ are taken to be related in some manner. Usually, but not always, the relation

$$\nabla \cdot \boldsymbol{\psi}^T = 0 \tag{3.8.5}$$

is taken as the additional constraint. This relation has the advantage that it is consistent with the Helmholtz decomposition of the displacement vector.

3.9 Wave motion generated by a point load

Distributions of body forces are an important source mechanism for wave motion. Of particular interest is the wave motion generated by a single point force. For the purposes of this section we define the components of a point force by

$$f_i(\boldsymbol{x}, t) = \delta(\boldsymbol{x}) g(t)\, a_i \tag{3.9.1}$$

where $\delta(\boldsymbol{x})$ is the three-dimensional Dirac delta function and $g(t)$ defines the time dependence of the force. The force is applied at the origin of the coordinate system, and it acts in the direction of the unit vector $\boldsymbol{a}$.

The stresses and displacements generated by a point force of time dependence $g(t) = \delta(t)$ comprise the basic singular solution of the field equations for elastodynamics. The displacements generated by a point force of arbitrary time dependence were first worked out by Stokes (1849); see also Love (1892). Explicit expressions can also be found elsewhere; see, e.g., Achenbach (1973).

Here we consider the special case of steady-state time-harmonic motions, i.e.,

$$g(t) = F_0 e^{-i\omega t}. \tag{3.9.2}$$

The magnitude of F_0 is generally taken as unity; this symbol is introduced to keep track of the dimensions. The dimension of F_0 is that of force.

The displacement equations of motion, in the presence of a system of body forces per unit volume, are given by Eq. (2.2.20). For the steady-state case, they become, in vector notation,

$$\mu\nabla^2\boldsymbol{u} + (\lambda+\mu)\nabla\nabla\cdot\boldsymbol{u} + \boldsymbol{f} + \rho\omega^2\boldsymbol{u} = 0.$$

In accordance with Eq. (3.7.1) we will decompose the body-force vector into a scalar potential $F(\boldsymbol{x})$ and a vector potential $G(x)$:

$$f(\boldsymbol{x}) = (\lambda+2\mu)\nabla F(\boldsymbol{x}) + \mu\nabla\wedge\boldsymbol{G}(\boldsymbol{x}).$$

The decomposition can be achieved by using the Helmholtz decomposition. By employing the results of the previous section, and because the volume integral over the Dirac delta function is unity, we find

$$(\lambda+2\mu)F(\boldsymbol{x}) = -F_0\nabla\cdot\left(\frac{\boldsymbol{a}}{4\pi R}\right),$$

$$\mu\boldsymbol{G}(\boldsymbol{x}) = F_0\nabla\wedge\left(\frac{\boldsymbol{a}}{4\pi R}\right),$$

where

$$R^2 = x_1^2 + x_2^2 + x_3^2. \tag{3.9.3}$$

Writing the displacement vector in the form given by Eq. (3.8.2), it follows analogously to the development of Eqs. (3.8.3), (3.8.4) that $\psi^L(x)$ and $\boldsymbol{\psi}^T(x)$ must satisfy the inhomogeneous wave equations

$$\nabla^2\psi^L + k_L^2\psi^L = \frac{F_0}{\lambda+2\mu}\nabla\cdot\left(\frac{\boldsymbol{a}}{4\pi R}\right), \tag{3.9.4}$$

$$\nabla^2\boldsymbol{\psi}^T + k_T^2\boldsymbol{\psi}^T = -\frac{F_0}{\mu}\nabla\wedge\left(\frac{\boldsymbol{a}}{4\pi R}\right), \tag{3.9.5}$$

where

$$k_L = \omega/c_L \qquad \text{and} \qquad k_T = \omega/c_T. \tag{3.9.6}$$

For convenience we define

$$\psi^L = \nabla\cdot\boldsymbol{\Psi}^L \qquad \text{and} \qquad \boldsymbol{\psi}^T = -\nabla\wedge\boldsymbol{\Psi}^T. \tag{3.9.7}$$

By setting

$$\boldsymbol{\Psi}^L = \Psi^L\boldsymbol{a} \qquad \text{and} \qquad \boldsymbol{\Psi}^T = \Psi^T\boldsymbol{a}, \tag{3.9.8}$$

Eqs. (3.9.4) and (3.9.5) are satisfied if Ψ^L and Ψ^T are solutions of the following inhomogeneous Helmholtz equations:

$$\nabla^2\Psi^L + k_L^2\Psi^L = \frac{F_0}{\lambda + 2\mu}\frac{1}{4\pi R}, \tag{3.9.9}$$

$$\nabla^2\Psi^T + k_T^2\Psi^T = \frac{F_0}{\mu}\frac{1}{4\pi R}, \tag{3.9.10}$$

where R is defined by Eq. (3.9.3).

Appropriate solutions of these equations are readily derived. Let us first consider Eq. (3.9.9) for the case where the right-hand side is a delta function:

$$\nabla^2\Psi^L + k_L^2\Psi^L = -\delta(\boldsymbol{x}).$$

The solution to this equation has been derived in Section 3.6 as

$$\Psi^L = \frac{e^{ik_L R}}{4\pi R}. \tag{3.9.11}$$

When $k_L \equiv 0$, i.e., when

$$\nabla^2\Psi^L = -\delta(\boldsymbol{x}), \tag{3.9.12}$$

we have

$$\Psi^L = \frac{1}{4\pi R}. \tag{3.9.13}$$

This information is sufficient to construct an outgoing wave solution of (3.9.9). First we consider a particular solution of the form

$$\Psi^L = \frac{1}{k_L^2}\frac{F_0}{\lambda + 2\mu}\frac{1}{4\pi R}.$$

However, according to (3.9.12) and (3.9.13) this expression would generate a term of the form

$$-\frac{1}{k_L^2}\frac{F_0}{\lambda + 2\mu}\delta(\boldsymbol{x})$$

on the right-hand side of Eq. (3.9.9). Adding a term of the form given by Eq. (3.9.11), namely,

$$\Psi^L = -\frac{1}{k_L^2}\frac{F_0}{\lambda + 2\mu}\frac{e^{ik_L R}}{4\pi R},$$

will eliminate this delta function. Thus, the expression that satisfies Eq. (3.9.9) is

$$\Psi^L = \frac{1}{k_L^2}\frac{F_0}{\lambda + 2\mu}\left(-\frac{e^{ik_L R}}{4\pi R} + \frac{1}{4\pi R}\right).$$

In general terms we write the solutions of (3.9.9) and (3.9.10) as

$$\Psi^\gamma = \frac{F_0}{\rho\omega^2}\left[-G(k_\gamma R) + \frac{1}{4\pi R}\right], \tag{3.9.14}$$

where

$$G(k_\gamma R) = \frac{e^{ik_\gamma R}}{4\pi R}, \tag{3.9.15}$$

and γ is either L or T.

In view of (3.8.2) and (3.9.7) the displacement vector can now be written as

$$\boldsymbol{u} = \nabla\nabla\cdot\boldsymbol{\Psi}^L - \nabla\wedge\nabla\wedge\boldsymbol{\Psi}^T.$$

By virtue of the vector identity

$$\nabla^2\boldsymbol{\Psi}^T = \nabla\nabla\cdot\boldsymbol{\Psi}^T - \nabla\wedge\nabla\wedge\boldsymbol{\Psi}^T,$$

the displacement vector can also be expressed as

$$\boldsymbol{u} = \nabla\nabla\cdot(\boldsymbol{\Psi}^L - \boldsymbol{\Psi}^T) + \nabla^2\boldsymbol{\Psi}^T. \tag{3.9.16}$$

Using (3.9.8) and (3.9.14) we find

$$\boldsymbol{\Psi}^L - \boldsymbol{\Psi}^T = \frac{F_0}{\rho\omega^2}[-G(k_L R) + G(k_T R)]\,\boldsymbol{a}, \tag{3.9.17}$$

while (3.9.10), (3.9.8) and (3.9.14) yield

$$\nabla^2\boldsymbol{\Psi}^T = \frac{F_0}{\mu}G(k_T R)\,\boldsymbol{a}. \tag{3.9.18}$$

It follows from (3.9.17) that

$$\nabla\cdot(\boldsymbol{\Psi}^L - \boldsymbol{\Psi}^T) = \frac{F_0}{\rho\omega^2}\frac{\partial}{\partial x_k}[-G(k_L R) + G(k_T R)]a_k,$$

and

$$\nabla\nabla\cdot(\boldsymbol{\Psi}^L - \boldsymbol{\Psi}^T) = \frac{F_0}{\rho\omega^2}\frac{\partial}{\partial x_i}\frac{\partial}{\partial x_k}[-G(k_L R) + G(k_T R)]a_k. \tag{3.9.19}$$

Thus, substituting (3.9.18) and (3.9.19) into (3.9.16), we see that the components of the displacement field generated by the point force depend linearly on the components of $\boldsymbol{a}$. To express this dependence we introduce a tensor of rank two, $u^G_{i;k}$, and a tensor of rank three, $\tau^G_{ij;k}$, that relate u^G_i and τ^G_{ij} to a_k through

$$u^G_i = u^G_{i;k} a_k \qquad \text{and} \qquad \tau^G_{ij} = \tau^G_{ij;k} a_k. \tag{3.9.20}$$

The superscript G is included to indicate that the displacement $u^G_{i;k}$ is often called the Green displacement tensor. Using this notation, $u^G_{i;k}$ is the displacement in the i-direction at position $\boldsymbol{x}$ due to a force in the k-direction applied at position $\boldsymbol{X}$. We rewrite Eq. (3.9.16) as

$$\frac{\mu}{F_0} u^G_{i;k} = \frac{1}{k_T^2}[-G(k_L R) + G(k_T R)],_{ik} + G(k_T R)\delta_{ik} \tag{3.9.21}$$

where $G(k_\gamma R)$ is defined by Eq. (3.9.15) and R is now defined as

$$R^2 = (x_1 - X_1)^2 + (x_2 - X_2)^2 + (x_3 - X_3)^2. \tag{3.9.22}$$

The result given by Eq. (3.9.21) can be found in the book by Kupradze (1963). The corresponding expression for $\tau^G_{ij;k}$ follows from (3.9.21) and Hooke's law:

$$\frac{1}{F_0}\tau^G_{ij;k} = \left(1 - \frac{2}{\kappa^2}\right)[G(k_L R)],_k \delta_{ij} - \frac{2}{k_T^2}[G(k_L R) - G(k_T R)],_{ijk}$$
$$+ [G(k_T R)],_j \delta_{ik} + [G(k_T R)],_i \delta_{jk} \tag{3.9.23}$$

where $\kappa = c_L/c_T$ is defined by (3.2.6).

An alternate way of deriving the expression for $u^G_{i;k}$ by use of the exponential Fourier transform can be found in the book by Achenbach, Gautesen and McMaken (1982).

For future reference we provide the following detailed expressions for $u^G_{i;k}$ and $\tau^G_{ij;k}$:

$$\frac{\mu}{F_0} u^G_{i;k}(\boldsymbol{x}; \boldsymbol{X}; \omega) = (U_1 \delta_{ik} - U_2 R,_i R,_k) \tag{3.9.24}$$

where R is defined by Eq. (3.9.22) and $R,_i = \partial R/\partial x_i$. Also,

$$U_1 = G(k_T R) + \left(\frac{1}{k_T R} - \frac{1}{(k_T R)^2}\right) G(k_T R)$$
$$- \left(\frac{k_L}{k_T}\right)^2 \left(\frac{i}{k_T R} - \frac{1}{(k_T R)^2}\right) G(k_L R), \tag{3.9.25}$$

$$U_2 = \left(1 + \frac{3i}{k_T R} - \frac{3}{(k_T R)^2}\right) G(k_T R),$$
$$- \left(\frac{k_L}{k_T}\right)^2 \left(1 + \frac{3i}{k_L R} - \frac{3}{(k_L R)^2}\right) G(k_L R), \tag{3.9.26}$$

where $G(k_\gamma R)$ is defined by Eq. (3.9.15). The corresponding expression for $(1/\mu)\tau^G_{ij;k}$ can be written as

$$\begin{aligned}\frac{1}{F_0}\tau^G_{ij;k}(x;X;\omega) = &\left(\frac{\lambda}{\mu}R_{,k}\,\delta_{ij} + R_{,j}\,\delta_{ik} + R_{,i}\,\delta_{jk}\right)\frac{dU_1}{dR}\\ &-\left(\frac{\lambda}{\mu}R_{,k}\,\delta_{ij} + 2R_{,i}\,R_{,j}\,R_{,k}\right)\frac{dU_2}{dR}\\ &-\left[2\frac{\lambda}{\mu}R_{,k}\,\delta_{ij} + 2(\delta_{ij} - 2R_{,i}\,R_{,j})R_{,k}\right.\\ &\left.+\ (R_{,i}\,\delta_{jk} + R_{,j}\,\delta_{ik})\right]\frac{U_2}{R}. \end{aligned} \tag{3.9.27}$$

In deriving Eq. (3.9.27) we have taken into account that

$$\frac{\partial^2 R}{\partial x_j x_k} = \frac{1}{R}(\delta_{jk} - R_{,j}\,R_{,k}).$$

Expressions (3.9.24)–(3.9.26) agree with expressions listed by Kobayashi (1987).

3.10 Expressions for the far field

For many applications it is useful to have far-field expressions for the displacements and stresses generated by a point force. These can be obtained by approximations of the following form:

$$|\boldsymbol{x} - \boldsymbol{X}| \approx |\boldsymbol{x}| - (\hat{\boldsymbol{x}} \cdot \boldsymbol{X}),$$

where $\hat{\boldsymbol{x}}$ is a unit vector in the direction of $\boldsymbol{x}$, i.e., $\hat{\boldsymbol{x}} = \boldsymbol{x}/\,|\boldsymbol{x}|$, and

$$|\boldsymbol{x}| = \left(x_1^2 + x_2^2 + x_3^2\right)^{1/2}.$$

Based on these approximations the following simplifications may be used in the far field:

$$\begin{aligned}\bar{G}(k_\gamma R) &\approx \frac{e^{ik_\gamma|\boldsymbol{x}|}}{4\pi\,|\boldsymbol{x}|}e^{-ik_\gamma\hat{\boldsymbol{x}}\cdot\boldsymbol{X}} = G(k_\gamma\,|x|)e^{-ik_\gamma\hat{\boldsymbol{x}}\cdot\boldsymbol{X}},\\ R_{,i} &\approx \hat{\boldsymbol{x}}_i,\\ U_1 &\approx \bar{G}(k_T R),\\ U_2 &\approx \bar{G}(k_T R) - \left(\frac{k_L}{k_T}\right)^2\bar{G}(k_L R),\\ \frac{dU_1}{dR} &\approx ik_T\bar{G}(k_T R),\end{aligned}$$

$$\frac{U_2}{R} \approx 0,$$

$$\frac{dU_2}{dR} \approx ik_T \bar{G}(k_T R) - \left(\frac{k_L}{k_T}\right)^2 ik_L \bar{G}(k_L R).$$

By using these simplified expressions we obtain

$$u_{i;k}^{G} = u_{i;k}^{G;L} + u_{i;k}^{G;T} \tag{3.10.1}$$

where

$$u_{i;k}^{G;L} \approx A_{i;k}^{G;L} \frac{\exp(ik_L\,|\boldsymbol{x}|}{4\pi\,|\boldsymbol{x}|} \exp(-ik_L \hat{\boldsymbol{x}} \cdot \boldsymbol{x}), \tag{3.10.2}$$

$$u_{i;k}^{G;T} \approx A_{i;k}^{G;T} \frac{\exp(ik_T\,|\boldsymbol{x}|}{4\pi\,|\boldsymbol{x}|} \exp(-ik_T \hat{\boldsymbol{x}} \cdot \boldsymbol{x}). \tag{3.10.3}$$

In these expressions, we have for the amplitudes

$$A_{i;k}^{G;L} = \frac{F_0}{\lambda + 2\mu} \hat{x}_i \hat{x}_k, \tag{3.10.4}$$

$$A_{i;k}^{G;T} = \frac{F_0}{\mu} (\delta_{ik} - \hat{x}_i \hat{x}_k). \tag{3.10.5}$$

It should be recalled that the subscript i refers to the displacement direction, while the subscript k refers to the direction of the point force. It can be verified that the tensors $A_{i;k}^{G;L}$ and $A_{i;k}^{G;T}$ satisfy the following properties:

$$A_{i;k}^{G;L} = \left[A_{j;k}^{G;L} \hat{x}_j \right] \hat{x}_i,$$

$$A_{i;k}^{G;T} \hat{x}_i = 0.$$

These results imply that the first term in Eq. (3.10.1) represents a longitudinal wave in the direction of $\boldsymbol{x}$, while the second term represents a transverse wave in the direction of $\boldsymbol{x}$ with displacement normal to $\boldsymbol{x}$.

The far-field stresses corresponding to (3.9.1) are obtained from (3.9.27). We find

$$\tau_{ij;k}^{G} = \tau_{ij;k}^{G;L} + \tau_{ij;k}^{G;T} \tag{3.10.6}$$

where

$$\tau_{ij;k}^{G;L} \approx ik_L B_{ij;k}^{G;L} \frac{\exp(ik_L\,|\boldsymbol{x}|)}{4\pi\,|\boldsymbol{x}|} \exp(-ik_L \hat{\boldsymbol{x}} \cdot \boldsymbol{X}), \tag{3.10.7}$$

$$\tau_{ij;k}^{G;T} \approx ik_T B_{ij;k}^{G;T} \frac{\exp(ik_T\,|\boldsymbol{x}|)}{4\pi\,|\boldsymbol{x}|} \exp(-ik_T \hat{\boldsymbol{x}} \cdot \boldsymbol{X}) \tag{3.10.8}$$

and

$$B_{ij;k}^{G;L} = F_0[2\kappa^{-2}\hat{x}_i\hat{x}_j + (1 - 2\kappa^{-2})\delta_{ij}]\hat{x}_k, \tag{3.10.9}$$

$$B_{ij;k}^{G;T} = F_0(\delta_{ik}\hat{x}_j + \delta_{jk}\hat{x}_i - 2\hat{x}_i\hat{x}_j\hat{x}_k). \tag{3.10.10}$$

It can also be verified that

$$B_{ij;k}^{G;L}\hat{x}_j = \hat{x}_i\hat{x}_k F_0 = (\lambda + 2\mu)A_{i;k}^{G;L}, \tag{3.10.11}$$

$$B_{ij;k}^{G;T}\hat{x}_j = (\delta_{ik} - \hat{x}_i\hat{x}_k)F_0 = \mu A_{i;k}^{G;T} \tag{3.10.12}$$

It then follows from Eqs. (3.10.1)–(3.10.3) and (3.10.7), (3.10.8) that

$$\tau_{ij;k}^{G;L}\hat{x}_j = ik_L(\lambda + 2\mu)u_{i;k}^{G;L}, \tag{3.10.13}$$

$$\tau_{ij;k}^{G;T}\hat{x}_j = ik_T\mu u_{i;k}^{G;T} \tag{3.10.14}$$

These relations are just the same as for plane waves.

Thus, at distances from the point of load application that are large compared with the wavelength, the point-force solution breaks up into two separate waves that locally show the same general form as plane longitudinal and transverse waves, respectively.

3.11 Wave motion generated by a line load

The two-dimensional basic singular solution is the wave field generated by a line load. A time-harmonic line load applied at $X = (X_1, X_2)$ and over the whole range of x_3 may be expressed in the form

$$f_i = a_i\bar{F}_0 e^{-i\omega t}\delta(x_1 - X_1)\delta(x_2 - X_2). \tag{3.11.1}$$

It may be noted that $\bar{F}_0$ has the dimension of force/length.

The wave field generated by f_i is of course independent of the x_3- coordinate. Even though the line load may be applied in an arbitrary direction, it can be verified that the fields generated by the in-plane and the out-of-plane components of the line load are uncoupled. Thus, it is convenient to consider these cases separately. The direction of the line load for the in-plane case is defined by

$$\mathbf{a} = a_1\mathbf{i}_1 + a_2\mathbf{i}_2, \qquad \text{where} \qquad a_1^2 + a_2^2 = 1.$$

For the anti-plane case we have

$$\mathbf{a} = \mathbf{i}_3.$$

Since a line load can be considered as a superposition of point forces, the wave field generated by the line load can be expressed as a superposition integral of the form

$$\int_{-\infty}^{\infty} u_{i;k}^{G}(\boldsymbol{x}, \boldsymbol{X})\, dX_3, \tag{3.11.2}$$

where $u_{i;k}^{G}(\boldsymbol{x}, \boldsymbol{X})$ is the solution for the point-force problem given by (3.9.21). A typical integral in (3.11.2) is

$$J_\gamma = \int_{-\infty}^{\infty} G(k_\gamma R)\, dX_3,$$

where $G(k_\gamma R)$ is defined by (3.9.15),

$$R^2 = r^2 + (x_3 - X_3)^2, \tag{3.11.3}$$

and

$$r^2 = (x_1 - X_1)^2 + (x_2 - X_2)^2. \tag{3.11.4}$$

By introducing the new variable $s = x_3 - X_3$, J_γ can be rewritten as

$$J_\gamma = \frac{1}{4\pi} \int_{-\infty}^{\infty} \frac{\exp\left[ik_\gamma (r^2 + s^2)^{1/2}\right]}{(r^2 + s^2)^{1/2}} ds.$$

This integral can be recognized as a Hankel function:

$$J_\gamma = \frac{i}{4} H_0^{(1)}(k_\gamma r).$$

It follows that the solution to the line-load problem can be obtained from the solution for a point force on replacing $G(k_\gamma R)$ by $\frac{1}{4} i H_0^{(1)}(k_\gamma r)$.
Hence, for $k = 1, 2$ we find that

$$\begin{aligned} -(4i\mu/\bar{F}_0) u_{i;k}^{G} &= k_T^{-2}\left[-H_0^{(1)}(k_L r) + H_0^{(1)}(k_T r)\right]_{,ik} \\ &\quad + H_0^{(1)}(k_T r)\delta_{ik}, \end{aligned} \tag{3.11.5}$$

$$\begin{aligned} -(4i/\bar{F}_0)\tau_{ij;k}^{G} &= (1 - 2\kappa^{-2})\left[H_0^{(1)}(k_L r)\right]_{,k} \delta_{ij} \\ &\quad - 2k_T^{-2}\left[H_0^{(1)}(k_L r) - H_0^{(1)}(k_T r)\right]_{,ijk} \\ &\quad + \left[H_0^{(1)}(k_T r)\right]_{,j} \delta_{ik} + \left[H_0^{(1)}(k_T r)\right]_{,i} \delta_{jk}. \end{aligned} \tag{3.11.6}$$

It may be recalled that $\bar{F}_0$ has the dimension of force/length.

When the line-load is acting in the x_3-direction, i.e., $\boldsymbol{a} = \boldsymbol{i}_3$, we find

$$-(4i\mu/\bar{F}_0) u_{3;3}^{G} = H_0^{(1)}(k_T r), \tag{3.11.7}$$

$$-(4i/\bar{F}_0)\tau_{ij;3}^{G} = \left[H_0^{(1)}(k_T r)\right]_{,j} \delta_{i3} + \left[H_0^{(1)}(k_T r)\right]_{,i} \delta_{j3}. \tag{3.11.8}$$

We note that as expected $u^G_{3;k} \equiv 0$ for $k = 1, 2$ and $u^G_{i;3} \equiv 0$ for $i = 1, 2$.

It is of interest to list the far-field expressions, which are valid when $|\boldsymbol{x}| \gg |\boldsymbol{X}|$. They are of the forms (3.10.1) and (3.10.6), where

$$u^{G;L}_{i;k} = A^{G;L}_{i;k}(\hat{\boldsymbol{x}}) \frac{\exp(ik_L\,|\boldsymbol{x}|)}{(8\pi k_L\,|\boldsymbol{x}|)^{1/2}} \exp\left[i\left(\frac{\pi}{4} - k_L \hat{\boldsymbol{x}} \cdot \boldsymbol{X}\right)\right], \tag{3.11.9}$$

$$u^{G;T}_{i;k} = A^{G;T}_{i;k}(\hat{\boldsymbol{x}}) \frac{\exp(ik_T\,|\boldsymbol{x}|)}{(8\pi k_T\,|\boldsymbol{x}|)^{1/2}} \exp\left[i\left(\frac{\pi}{4} - k_T \hat{\boldsymbol{x}} \cdot \boldsymbol{X}\right)\right] \tag{3.11.10}$$

$$\tau^{G;L}_{ij;k} = ik_L B^{G;L}_{ij;k}(\hat{\boldsymbol{x}}) \frac{\exp(ik_L\,|\boldsymbol{x}|)}{(8\pi k_L\,|x|)^{1/2}} \exp\left[i\left(\frac{\pi}{4} - k_L \hat{\boldsymbol{x}} \cdot \boldsymbol{X}\right)\right] \tag{3.11.11}$$

$$\tau^{G;T}_{ij;k} = ik_T B^{G;T}_{ij;k}(\hat{\boldsymbol{x}}) \frac{\exp(ik_T\,|\boldsymbol{x}|)}{(8\pi k_T\,|\boldsymbol{x}|)^{1/2}} \exp\left[i\left(\frac{\pi}{4} - k_T \hat{\boldsymbol{x}} \cdot \boldsymbol{X}\right)\right] \tag{3.11.12}$$

In these expressions $k = 1, 2, 3$,

$$|x| = \left(x_1^2 + x_2^2\right)^{1/2}$$

and

$$\hat{\boldsymbol{x}} = \frac{(x_1, x_2, 0)}{|\boldsymbol{x}|}$$

is a unit vector in the direction of $\boldsymbol{x}$. The constants $A^{G;\gamma}_{i;k}(\hat{\boldsymbol{x}})$ and $B^{G;\gamma}_{i;k}(\hat{\boldsymbol{x}})$ follow from (3.10.4), (3.10.5) and (3.10.9), (3.10.10). Since $\hat{\boldsymbol{x}}_3 \equiv 0$, we find for the in-plane case ($k = 1, 2$)

$$A^{G;\gamma}_{3;k} = 0, \qquad \gamma = L, T,$$

$$B^{G;L}_{i3;k} = B^{G;L}_{3i;k} = (1 - 2\kappa^{-2})\bar{F}_0 \delta_{3i} \hat{x}_k$$

$$B^{G;T}_{i3;k} = B^{G;T}_{3i;k} = 0,$$

while for the anti-plane case ($k = 3$) we find

$$A^{G;L}_{i;3} = B^{G;L}_{ij;3} = 0,$$

$$A^{G;T}_{i;3} = \mu^{-1} \bar{F}_0 \delta_{3i},$$

$$B^{G;T}_{ij;3} = \bar{F}_0(\delta_{3i}\hat{x}_j + \delta_{3j}\hat{x}_i).$$

4

Reciprocity in acoustics

4.1 Introduction

The first reciprocity relation specifically for acoustics was stated by von Helmholtz (1859). This relation caught the attention of and was elaborated by Rayleigh (1873) and Lamb (1888). Rayleigh (1873) briefly discussed the reciprocal theorem for acoustics in his paper "Some general theorems relating to vibrations." In *The Theory of Sound* (1878, Dover reprint 1945, Vol. II, pp. 145–8), Rayleigh paraphrased this theorem as follows: "If in a space filled with air which is partly bounded by finitely extended fixed bodies and is partly unbounded, sound waves may be excited at any point A, the resulting velocity potential at a second point B is the same both in magnitude and phase, as it would have been at A, had B been the source of sound." In this statement it is implicitly assumed that sources of the same strength would be applied at both places. In Rayleigh's book the statement is accompanied by a simple proof. A similar statement of the Helmholtz reciprocity theorem for acoustics can be found in the paper by Lamb (1888). Both Rayleigh and Lamb generalized the theorem to more complicated configurations, and in time the reciprocity theorem became known as Rayleigh's reciprocity theorem.

Most books on acoustics devote attention to the reciprocity theorem; see for example Pierce (1981), Morse and Ingard (1968), Jones (1986), Dowling and Ffowcs Williams (1983) and Crighton *et al.* (1992). A book by Fokkema and van den Berg (1993) is exclusively concerned with acoustic reciprocity.

This chapter starts with a derivation of the local form of the time-domain reciprocity theorem. More useful forms of the theorem, which are derived in Sections 4.2 and 4.3, are in the frequency and Laplace transform domains, respectively. In Section 4.4 it is shown that the global form of the reciprocity theorem is the same whether or not a compact inhomogeneity is present in its domain. Direct reciprocity for the wave equation is discussed in Section 4.5.

Reciprocity theorems in addition to those in the frequency and Laplace transform domains are stated in Section 4.6. The last three sections of the chapter present applications of reciprocity relations, including an inverse problem.

The volume density of mass, ρ, and the compressibility, κ, characterize the mechanical behavior of the acoustic medium. The equations that relate the pressure and particle velocity perturbations are given by (2.6.4) and (2.6.7). For simplicity we will omit the prime, and rewrite these equations as

$$p,_i + \rho_0 \dot{v}_i = f_i, \tag{4.1.1}$$

$$v_{i,i} + \kappa \dot{p} = q. \tag{4.1.2}$$

It should be noted that the derivation of these equations as given in Section 2.6 does not change if ρ_0 and κ depend on position. Hence Eqs. (4.1.1) and (4.1.2) are also valid for an inhomogeneous acoustic medium. In the present work, however, we will generally consider the case where ρ_0 and κ are constants.

For an acoustic medium, reciprocity theorems provide a relation between $p(\boldsymbol{x}, t)$ and $v_i(\boldsymbol{x}, t)$ for two different acoustic states of the same region, or for two regions having the same geometry but different values of ρ_0 and κ. Here we will first consider the case of two regions having different material constants. The two states will be denoted state A and state B, and so the field variables for the two states are indicated by superscripts A and B, respectively. Hence we have for state A

$$p,_i^A + \rho_0^A \dot{v}_i^A = f_i^A, \tag{4.1.3}$$

$$v_{i,i}^A + \kappa^A \dot{p}^A = q^A. \tag{4.1.4}$$

Similarly, for state B

$$p,_i^B + \rho_0^B \dot{v}_i^B = f_i^B \tag{4.1.5}$$

$$v_{i,i}^B + \kappa^B \dot{p}^B = q^B. \tag{4.1.6}$$

To derive the reciprocity theorem we multiply Eq. (4.1.3) by v_i^B and Eq. (4.1.4) by p^B. Taking the difference of the two results yields

$$v_i^B p,_i^A - p^B v_{i,i}^A = -\rho_0^A \dot{v}_i^A v_i^B + f_i^A v_i^B + \kappa^A \dot{p}^A p^B - q^A p^B. \tag{4.1.7}$$

A similar procedure, i.e., multiplying (4.1.5) by v_i^A and (4.1.6) by p_i^A and taking the difference of the two results yields

$$v_i^A p,_i^B - p^A v_{i,i}^B = -\rho_0^B \dot{v}_i^B v_i^A + f_i^B v_i^A + \kappa^B \dot{p}^B p^A - q^B p^A \tag{4.1.8}$$

By subtracting (4.1.8) from (4.1.7) we obtain

$$\begin{aligned}(p^A v_i^B - p^B v_i^A)_{,i} &= \rho_0^B v_i^A \dot{v}_i^B - \rho_0^A \dot{v}_i^A v_i^B + \kappa^A \dot{p}^A p^B - \kappa^B p^B \dot{p}^A \\ &\quad + f_i^A v_i^B - f_i^B v_i^A - q^A p^B + q^B p^A. \end{aligned} \tag{4.1.9}$$

Equation (4.1.9), which is the local form of the time-domain reciprocity theorem, has limited use. We will therefore consider other forms of the theorem in the frequency and Laplace transform domains.

4.2 Reciprocity in the frequency domain

It is often useful to determine a time-harmonic, steady-state solution to a problem. Such a solution in the frequency domain has value for its own sake and it can also be used to obtain by superposition a general time-domain solution. Solutions in the frequency domain are chosen in the form

$$p(\mathbf{x}, t) = \mathcal{R}[p(\mathbf{x}, \omega)e^{-i\omega t}], \tag{4.2.1}$$

$$v_i(\mathbf{x}, t) = \mathcal{R}[v_i(\mathbf{x}, \omega)e^{-i\omega t}], \tag{4.2.2}$$

$$f_i(\mathbf{x}, t) = \mathcal{R}[f_i(\mathbf{x}, \omega)e^{-i\omega t}], \tag{4.2.3}$$

$$q(\mathbf{x}, t) = \mathcal{R}[q(\mathbf{x}, \omega)e^{-i\omega t}], \tag{4.2.4}$$

where $\mathcal{R}$ denotes the real part of the function. Substitution of these complex representations into the time-domain equations (4.1.1), (4.1.2) yields

$$p_{,i} - i\omega\rho_0 v_i = f_i \tag{4.2.5}$$

$$v_{i,i} - i\omega\kappa p = q. \tag{4.2.6}$$

For time-harmonic solutions the reciprocity theorem given by Eq. (4.1.9) simplifies considerably to

$$\begin{aligned}(p^A v_i^B - p^B v_i^A)_{,i} &= -i\omega(\rho_0^B - \rho_0^A) v_i^A v_i^B - i\omega(\kappa^A - \kappa^B) p^A p^B \\ &\quad + f_i^A v_i^B - f_i^B v_i^A - q^A p^B + q^B p^A. \end{aligned} \tag{4.2.7}$$

Usually the reciprocity theorem is considered for the case where the material constants are the same for the two states A and B. We then have

$$(p^A v_i^B - p^B v_i^A)_{,i} = f_i^A v_i^B - f_i^B v_i^A - q^A p^B + q^B p^A. \tag{4.2.8}$$

Equations (4.2.7), (4.2.8) are local reciprocity theorems. The corresponding global reciprocity theorems are obtained by integrating these equations over a body with volume V and bounding surface S. For Eq. (4.2.8) the result is

$$\int_S \left(p^A v_i^B - p^B v_i^A\right) n_i \, dS = \int_V \left(f_i^A v_i^B - f_i^B v_i^A - q^A p^B + q^B p^A\right) dV, \tag{4.2.9}$$

where Gauss' integral theorem has been used.

For a good many cases we wish to apply a reciprocity theorem to an unbounded domain. For such cases the domain V is defined as being interior to a sphere of surface S_0 and of radius R_0, where $R_0 \to \infty$. We need to examine what happens to the integral on the left-hand side of Eq. (4.2.9) as $S \to S_0$. From the solutions for the acoustic variables given in Section 3.3 it follows that both the pressure distribution and the particle velocity are of order $O(R_0^{-1})$ as $R_0 \to \infty$. Hence $p^A v^B$ is of order $O(R_0^{-2})$. However, in $p^A v^B - p^B v^A$ the term of order $O(R_0^{-2})$ cancels out, and hence the difference of terms is of order $O(R_0^{-3})$. Because the area of S_0 is $4\pi R_0^2$, we then conclude that

$$\int_{S_0} \left(p^A v_i^B - p^B v_i^A\right) n_i \, dS = O\left(R_0^{-1}\right) \quad \text{as} \quad R_0 \to \infty. \tag{4.2.10}$$

Thus the left-hand side of Eq. (4.2.9) vanishes as the domain becomes unbounded.

4.3 Reciprocity in the Laplace transform domain

It will be useful to review briefly some results from Laplace transform theory. The one-sided Laplace transform, $\overline{f}(\boldsymbol{x}, s)$, of a function $f(\boldsymbol{x}, t)$ is defined as

$$\overline{f}(\boldsymbol{x}, s) = \mathcal{L}[f(\boldsymbol{x}, t)] = \int_0^\infty e^{-st} f(\boldsymbol{x}, t) \, dt. \tag{4.3.1}$$

If the integral of Eq. (4.3.1) converges for $s = s_1$ then it also converges for any value of s satisfying $\mathcal{R}(s) > \mathcal{R}(s_1)$, where $\mathcal{R}(s)$ denotes the real part of s. In general the function $\overline{f}(\boldsymbol{x}, s)$ is a regular function of the complex variable s for $\mathcal{R}(s) > \mathcal{R}(s_1)$. For many functions the inverse Laplace transform $\overline{f}(\boldsymbol{x}, s)$ can be found in tables of Laplace transforms or can be constructed by using certain simple properties of Laplace transforms. A formal expression for the inverse Laplace transform is

$$f(\boldsymbol{x}, t) = \mathcal{L}^{-1}[\overline{f}(\boldsymbol{x}, s)] = \frac{1}{2\pi i} \int_{\gamma - i\infty}^{\gamma + i\infty} \overline{f}(\boldsymbol{x}, s) \, e^{st} \, ds, \tag{4.3.2}$$

where $\mathcal{R}(\gamma) > \mathcal{R}(s_1)$.

The Laplace transform of the derivatives of a function can be obtained from Eq. (4.3.1) by integration by parts:

$$\mathcal{L}\left[\frac{\partial f}{\partial t}\right] = s\overline{f}(\boldsymbol{x}, s) - f(\boldsymbol{x}, 0^+), \tag{4.3.3}$$

$$\mathcal{L}\left[\frac{\partial^2 f}{\partial t^2}\right] = s^2\overline{f}(\boldsymbol{x}, s) - sf(\boldsymbol{x}, 0^+) - \dot{f}(\boldsymbol{x}, 0^+), \tag{4.3.4}$$

where $\dot{f}(\boldsymbol{x}, 0^+)$ and $f(\boldsymbol{x}, 0^+)$ are the limits of $\partial f/\partial t$ and $f(\boldsymbol{x}, t)$ as $t \to 0$ for $t > 0$.

Another useful property of Laplace transforms is the convolution theorem. The temporal convolution $C_t\{f, g\}$ of two functions $f(x, t)$ and $g(x, t)$ is defined as

$$\begin{aligned} C_t\{f, g\} &= \int_0^t f(\boldsymbol{x}, t - \tau)g(\boldsymbol{x}, \tau)\, d\tau \\ &= \int_0^t f(\boldsymbol{x}, \tau)g(\boldsymbol{x}, t - \tau)\, d\tau = C_t\{g, f\}. \end{aligned} \tag{4.3.5}$$

Another notation for the temporal convolution is defined by

$$C_t\{f, g\} \equiv f(\boldsymbol{x}, t) * g(\boldsymbol{x}, t) \equiv f * g. \tag{4.3.6}$$

This somewhat more convenient notation will be used in what follows.

It can be shown that the Laplace transform of the temporal convolution is of the simple form

$$\overline{C}_t\{f, g\} = \overline{f}(\boldsymbol{x}, s)\overline{g}(\boldsymbol{x}, s). \tag{4.3.7}$$

We now return to Eqs. (4.1.1) and (4.1.2) and apply the Laplace transform to these equations to obtain

$$\overline{p}_{,i} + \rho_0 s \overline{v}_i = \overline{f}_i + \rho_0 v_i(\boldsymbol{x}, 0^+), \tag{4.3.8}$$

$$\overline{v}_{i,i} + \kappa s \overline{p} = \overline{q} + \kappa p(\boldsymbol{x}, 0^+). \tag{4.3.9}$$

It is convenient to define two new functions

$$\overline{g}_i = \overline{f}_i + \rho_0 v_i(\boldsymbol{x}, 0^+), \tag{4.3.10}$$

$$\overline{h} = \overline{q} + \kappa p(\boldsymbol{x}, 0^+). \tag{4.3.11}$$

Then, using the same manipulations that led to Eq. (4.1.9) we obtain the local reciprocity relation in the form

$$\begin{aligned} \left(\overline{p}^A \overline{v}_i^B - \overline{p}^B \overline{v}_i^A\right)_{,i} &= s(\rho^B - \rho^A)\overline{v}_i^A \overline{v}_i^B - s(\kappa^B - \kappa^A)\overline{p}^A \overline{p}^B \\ &\quad + \overline{g}_i^A \overline{v}_i^B + \overline{h}^B \overline{p}^A - \overline{g}_i^B \overline{v}_i^A - \overline{h}^A \overline{p}^B. \end{aligned} \tag{4.3.12}$$

It is noted that by setting $\overline{g}_i = f_i$, $\overline{h} = q$ and $s = -i\omega$, Eq. (4.3.12) reduces to the corresponding equation in the frequency domain.

The global form of the reciprocity theorem follows from Eq. (4.3.12) as

$$\int_S (\overline{p}^A \overline{v}_i^B - \overline{p}^B \overline{v}_i^A) n_i \, dS = \int_V \left[s(\rho^B - \rho^A) \overline{v}_i^A \overline{v}_i^B - s(\kappa^B - \kappa^A) \overline{p}^A \overline{p}^B \right] dV + \int_V \left(\overline{g}_i^A \overline{v}_i^B + \overline{h}^B \overline{p}^A - \overline{g}_i^B \overline{v}_i^A - \overline{h}^A \overline{p}^B \right) dV. \tag{4.3.13}$$

4.4 Reciprocity in the presence of a compact inhomogeneity

The configuration that will be considered consists of a semicircular domain D of radius R_0, located in the half-space $x_3 \geq 0$. The domain contains a compact inhomogeneity defined by $\hat{V}$, with boundary $\hat{S}$. The geometry is shown in Fig. 4.1. In the domain $V = D - \hat{V}$ the medium is homogeneous with mass density ρ and compressibility κ. The compact inhomogeneity is itself also homogeneous with mass density $\hat{\rho}$ and compressibility $\hat{\kappa}$. There may be source distributions f_i and q in V but not in $\hat{V}$.

We now write the reciprocity theorem (4.3.13) for the domain $V = D - \hat{V}$. As shown in Fig. 4.1, the integrations along S_0 and S are counter-clockwise. The integration along $\hat{S}_{\text{ext}}$ is, however, clockwise. We have

$$\int_{S+S_0} (\overline{p}^A \overline{v}_i^B - \overline{p}^B \overline{v}_i^A) n_i \, dS + \int_{\hat{S}_{\text{ext}}} (\overline{p}^A \overline{v}_i^B - \overline{p}^B \overline{v}_i^A) n_i \, dS = \int_V \left(\overline{g}_i^A \overline{v}_i^B + \overline{h}^B \overline{p}^A - \overline{g}_i^B \overline{v}_i^A - \overline{h}^A \overline{p}^B \right) dV. \tag{4.4.1}$$

The boundary of the inhomogeneity is represented by both $\hat{S}_{\text{ext}}$ and $\hat{S}_{\text{int}}$. For the inhomogeneity we have

$$\int_{\hat{S}_{\text{int}}} (\overline{p}^A \overline{v}_i^B - \overline{p}^B \overline{v}_i^A) n_i \, dS = 0, \tag{4.4.2}$$

where the integration along $\hat{S}_{\text{int}}$ is counter-clockwise. The right-hand side of this equation is zero because there are no source distributions in $\hat{V}$. We emphasize that the fields $\overline{p}^A(\boldsymbol{x}, s)$, $\overline{v}_i^A(\boldsymbol{x}, s)$, $\overline{p}^B(\boldsymbol{x}, s)$ and $\overline{v}_i^B(\boldsymbol{x}, s)$ represent the pressure perturbations and particle velocities in both V and $\hat{V}$. Now, since the pressure perturbation and the normal component of the particle velocity are continuous along the interface of the inhomogeneity and the surrounding medium, the

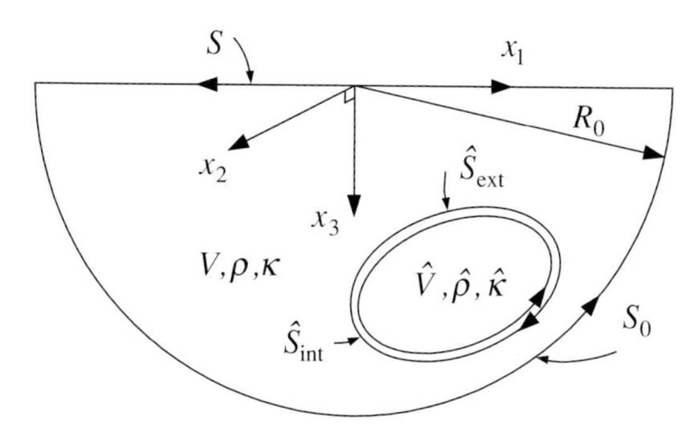

Figure 4.1 Compact inhomogeneity in a half-space.

integral along $\hat{S}_{\text{ext}}$ in Eq. (4.4.1) vanishes by virtue of Eq. (4.4.2). Also, letting $R_0 \to \infty$, Eq. (4.4.1) yields the expected result that

$$\int_S (\overline{p}^A \overline{v}_i^B - \overline{p}^B \overline{v}_i^A)\, dS = \int_V (\overline{f}_i^A \overline{v}_i^B + \overline{q}^B \overline{p}^A - \overline{f}_i^B \overline{v}_i^A - \overline{q}^A \overline{p}^B)\, dV. \quad (4.4.3)$$

Thus the reciprocity theorem has the same form as for a domain without an inhomogeneity, given a proper interpretation of the field quantities. For future reference, it has also been assumed in Eq. (4.4.3) that the body is at rest prior to $t = 0$.

4.5 Direct reciprocity for the wave equation

There are some advantages to a reciprocity theorem that relates just one of the field variables for states A and B, the particle velocity, the pressure perturbation or the velocity potential. Here we will consider the frequency-domain reciprocity theorem for the velocity potential. The wave equation for $\varphi(x, t)$ is given by Eq. (2.6.26). For state A we have in the frequency domain

$$\nabla^2 \varphi^A + \frac{\omega^2}{c^2} \varphi^A = q^A(\boldsymbol{x}). \quad (4.5.1)$$

For state B we take

$$\nabla^2 \varphi^B + \frac{\omega^2}{c^2} [1 + \alpha(\boldsymbol{x})] \varphi^B = q^B(\boldsymbol{x}). \quad (4.5.2)$$

Thus states A and B are defined in different acoustic media; the acoustic medium for state B is inhomogeneous, as indicated by the presence of the function $\alpha(\boldsymbol{x})$.

Now Eqs. (4.5.1) and (4.5.2) are multiplied by $\varphi^B(x)$ and $\varphi^A(x)$, respectively; the second equation obtained in this manner is subtracted from the first, and the result is integrated over a volume V with boundary S. To simplify the resulting equation we use Green's theorem,

$$\int_V \left(\varphi^B \nabla^2 \varphi^A - \varphi^A \nabla^2 \varphi^B\right) dV = \int_S \left(\varphi^B \frac{\partial \varphi^A}{\partial n} - \varphi^A \frac{\partial \varphi^B}{\partial n}\right) dS, \quad (4.5.3)$$

where $\partial/\partial n$ is the derivative in the direction normal to the surface S. These manipulations result in the following reciprocity theorem:

$$\int_V \left(\varphi^A q^B - \varphi^B q^A\right) dV - \frac{\omega^2}{c^2} \int_V \alpha(x) \varphi^A \varphi^B dV + \int_S \left(\varphi^B \frac{\partial \varphi^A}{\partial n} - \varphi^A \frac{\partial \varphi^B}{\partial n}\right) dS = 0. \quad (4.5.4)$$

It should be noted that replacing ω^2 by $-s^2$, φ^B by $\overline{\varphi}^B$ and φ^A by $\overline{\varphi}^A$ yields the corresponding theorem in the Laplace transform domain. As we shall see in what follows the theorem given by Eq. (4.5.4) and other theorems of this type play a role in the formulation of inverse problems.

4.6 Other reciprocity theorems

The reciprocity theorems in the frequency and Laplace transform domains are best known and most frequently used. There are, however, other reciprocity theorems, stated here for completeness even though in this book they are not used for specific applications. The theorems stated in this section are the time-domain reciprocity theorem of the convolution type and the power reciprocity theorems in the Laplace transform and frequency domains. For more extensive discussions of these theorems we refer to Fokkema and van den Berg (1993).

Time-domain reciprocity of the convolution type

The reciprocity theorem in the time domain can be obtained directly by Laplace transform inversion of the reciprocity theorem in the Laplace transform domain given by Eq. (4.3.13). By using the definition of the temporal convolution given by Eq. (4.3.6) and the result for the Laplace transform inversion of the product of two Laplace transforms that follows from Eq. (4.3.7), and also using Eq. (4.3.3), we obtain for quiescent initial conditions

$$\begin{aligned} &\int_S \left(p^A * v_i^B - p^B * v_i^A\right) n_i \, dS \\ &= \int_V \frac{\partial}{\partial t}\left[(\rho^B - \rho^A) v_i^A * v_i^B - (\kappa^B - \kappa^A) p^A * p^B\right] dV \\ &\quad + \int_V \left(f_i^A * v_i^B + q^B * p^A - f_i^B * v_i^A - q^A * p^B\right) dV. \quad (4.6.1) \end{aligned}$$

Power reciprocity in the time domain

For this case, state B is the anti-causal wave field in the Laplace transform domain. This field differs from the causal field in that s has been replaced by $-s$. Thus for state B we have

$$\{\overline{p}^B, \overline{v}_i^B\} = \{\hat{p}^B, \hat{v}_i^B\}(\boldsymbol{x}, -s), \tag{4.6.2}$$

$$\{\overline{q}^B, \overline{f}_i^B\} = \{\hat{q}^B, \hat{f}_i^B\}(\boldsymbol{x}, -s), \tag{4.6.3}$$

where the caret indicates the anti-causal nature of the quantities. Instead of Eqs. (4.3.8), (4.3.9) we now have

$$\hat{p}^B_{,i} - \rho_0^B s \hat{v}_i^B = \hat{g}_i, \tag{4.6.4}$$

$$\hat{v}^B_{i,i} - \kappa^B s \hat{p}^B = \hat{h}. \tag{4.6.5}$$

Following the same manipulations that led to Eq. (4.3.13) we obtain

$$\int_V (\overline{p}^A \hat{v}_i^B + \hat{p}^B \overline{v}_i^A) n_i \, dS = \int_V \left[s(\rho^B - \rho^A) \overline{v}_i^A \hat{v}_i^B + s(\kappa^B - \kappa^A) \overline{p}^A \hat{p}^B \right] dV$$
$$+ \int_V \left(\overline{g}_i^A \hat{v}_i^B + \hat{h}^B \overline{p}^A + \hat{g}_i^B \overline{v}_i^A + h^A \hat{p}^B \right) dV. \tag{4.6.6}$$

This relation is sometimes referred to as the power reciprocity theorem.

Power reciprocity theorem in the frequency domain

In the frequency domain we now take for state B the complex conjugate of state A. Instead of Eqs. (4.2.5) and (4.2.6) we then have

$$p_{,i}^{B*} + i\omega\rho_0^B v_i^{B*} = f_i^{B*}, \tag{4.6.7}$$

$$v_{i,i}^{B*} + i\omega\kappa^B p^{B*} = q^*. \tag{4.6.8}$$

Then, following the same manipulations that led to the reciprocity theorem in the frequency domain, we obtain

$$\int_V (p v_i^* + p^* v_i) n_i \, dS = \int_V (f_i v_i^* + p q^* + f_i^* v_i + q p^*) \, dV. \tag{4.6.9}$$

Just as for the regular reciprocity theorem in the frequency domain, Eq. (4.2.9), we have considered the case for which $\rho_0^A = \rho_0^B$ and $\kappa_0^A = \kappa_0^B$. We have also left out the superscripts A and B altogether because the two states are now distinguished from each other in that one is the complex conjugate of the other.

4.7 Reciprocity between source solutions

We consider the configuration of two time-harmonic point sources applied at positions $\boldsymbol{x}^A$ and $\boldsymbol{x}^B$, respectively, in an unbounded acoustic medium. Either source can be of the volume-force or the volume-injection type. The source functions are of the general forms

$$\{q^A(\boldsymbol{x}), f_i^A(\boldsymbol{x})\} = \{Q^A, F_i^A\}\delta(\boldsymbol{x} - \boldsymbol{x}^A), \tag{4.7.1}$$

$$\{q^B(\boldsymbol{x}), f_i^B(\boldsymbol{x})\} = \{Q^B, F_i^B\}\delta(\boldsymbol{x} - \boldsymbol{x}^B). \tag{4.7.2}$$

Since we are considering the reciprocity theorem for an unbounded domain, the integral over S on the left-hand side of Eq. (4.2.9) vanishes, and thus

$$\int_{V_\infty} \left(f_i^A v_i^B - f_i^B v_i^A - q^A p^B + q^B p^A\right) dV = 0, \tag{4.7.3}$$

where V_∞ defines the unbounded domain. Substitution of the source functions given by Eqs. (4.7.1), (4.7.2) into this equation yields, by virtue of the sifting property of the delta function

$$\begin{aligned} F_i^A v_i^B(\boldsymbol{x}^A - \boldsymbol{x}^B) - F_i^B v_i^A(\boldsymbol{x}^B - \boldsymbol{x}^A) - Q^A p^B(\boldsymbol{x}^A - \boldsymbol{x}^B) \\ + Q^B p^A(\boldsymbol{x}^B - \boldsymbol{x}^A) = 0. \end{aligned} \tag{4.7.4}$$

This reciprocity relation includes three special cases, namely: both sources are of the volume-force type; both are of the volume-injection type; or they are of different types. For example in the latter case we have

$$F_i^A = 0, \quad Q^B = 0, \tag{4.7.5}$$

and it follows from Eq. (4.7.4) that

$$-F_i^B v_i^A(\boldsymbol{x}^B - \boldsymbol{x}^A) = Q^A p^B(\boldsymbol{x}^A - \boldsymbol{x}^B). \tag{4.7.6}$$

The other cases follow immediately in the same way from Eq. (4.7.4).

Similar results can be obtained in the Laplace transform domain. Instead of Eqs. (4.7.1), (4.7.2) we now have

$$\{q^A(\boldsymbol{x}, t), f_i^A(\boldsymbol{x}, t)\} = \{Q^A(t), F_i^A(t)\}\delta(\boldsymbol{x} - \boldsymbol{x}^A), \tag{4.7.7}$$

$$\{q^B(\boldsymbol{x}, t), f_i^B(\boldsymbol{x}, t)\} = \{Q^B(t), F_i^B(t)\}\delta(\boldsymbol{x} - \boldsymbol{x}^B). \tag{4.7.8}$$

For this case Eq. (4.7.4) becomes a relation between Laplace transforms, namely,

$$\begin{aligned} \overline{F}_i^A \overline{v}_i^B(\boldsymbol{x}^A - \boldsymbol{x}^B, s) - \overline{F}_i^B \overline{v}_i^A(\boldsymbol{x}^B - \boldsymbol{x}^A, s) - \overline{Q}^A p^B(\boldsymbol{x}^A - \boldsymbol{x}^B, s) \\ + \overline{Q}^B p^A(\boldsymbol{x}^B - \boldsymbol{x}^A, s) = 0. \end{aligned} \tag{4.7.9}$$

In the time domain the equivalent of Eq. (4.7.6) then becomes

$$-F_i^B(t) * v_i^A(\boldsymbol{x}^B - \boldsymbol{x}^A, t) = Q^A(t) * p^B(\boldsymbol{x}^A - \boldsymbol{x}^B, t). \quad (4.7.10)$$

4.8 Distribution of sources

Now let us consider the case where acoustic wave motion is generated by a distribution of sources over the region $\hat{V}$. In the frequency domain the relevant equation is given by Eq. (4.5.1):

$$\nabla^2 \varphi^A + \frac{\omega^2}{c^2} \varphi^A = q^A(\boldsymbol{x}). \quad (4.8.1)$$

For state B we take a concentrated source applied at $x = x^B$:

$$\nabla^2 \varphi^B + \frac{\omega^2}{c^2} \varphi^B = Q^B \delta(\boldsymbol{x} - \boldsymbol{x}^B). \quad (4.8.2)$$

The solution to this equation has been given earlier in Section 3.6 as

$$\varphi^B(\boldsymbol{x} - \boldsymbol{x}^B) = -\frac{Q^B}{4\pi\,|\boldsymbol{x} - \boldsymbol{x}^B|} e^{i(\omega/c)|\boldsymbol{x} - \boldsymbol{x}^B|}. \quad (4.8.3)$$

In the next step the reciprocity theorem is applied. The appropriate form follows immediately from Eq. (4.5.4) since the second term in that equation vanishes because the media for the two states are the same, and the third term is zero as S becomes the spherical surface S_o when $R \to R_0$. Thus we have

$$\int_{V_\infty} (\varphi^A q^B - \varphi^B q^A)\, dV = 0. \quad (4.8.4)$$

Substitution of $q^B = Q^B \delta(\boldsymbol{x} - \boldsymbol{x}^B)$, and $\varphi^B(\boldsymbol{x} - \boldsymbol{x}^B)$ from Eq. (4.8.3), yields

$$\varphi^A(\boldsymbol{x}^B) = -\frac{1}{4\pi} \int_{\hat{V}} q^A(\boldsymbol{x}) \frac{e^{i(\omega/c)|\boldsymbol{x} - \boldsymbol{x}^B|}}{|\boldsymbol{x} - \boldsymbol{x}^B|}\, dV_x. \quad (4.8.5)$$

Clearly this expression can also be written as

$$\varphi^A(\boldsymbol{x}^B) = -\frac{1}{4\pi} \int_{\hat{V}} q^A(\boldsymbol{x}) \frac{e^{i(\omega/c)|\boldsymbol{x}^B - \boldsymbol{x}|}}{|\boldsymbol{x}^B - \boldsymbol{x}|}\, dV_x. \quad (4.8.6)$$

4.9 Inverse problem for scattering by a compact inhomogeneity

In inverse scattering problems the objective is to determine the shape and/or material constitution of an interior domain that is not accessible to direct

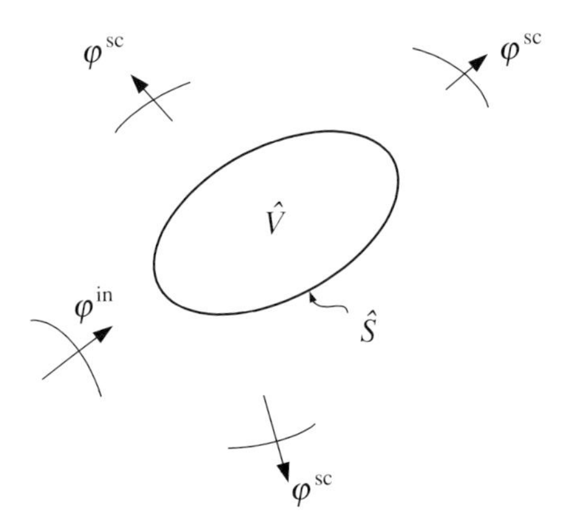

Figure 4.2 Compact inhomogeneity in homogeneous embedding.

inspection and measurement. For an inverse scattering problem in acoustics the interior domain of interest, here referred to as a compact inhomogeneity, is sonified by incident acoustic wave motion, and the sound scattered by the inclusion is detected by a number of detectors. The objective of the inverse problem is to obtain information on the inhomogeneity from the detected scattered field at several locations. There is a considerable literature on inverse problems. For acoustics a useful and clear discussion can be found in the paper by Blok and Zeylmans (1987).

There are several ways to formulate inverse scattering problems. Here we present a formulation in terms of domain integral equations for the velocity potential in the frequency domain. The geometrical configuration is shown in Fig. 4.2. The domain of the inhomogeneity is denoted by $\hat{V}$ with boundary $\hat{S}$.

A steady-state field of the velocity potential, $\varphi^{in}(\boldsymbol{x})\exp(-i\omega t)$, is incident on the inhomogeneity and gives rise to a scattered field. The resulting total field is written as

$$\varphi = \varphi^{in} + \varphi^{sc}. \tag{4.9.1}$$

The incident field is defined in the whole domain, including the part occupied by the inhomogeneity. Away from its source region, the incident wave potential is governed by the homogeneous Helmholtz equation:

$$\nabla^2\varphi^{in} + \frac{\omega^2}{c^2}\varphi^{in} = 0, \tag{4.9.2}$$

where c^2 is the phase velocity in the embedding. Inside the inhomogeneity the total field must satisfy

$$\nabla^2\varphi + \frac{\omega^2}{c^2}[1 + \alpha(\boldsymbol{x})]\varphi = 0. \tag{4.9.3}$$

Thus $\alpha(\boldsymbol{x})$ is only different from zero when $\boldsymbol{x}$ is in $\hat{V}$. By subtracting Eq. (4.9.2) from Eq. (4.9.3) we obtain

$$\nabla^2\varphi^{\mathrm{sc}} + (\omega^2/c^2)\varphi^{\mathrm{sc}} = \begin{cases} -(\omega^2/c^2)\alpha(\boldsymbol{x})\varphi(\boldsymbol{x}) & \text{in } \hat{V}, \\ 0 & \text{outside } \hat{V}. \end{cases} \tag{4.9.4}$$

The right-hand side of this equation can be considered as a distribution of sources. By using the results of the previous section, the solution to Eq. (4.9.4) may then be written formally as

$$\varphi^{\mathrm{sc}}(\boldsymbol{x};\omega) = \frac{1}{4\pi}\frac{\omega^2}{c^2}\int_{\hat{V}} \alpha(\boldsymbol{y})e^{i(\omega/c)|\boldsymbol{x}-\boldsymbol{y}|}[\varphi^{\mathrm{in}}(\boldsymbol{y}) + \varphi^{\mathrm{sc}}(\boldsymbol{y})]\,dV_y. \tag{4.9.5}$$

This is an exact integral equation. Unfortunately, this equation contains two unknown functions, $\alpha(\boldsymbol{y})$ and $\varphi^{\mathrm{sc}}(\boldsymbol{y},\omega)$, and therefore does not directly lend itself to exact solution techniques. The equation is also non-linear in the unknowns through the product $\alpha(\boldsymbol{y})\varphi^{\mathrm{sc}}(\boldsymbol{y};\omega)$.

Bleistein (1984) briefly discussed an approximation procedure for the solution of Eq. (4.9.5). He considered the case where $\alpha(\boldsymbol{y})$ is small. Then, the source term in Eq. (4.9.5) is also small. It would be reasonable, therefore, to expect that the solution to the problem is small as well, of the same order, $O(\alpha(\boldsymbol{y}))$, as the source term itself. Consequently, it is reasonable to expect that the product $\alpha(\boldsymbol{y})\varphi^{\mathrm{sc}}(\boldsymbol{y};\omega)$ on the right-hand side is quadratic in $\alpha(\boldsymbol{y})$ while the product $\alpha(\boldsymbol{y})\varphi^{\mathrm{in}}(\boldsymbol{y};\omega)$ is linear in $\alpha(\boldsymbol{y})$. As a first-order approximation, then, we neglect the higher-order term in $\alpha(\boldsymbol{y})$ in the source; that is, we replace the sum $\varphi^{\mathrm{in}}(\boldsymbol{y},\omega) + \varphi^{\mathrm{sc}}(\boldsymbol{y},\omega)$ by the first term $\varphi^{\mathrm{in}}(\boldsymbol{y},\omega)$ alone. Mathematically, we are assuming that the solution $\varphi^{\mathrm{sc}}(\boldsymbol{y},\omega)$ can be derived as a regular perturbation series in $\alpha(\boldsymbol{y})$. This technique is equivalent to the Born approximation for potential scattering in theoretical physics, and Born's name has been attached to this application of the regular perturbation method as well.

The effect of modifying the source in this manner is to recast the integral equation (4.9.5) as

$$\varphi^{\mathrm{sc}}(\boldsymbol{x};\omega) = \frac{1}{4\pi}\frac{\omega^2}{c^2}\int_{\hat{V}} \alpha(\boldsymbol{y})e^{i(\omega/c)|\boldsymbol{x}-\boldsymbol{y}|}\varphi^{\mathrm{in}}(\boldsymbol{y})\,dV_y. \tag{4.9.6}$$

This is a linear integral equation relating the values (assumed known) of the backscattered field $\varphi^{\mathrm{sc}}(\boldsymbol{y},\omega)$ to $\alpha(\boldsymbol{y})$, which is now the only unknown in the equation.

As pointed out in Bleistein (1984), Eq. (4.9.6) is a Fredholm integral equation of the first kind for $\alpha(\boldsymbol{y})$. Bleistein noted that when the kernel of such an integral equation – in this case $e^{i(\omega/c)|\boldsymbol{x}-\boldsymbol{y}|}\varphi^{\mathrm{in}}(\boldsymbol{y})$ – is such that its modulus has a bounded square integral in all its variables, here $\boldsymbol{y}$ and ω, then it is known that the solution to this type of integral equation is ill conditioned, with eigenvalues that have a limit point at zero. However, the kernel in the present problem is not square integrable in all its variables, and hence that theory does not apply. Indeed, as Bleistein noted, a prototypical one-dimensional analog of Eq. (4.9.6) has as its kernel the square of the Fourier kernel $\exp\{2i\omega z/c\}$, which is known to have all its (complex) eigenvalues on a circle of non-zero radius. If we think of the kernel as always being the ray-method generalization of this one-dimensional kernel, then it is reasonable to expect that it never has a bounded square integral in all its variables.

Alternatively, a formulation can be given based directly on reciprocity. In that approach state A is the field in the actual configuration of the embedding and the inhomogeneity, as defined by the inhomogeneous form of Eq. (4.9.3). State B, which is the virtual-wave solution, is taken as satisfying

$$\nabla^2\varphi^B + \frac{\omega^2}{c^2}\varphi^B = q^B(\boldsymbol{x}). \tag{4.9.7}$$

If the domain of the embedding is taken as V_∞, the reciprocity theorem follows as a special case from Eq. (4.5.4) as

$$\int_{V_\infty} (\varphi^A q^B - \varphi^B q^A) dV - \frac{\omega^2}{c^2}\int_{\hat{V}} \alpha(\boldsymbol{x})\varphi^A\varphi^B dV = 0. \tag{4.9.8}$$

As an example we consider the case where sound is generated by a point source at the point $\boldsymbol{x}^A$ and received at the point $\boldsymbol{x}^B$. This arrangement implies that

$$q^A(\boldsymbol{x};\boldsymbol{x}^A) = Q^A\delta(\boldsymbol{x}-\boldsymbol{x}^A), \tag{4.9.9}$$
$$q^B(\boldsymbol{x};\boldsymbol{x}^B) = Q^B\delta(\boldsymbol{x}-\boldsymbol{x}^B). \tag{4.9.10}$$

The solutions that are obtained by substituting $q^A(\boldsymbol{x};\boldsymbol{x}^A)$ and $q^B(\boldsymbol{x};\boldsymbol{x}^B)$ in Eqs. (4.9.3) and (4.9.7), respectively, are denoted by $G^A(\boldsymbol{x};\boldsymbol{x}^A)$ and $G^B(\boldsymbol{x};\boldsymbol{x}^B)$. The expression for $G^B(\boldsymbol{x};\boldsymbol{x}^B)$ is simple and has already been given by Eq. (4.8.3) as

$$\varphi^B(\boldsymbol{x};\boldsymbol{x}^B) = G^B(\boldsymbol{x};\boldsymbol{x}^B) = -\frac{Q^B}{4\pi|\boldsymbol{x}-\boldsymbol{x}^B|}e^{i(\omega/c)|\boldsymbol{x}-\boldsymbol{x}^B|}. \tag{4.9.11}$$

The expression for $G^A(\boldsymbol{x};\boldsymbol{x}^A)$ is much more complicated and generally requires a complicated calculation, because the domain for $G^A(\boldsymbol{x};\boldsymbol{x}^A)$ includes the compact inhomogeneity. The reciprocity relation given by Eq. (4.9.8) now yields

$$Q^B G^A(\boldsymbol{x}^B;\boldsymbol{x}^A) - Q^A G^B(\boldsymbol{x}^A;\boldsymbol{x}^B) = -\frac{\omega^2}{c^2}\int_{\hat{V}} \alpha(\boldsymbol{x})G(\boldsymbol{x};\boldsymbol{x}^A)G(\boldsymbol{x};\boldsymbol{x}^B)\,dV_{\boldsymbol{x}}. \tag{4.9.12}$$

Equation (4.9.12) is a well-known relation and is the basis for many inverse problems. For an actual inverse problem there would be many points of observation of the scattered field. Still, even if $\hat{V}$ should be known, the integral would contain the unknowns $\alpha(\boldsymbol{x})$ and $G(\boldsymbol{x};\boldsymbol{x}^A)$. Problems connected with this equation have been discussed by Weston (1984). A simple and often useful approximation can again be obtained by the Born approximation, for which $G(\boldsymbol{x};\boldsymbol{x}^A)$ is taken as Eq. (4.9.11) with subscripts B replaced by A, i.e., as the solution for the concentrated force in the domain V_∞ when the inhomogeneity is absent.

5

Reciprocity in one-dimensional elastodynamics

5.1 Introduction

For practical purposes much interest exists in the analysis of wave motion in elastic bodies that are geometrically defined by one or two large length parameters and at least one small length parameter. Examples are plates, beams and rods. The exact treatment by analytical methods of wave motion in such structural components is often very difficult, if not impossible. For that reason several one- or two-dimensional models that provide approximate descriptions have been developed. These models are based on a priori assumptions with regard to the form of the displacements across the smaller dimension(s) of the component, generally in the cross-sectional area. For beams and rods the assumptions simplify the description of the kinematics to such an extent that the wave motions can be described by one-dimensional approximate theories. For the propagation of time-harmonic waves it was found that the approximate theories can account adequately for the dispersive behavior of at least the lowest mode of the exact solution over a limited but significant range of wavenumbers and frequencies.

One of the best-known examples is the Bernoulli–Euler beam theory. In this simplest model for the description of flexural motions of beams of arbitrary but small uniform cross section with a plane of symmetry, it is assumed that the dominant displacement component is parallel to the plane of symmetry. It is also assumed that the deflections are small and that the cross-sectional area remains plane and normal to the neutral axis. The equations governing Bernoulli–Euler theory are summarized in Section 5.2. Reciprocity relations for a Bernoulli–Euler beam are also established in Section 5.2.

Section 5.3 presents an application of reciprocity considerations to determine the wave motion in a Bernoulli–Euler beam due to the application of a time-harmonic force. Forced vibrations of a simply supported beam are analyzed by means of reciprocity considerations in Section 5.4. For the one-dimensional case, reflection and transmission by an interface are discussed in Section 5.5. The remaining sections of this chapter are concerned with applications of reciprocity to bodies in a state of one-dimensional strain. By appropriate choice of the elastic constant relating longitudinal stress and strain, the results are, however, also valid for one-dimensional stress, i.e., for a thin rod. The results include the wave motion generated by a longitudinal time-harmonic force, the determination of reflection and transmission coefficients in the presence of an interface, and reciprocity between two one-dimensional solutions for concentrated forces.

An inverse problem for a layer of unknown elastic properties embedded in between two homogeneous elastic domains is also investigated. The objective of the inverse problem is to determine the function or constant that governs the elastic behavior of the layer resulting from the reflection and/or transmission of an incident wave. Reciprocity considerations lead to an integral equation for that function or constant. A well-known approximate method, known as the Born approximation, is used to solve the inverse problem. In the last section of the chapter the accuracy of the Born approximation is examined for the special case where the inverse problem consists of determining the unknown elastic constant of a homogeneous layer located in between two homogeneous elastic domains.

5.2 Reciprocity for the Bernoulli–Euler beam

In the geometry shown in Fig. 5.1, the bending moment and the beam deflection of a Bernoulli–Euler beam are related by

$$M = -EI\frac{\partial^2 w}{\partial x^2}, \tag{5.2.1}$$

where $w(x,t)$ is the deflection of the beam, E is Young's modulus and I is the principal moment of inertia of the cross section with respect to the neutral axis. Since the transverse force V is related to M by $V = \partial M/\partial x$, the equation of motion in the z-direction is obtained as

$$\frac{\partial^2 M}{\partial x^2} + q(x,t) = \rho A\frac{\partial^2 w}{\partial t^2}. \tag{5.2.2}$$

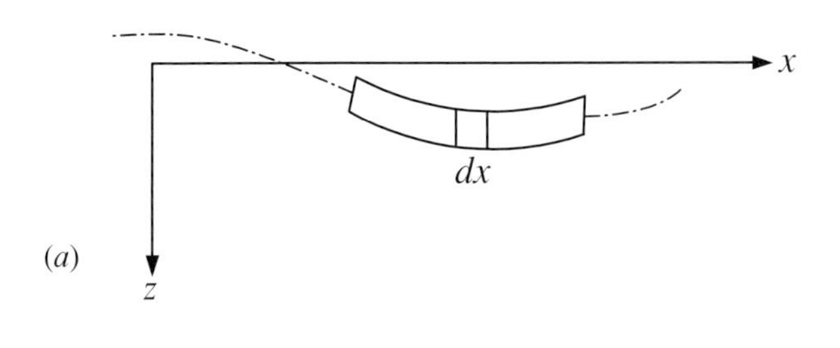

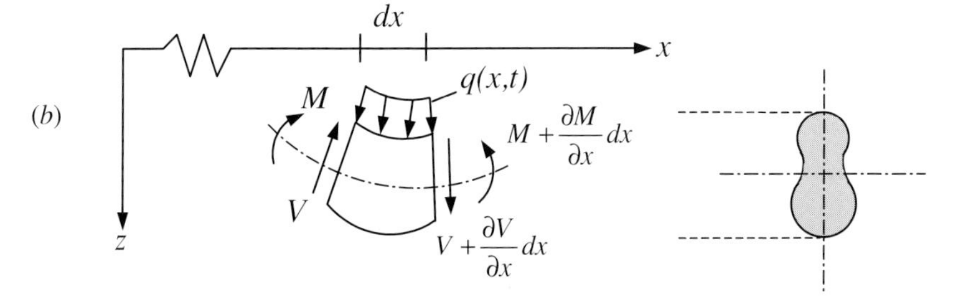

Figure 5.1 Element of a Bernoulli–Euler beam.

Here $q(x, t)$ is the transverse loading per unit length of the beam, ρ is the mass density and A is the cross-sectional area of the beam.

To apply elastodynamic reciprocity to the beam, we consider two elastodynamic states, indicated by superscripts A and B. For these two states Eq. (5.2.2) becomes

$$-\frac{\partial^2 M^A}{\partial x^2} + \rho A \frac{\partial^2 w^A}{\partial t^2} = q^A, \tag{5.2.3}$$

$$-\frac{\partial^2 M^B}{\partial x^2} + \rho A \frac{\partial^2 w^B}{\partial t^2} = q^B. \tag{5.2.4}$$

In the usual manner we multiply (5.2.3) by w^B, (5.2.4) by w^A and subtract the resulting equations, to obtain

$$\frac{\partial^2 M^A}{\partial x^2} w^B + \left(q^A - \rho A \frac{\partial^2 w^A}{\partial t^2} \right) w^B - \frac{\partial^2 M^B}{\partial x^2} w^A - \left(q^B - \rho A \frac{\partial^2 w^B}{\partial t^2} \right) w^A = 0. \tag{5.2.5}$$

This equation is a local reciprocity relation. To obtain a global reciprocity relation the equation is integrated over x, between the limits $x = a$ and $x = b$.

After integration by parts the following global relation is obtained:

$$\left[\frac{\partial M^A}{\partial x}w^B - \frac{\partial M^B}{\partial x}w^A - M^A\frac{\partial w^B}{\partial x} + M^B\frac{\partial w^A}{\partial x}\right]_a^b + \int_a^b \left\{\left(q^A - \rho A\frac{\partial^2 w^A}{\partial t^2}\right)w^B - \left(q^B - \rho A\frac{\partial^2 w^B}{\partial t^2}\right)w^A\right\}dx = 0. \tag{5.2.6}$$

It is of interest to note that the two elastodynamic states do not have to be defined at the same time t.

Equation (5.2.6) simplifies further for time-harmonic motion induced by

$$q^A(x,t) = q^A(x)e^{-i\omega t} \quad \text{and} \quad q^B(x,t) = q^B(x)e^{-i\omega t}. \tag{5.2.7}$$

In the steady state, $M(x,t)$ and $w(x,t)$ will also be products of $\exp(-i\omega t)$ and functions depending on x and ω, indicated by $M(x)$ and $w(x)$. The global reciprocity theorem then becomes

$$\int_a^b (q^A w^B - q^B w^A)\,dx = -\left[\frac{dM^A}{dx}w^B - \frac{dM^B}{dx}w^A - M^A\frac{dw^B}{dx} + M^B\frac{dw^A}{dx}\right]_a^b. \tag{5.2.8}$$

5.3 Waves in a Bernoulli–Euler beam due to a time-harmonic force

As a preliminary step we investigate free waves in the beam. Thus we set $q(x,t) \equiv 0$ and substitute (5.2.1) into (5.2.2) to obtain

$$\frac{\partial^4 w}{\partial x^4} + \frac{1}{\alpha^2}\frac{\partial^2 w}{\partial t^2} = 0, \quad \text{where} \quad \alpha^2 = \frac{EI}{\rho A}. \tag{5.3.1}$$

A free time-harmonic wave of beam deflection is represented by

$$w(x,t) = Ce^{i(kx-\omega t)}, \tag{5.3.2}$$

where C is an arbitrary constant. Substitution of (5.3.2) into (5.3.1) yields

$$k^4 - \frac{\omega^2}{\alpha^2} = 0, \tag{5.3.3}$$

which implies that

$$k = \pm\left(\frac{\omega}{\alpha}\right)^{1/2} \quad \text{and} \quad \bar{k} = \pm i\left(\frac{\omega}{\alpha}\right)^{1/2}. \tag{5.3.4}$$

The first two wavenumbers k are associated with propagating waves in the positive and negative x-directions, respectively. The wavenumbers $\bar{k}$ define standing waves, whose amplitudes decay exponentially with distance from $x = 0$.

Once again the global reciprocity theorem for time-harmonic motion will be used to calculate the wave motion generated by a concentrated force. For an application to the beam based on Eq. (5.2.8), the elastodynamic state A corresponds to the application of a concentrated force

$$q^A(x) = F^A \delta(x - x_1), \qquad \text{where} \qquad a < x_1 < b. \tag{5.3.5}$$

On the basis of the free-wave solution defined by Eq. (5.3.2) it may then be stated that far away from the force the deflections are of the forms

$$x > x_1: \qquad w^A(x) = W^A e^{ik(x-x_1)}, \tag{5.3.6}$$

$$x < x_1: \qquad w^A(x) = W^A e^{-ik(x-x_1)}, \tag{5.3.7}$$

where symmetry with respect to the point of application of the load is implied. It is now shown that an appropriate virtual wave for the elastodynamic state B immediately leads to the solution for constant W^A. For the present case we take

$$w^B(x) = e^{ikx} \qquad \text{and} \qquad q^B(x) \equiv 0. \tag{5.3.8}$$

Thus, the virtual wave is a free wave propagating in the positive x-direction. Substitution of (5.3.5)–(5.3.8) into (5.2.8) yields

$$F^A = 4EI(ik)^3 W^A. \tag{5.3.9}$$

This result comes from the side of the counter-propagating waves ($x = a$) and from the term $q^A(x)w^B(x)$ in the integral. It follows that for $x > x_1$ we have

$$w^A(x, t) = \frac{iF^A}{4EIk^3} e^{i[k(x-x_1)-\omega t]}. \tag{5.3.10}$$

This expression can be derived in other ways, one of them being the application of the exponential Fourier transform with respect to x to the point-force loaded beam; see Graff (1975), pp. 151–3.

5.4 Forced vibrations of a simply supported beam

The global reciprocity theorem given by Eq. (5.2.8) can also be used conveniently to relate the vibratory fields for two cases of concentrated-force application to a Bernoulli–Euler beam of length l. The geometry is shown in Fig. 5.2.

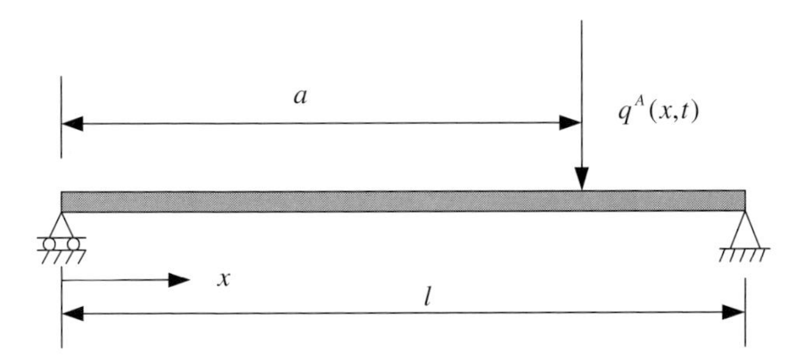

Figure 5.2 Time-harmonic load on a Bernoulli–Euler beam.

Two states are defined by time-harmonic concentrated forces applied at $x = a$ and $x = b$, respectively:

$$q^A(x,t) = F^A \delta(x - a)e^{-i\omega t}, \tag{5.4.1}$$

$$q^B(x,t) = F^B \delta(x - b)e^{-i\omega t}. \tag{5.4.2}$$

The corresponding steady-state deflections are

$$w^A(x,t) = W(x;a)e^{-i\omega t}, \tag{5.4.3}$$

$$w^B(x,t) = W(x;b)e^{-i\omega t}. \tag{5.4.4}$$

Note that a and b define the points of application of the concentrated loads. The boundary conditions on the beam are

$$W^A(0;a) = W^A(l;a) \equiv 0, \tag{5.4.5}$$

$$M^A(0;a) = M^A(l;a) \equiv 0, \tag{5.4.6}$$

$$W^B(0;b) = W^B(l;b) \equiv 0, \tag{5.4.7}$$

$$M^B(0;b) = M^B(l;b) \equiv 0. \tag{5.4.8}$$

When Eqs. (5.4.6)–(5.4.8) are substituted in the global reciprocity relation (5.2.8), the right-hand side vanishes since both the displacements and the bending moments are zero at the support points. Using the definitions of $q^A(x)$ and $q^B(x)$ given by (5.4.1) and (5.4.2) we obtain, by virtue of the sifting property of the delta function,

$$F^B W(b;a) = F^A W(a;b). \tag{5.4.9}$$

Equation (5.4.9) is the reciprocity relation for the two loading cases.

The reciprocity theorem given by Eq. (5.2.8) can also be used to determine the actual displacement solution for a simply supported beam subjected to a time-harmonic normal force. Since we are now concerned with a vibration problem, we employ virtual modes for state B rather than a virtual wave as in Section 5.3.

For time-harmonic motion the governing equation for forced vibrations of the beam follows from Eqs. (5.2.1) and (5.2.2) as

$$\frac{d^4 w}{dx^4} - \frac{\omega^2}{\alpha^2} w = \frac{q(x)}{EI}. \tag{5.4.10}$$

The modal solution that satisfies the boundary conditions is

$$w(x) = \sin\left(\frac{n\pi x}{l}\right), \tag{5.4.11}$$

where the natural frequencies are obtained from the condition

$$\left(\frac{\omega_n}{\alpha}\right)^{1/2} l = n\pi, \qquad n \quad \text{integer}. \tag{5.4.12}$$

For the application of the reciprocity theorem, state A is defined by

$$q^A = F^A \delta(x - a). \tag{5.4.13}$$

The displacement solution is taken as a superposition of natural modes

$$W^A = \sum_m W_m^A \sin\left(\frac{m\pi x}{l}\right), \tag{5.4.14}$$

where the summation is over all positive integers m. For state B we take

$$q^B = Q^B \sin\left(\frac{n\pi x}{l}\right). \tag{5.4.15}$$

The corresponding displacement follows from Eq. (5.4.10) as

$$W^B = \frac{1}{EI} \frac{Q^B \sin\left(\frac{n\pi x}{l}\right)}{\left(\frac{n\pi}{l}\right)^4 + \left(\frac{\omega}{\alpha}\right)^2}. \tag{5.4.16}$$

Substitution of the expressions for q^A, W^A, q^B and W^B into the global reciprocity theorem (5.2.8) with the integration taken over the range $x = 0$ to $x = l$ yields (since the right-hand side vanishes)

$$F^A \frac{1}{EI} \frac{Q^B \sin\left(\frac{n\pi a}{l}\right)}{\left(\frac{n\pi}{l}\right)^2 + \left(\frac{\omega}{\alpha}\right)^2} - Q^B W_m^A \int_0^l \sin\left(\frac{n\pi x}{l}\right) \sin\left(\frac{m\pi x}{l}\right) dx = 0. \tag{5.4.17}$$

Since

$$\int_0^l \sin\left(\frac{n\pi x}{l}\right) \sin\left(\frac{m\pi x}{l}\right) dx = \begin{cases} 0 & \text{for } n \neq m, \\ \frac{1}{2} l & \text{for } n = m, \end{cases} \tag{5.4.18}$$

we obtain from Eq. (5.4.16)

$$W_m^A = \frac{2}{l}\frac{F^A}{EI}\frac{\sin\left(\frac{m\pi a}{l}\right)}{\left(\frac{m\pi}{l}\right)^2 + \left(\frac{\omega}{\alpha}\right)^2}. \tag{5.4.19}$$

The solution for state A, Eq. (5.4.14), then becomes

$$W(x;a) = \frac{2}{l}\frac{F^A}{EI}\sum_m \frac{\sin\left(\frac{m\pi a}{l}\right)\sin\left(\frac{m\pi x}{l}\right)}{\left(\frac{m\pi}{l}\right)^2 + \left(\frac{\omega}{\alpha}\right)^2}. \tag{5.4.20}$$

The expressions for $W(b;a)$ and $W(a;b)$ follow immediately from Eq. (5.4.20), and the reciprocity relation given by Eq. (5.4.9) can then be directly verified.

5.5 Reciprocity for wave motion in one-dimensional strain

Next we consider the simple case of one-dimensional strain in a domain defined by $a \le x \le b$. As the point of departure we take the equation of motion given by Eq. (2.3.1). For the elastodynamic state A, we then have

$$\frac{\partial \tau^A}{\partial x} + f^A(x,t) = \rho \frac{\partial^2 u^A}{\partial t^2}. \tag{5.5.1}$$

Similarly we may write for state B, which is generated by a different distribution of body forces, $f^B(x,t)$,

$$\frac{\partial \tau^B}{\partial x} + f^B(x,t) = \rho \frac{\partial^2 u^B}{\partial t^2}. \tag{5.5.2}$$

In the usual manner we multiply the first equation by $u^B(x,t)$ and the second equation by $u^A(x,t)$. The difference between the two resulting equations is subsequently integrated over the domain $a \le x \le b$. The result is

$$\int_a^b \left\{ \left(f^A(x,t) - \rho\frac{\partial^2 u^A}{\partial t^2}\right) u^B(x,t) - \left(f^B(x,t) - \rho\frac{\partial^2 u^B}{\partial t^2}\right) u^A(x,t)\right\} dx$$
$$+ \int_a^b \left(\frac{\partial \tau^A}{\partial x}u^B - \frac{\partial \tau^B}{\partial x}u^A\right) dx = 0. \tag{5.5.3}$$

The second integral can be simplified by integration by parts, which reduces the equation to

$$\int_a^b \left\{ \left(f^A - \rho \frac{\partial^2 u^A}{\partial t^2} \right) u^B - \left(f^B - \rho \frac{\partial^2 u^B}{\partial t^2} \right) u^A \right\} dx$$
$$+ \left[\tau^A u^B - \tau^B u^A \right]_{x=a}^{x=b} = 0, \tag{5.5.4}$$

where use has been made of the one-dimensional stress–strain relation. Equation (5.5.4), which relates the elastodynamic states A and B, is the one-dimensional version of the reciprocity theorem.

Now let us consider the case of steady-state time-harmonic motions generated by distributions of body forces

$$f^A(x,t) = f^A(x)e^{-i\omega t}, \quad f^B(x,t) = f^B(x)e^{-i\omega t}. \tag{5.5.5}$$

The displacements and stresses are taken in the forms

$$u^A(x,t) = u^A(x)e^{-i\omega t}, \quad \tau^A(x,t) = \tau^A(x)e^{-i\omega t}, \tag{5.5.6}$$

with similar expressions for $u^B(x,t)$ and $\tau^B(x,t)$. Substitution into Eq. (5.5.4) yields

$$\int_a^b \left(f^A u^B - f^B u^A \right) dx = -\left[\tau^A u^B - \tau^B u^A \right]_{x=a}^{x=b}. \tag{5.5.7}$$

Equation (5.5.7) has some interesting applications. For example, the equation can be used to determine the wave motion in $\infty < x < \infty$ by uniform loads, distributed over the plane $x = x_1$,

$$f^A(x) = F^A \delta(x - x_1), \tag{5.5.8}$$

where $a < x_1 < b$. For $x \geq x_1$ the wave motion is of the form

$$u^A(x) = U^A e^{ik(x-x_1)}, \qquad k = \omega/c_L, \tag{5.5.9}$$

where U^A is to be determined. Similarly we have for $x \leq x_1$

$$u^A(x) = U^A e^{-ik(x-x_1)}. \tag{5.5.10}$$

To determine U^A, we select (5.5.8), (5.5.9) for solution A, and we take a virtual wave in the form of a simple harmonic wave of unit amplitude for solution B:

$$u^B = e^{ikx} \qquad \text{and} \qquad f^B(x) \equiv 0. \tag{5.5.11}$$

As before, substitution of these two elastodynamic states into Eq. (5.5.7) yields an equation that can be solved for U^A. The result is

$$U^A = -\frac{F^A}{2ikC}, \qquad \text{where} \qquad C = \lambda + 2\mu. \tag{5.5.12}$$

This solution becomes valid for a thin rod on replacing $\lambda + 2\mu$ by Young's modulus E.

It is of interest to examine the reciprocity relation for the case where states A and B are for materials with different mass densities, ρ^A and ρ^B, and different elastic constants. For the time-harmonic case we obtain after some manipulation

$$\int_a^b \left(f^A u^B - f^B u^A\right) dx + \left(\rho^A - \rho^B\right)\omega^2 \int_a^b u^A u^B dx$$
$$+ \left(C^A - C^B\right) \int_a^b \frac{\partial u^A}{\partial x}\frac{\partial u^B}{\partial x} dx = -\left[\tau^A u^B - \tau^B u^A\right]_{x=a}^{x=b}. \tag{5.5.13}$$

5.6 The effect of an interface

The incidence of a wave on an interface between two bodies of different materials gives rise to reflected and transmitted wave motion. We will consider the details for the case of one-dimensional strain. The relevant equations follow from Eq. (2.3.2) as

$$\tau = C\frac{\partial u}{\partial x}, \qquad \text{where } C = (\lambda + 2\mu), \tag{5.6.1}$$

and

$$\frac{\partial^2 u}{\partial x^2} = \frac{1}{c_L^2}\frac{\partial^2 u}{\partial t^2} \tag{5.6.2}$$

where

$$c_L^2 = \frac{C}{\rho}. \tag{5.6.3}$$

The notation has been simplified by leaving out the subscripts. As discussed in Section 1.4, the general solution of Eq. (5.6.1) is

$$u = f(t - x/c_L) + g(t + x/c_L). \tag{5.6.4}$$

Consider a wave motion defined by the first term of Eq. (5.6.4). Using Eq. (5.6.1) we find

$$\tau = -\rho c_L \dot{u}(x, t),$$

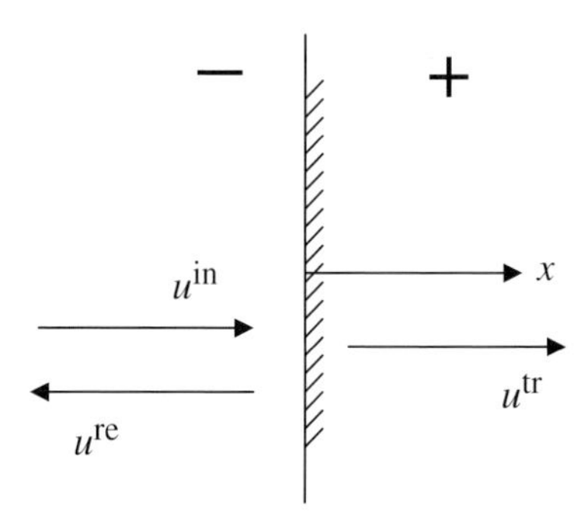

Figure 5.3 Reflection and transmission by an interface at $x = 0$.

where $\dot{u}(x, t) = \partial u(x, t)/\partial t$ is the particle velocity. The quantity $Z = \rho c_L$ is called the mechanical impedance or wave resistance.

Now we consider the reflection and transmission of a one-dimensional wave at an interface. Two solid half-spaces are perfectly bonded in the plane $x = 0$, as shown in Fig. 5.3. A wave defined by

$$u^{\text{in}}(x, t) = f(t - x/c_L^-), \qquad \text{where} \qquad f(t) \equiv 0 \text{ for } t \leq 0, \tag{5.6.5}$$

is incident from $x < 0$. The wave which arrives at the interface ($x = 0$) at time $t = 0$ is reflected and transmitted. The reflected and transmitted waves are taken to have the general forms

$$x \leq 0: \qquad u^{\text{re}}(x, t) = g(t + x/c_L^-), \qquad \text{where} \qquad g(t) \equiv 0 \quad \text{for} \quad t \leq 0, \tag{5.6.6}$$

and

$$x \geq 0: \qquad u^{\text{tr}}(x, t) = h(t - x/c_L^+), \qquad \text{where} \qquad h(t) \equiv 0 \quad \text{for} \quad t \leq 0. \tag{5.6.7}$$

The conditions of a perfect bond at $x = 0$ are the continuity of $\tau(x, t)$ and of $u(x, t)$. We find

$$-\frac{C^-}{c_L^-} f'(t) + \frac{C^-}{c_L^-} g'(t) = -\frac{C^+}{c_L^+} h'(t),$$
$$f(t) + g(t) = h(t), \tag{5.6.8}$$

where the primes denote differentiation with respect to the argument of the function. If the first equation is integrated with respect to t, and it is taken into account that the body is at rest prior to the arrival of the incident wave, we find

$$-\frac{C^-}{c_L^-}f(t)+\frac{C^-}{c_L^-}g(t)=-\frac{C^+}{c_L^+}h(t). \tag{5.6.9}$$

The functions $g(t)$ and $h(t)$ can be solved from Eqs. (5.6.8) and (5.6.9) as

$$g(t)=R_{-+}f(t),$$
$$h(t)=T_{-+}f(t),$$

where

$$R_{-+}=-\frac{\rho^+c_L^+-\rho^-c_L^-}{\rho^+c_L^++\rho^-c_L^-}=-\frac{Z^+-Z^-}{Z^++Z^-}, \tag{5.6.10}$$

$$T_{-+}=1-\frac{\rho^+c_L^+-\rho^-c_L^-}{\rho^+c_L^++\rho^-c_L^-}=\frac{2Z^-}{Z^++Z^-}. \tag{5.6.11}$$

Here the relations $C=\rho c_L^2$ and $Z=\rho c_L$ have been used. The constants R and T are the reflection and transmission coefficients, respectively. Thus, the reflected and transmitted waves are

$$x\le 0:\quad u^{\rm re}(x,t)=R_{-+}f(t+x/c_L^-), \tag{5.6.12}$$

$$x\ge 0:\quad u^{\rm tr}(x,t)=T_{-+}f(t-x/c_L^+). \tag{5.6.13}$$

5.7 Use of reciprocity to determine the reflection coefficient

In Section 5.6 the reflection and transmission coefficients for an interface were determined. In the present section we show that the same results can be obtained by reciprocity considerations. The wave motion to be considered here is time harmonic. An interesting new feature is that the virtual wave is taken as propagating in a homogeneous body of one of the materials.

The geometry is shown in Fig. 5.3. Similarly to Eqs. (5.6.5), (5.6.7) we write

$$x\le 0:\quad u^{\rm in}(x)=U^Ae^{ik^-x}, \tag{5.7.1}$$

$$x\le 0:\quad u^{\rm re}(x)=R_{-+}U^Ae^{-ik^-x}, \tag{5.7.2}$$

$$x\ge 0:\quad u^{\rm tr}(x)=T_{-+}U^Ae^{ik^+x}, \tag{5.7.3}$$

where R_{-+} and T_{-+} are the reflection and transmission coefficients. For the virtual wave we write

$$-\infty<x<\infty:\quad u^B(x)=e^{ik^+x}. \tag{5.7.4}$$

Thus the virtual wave propagates in a material defined by superscripts $+$ as if no interface exists.

Since for $a < x < 0$ the materials for states A and B are different, but for $0 < x < b$ they are the same, the usual integral over $a \leq x \leq b$ splits up into two parts. Considering the case where the two materials have the same mass densities we find, since there are no body forces,

$$\int_a^0 \left(\frac{\partial \tau^A}{\partial x} u^B - \frac{\partial \tau^B}{\partial x} u^A \right) dx + \int_0^b \left(\frac{\partial \tau^A}{\partial x} u^B - \frac{\partial \tau^B}{\partial x} u^A \right) dx = 0. \quad (5.7.5)$$

Instead of evaluating the integrals it is easier to note that the first integral must be independent of a and the second independent of b. This implies that the integrands must vanish over the respective ranges $a \leq x \leq 0$ and $0 \leq x \leq b$. Using the expressions for states A and B given above, we find from the first integral

$$R_{-+} = \frac{C^- k^- - C^+ k^+}{C^+ k^+ + C^- k^-} e^{2ik^- x}, \quad (5.7.6)$$

while the integrand for the second integral vanishes identically. For $x = 0$, Eq. (5.7.6) becomes the same expression for R_{-+} as given by Eq. (5.6.10), namely,

$$R_{-+} = -\frac{Z^+ - Z^-}{Z^+ + Z^-}, \quad (5.7.7)$$

where we have used that

$$C^- = \rho^- (c_L^-)^2 \qquad \text{and} \qquad k^- = \frac{\omega}{c_L^-}, \quad (5.7.8)$$

and thus

$$C^- k^- = \omega \rho^- c_L^- = \omega Z^-, \quad (5.7.9)$$

with a similar expression for $C^+ k^+$.

5.8 Reciprocity of two one-dimensional solutions for concentrated forces

We return to the case discussed in Section 5.5, but in Eq. (5.5.7) we now take $f^B(x)$ as a concentrated force also:

$$f^B(x) = F^B \delta(x - x_2).$$

The solutions for $x \geq x_2$ and $x \leq x_2$ are found from Eqs. (5.5.9) and (5.5.10) by replacing x_1 by x_2. Equation (5.5.7) then yields

$$F^A u^B(x_1) - F^B u^A(x_2) = -\left[\tau^A u^B - \tau^B u^A\right]_{x=a}^{x=b}. \tag{5.8.1}$$

The left-hand side of this equation depends only on x_1 and x_2, while the right-hand side depends on a and b. The equality can only be satisfied for arbitrary a and b if the right-hand side vanishes. This statement may also be verified by direct substitutions of the expressions for states A and B. It follows that

$$F^A u^B(x_1) = F^B u^A(x_2). \tag{5.8.2}$$

This result can be verified by substituting the actual solutions, which follow from (5.5.9), (5.5.10) and (5.5.12), into Eq. (5.8.1). We obtain

$$\frac{U^B}{F^B} e^{-ik(x_1 - x_2)} = \frac{U^A}{F^A} e^{ik(x_2 - x_1)}. \tag{5.8.3}$$

In words: the displacement at $x = x_1$ due to a unit load at $x = x_2$ ($x_2 > x_1$) is the same as the displacement at $x = x_2$ due to a unit load at $x = x_1$. This result was first noted by Lord Rayleigh (1873).

Equation (5.8.2) may appear to be obvious. It is perhaps less obvious that the same result also holds for the case where the body consists of two materials, namely, half-spaces of materials 1 and 2 that are joined at $x = x_i$, where $x_1 < x_i < x_2$. Equation (5.5.7) is still valid, except that the integration should be carried out from $x = a$ to $x = x_i$ and from $x = x_i$ to $x = b$. Adding the two integrals gives rise to twice the same term but of opposite signs, evaluated at $x = x_i$. Continuity of stress and displacement for both solutions does eliminate the terms at $x = x_i$, however, and the final result becomes just the same as Eq. (5.5.7) and subsequently as (5.8.1). Thus Eq. (5.8.2) is also valid for the case of two materials when the forces are applied at different sides of the interface.

We can verify the result on the basis of constructed solutions, using the transmission coefficient defined by Eq. (5.6.11). The transmission coefficient for a wave incident from the negative side is denoted by T_{-+}. From the expression for T_{-+} we conclude that

$$T_{+-} = \frac{2Z^+}{Z^+ + Z^-}. \tag{5.8.4}$$

The displacement at $x = x_1$ due to a point load of magnitude F^B at $x = x_2$, where $x_1 < x_i < x_2$, may then be expressed as

$$u^B(x_1) = \frac{F^B}{2i\omega Z^+} e^{-ik_+(x_i - x_2)} T_{+-} e^{-ik_-(x_1 - x_i)}. \tag{5.8.5}$$

Similarly the displacement at $x = x_2$ due to a unit load at $x = x_1$ is

$$u^A(x_2) = \frac{F^A}{2i\omega Z^-} e^{ik_-(x_i - x_1)} T_{-+} e^{ik_+(x_2 - x_i)}. \tag{5.8.6}$$

The displacements given by (5.8.5) and (5.8.6) satisfy the reciprocity relation Eq. (5.8.2).

5.9 An inverse problem

In the discussion of reciprocity in acoustics presented in Chapter 4, it was pointed out that reciprocity theorems can be used to derive an integral equation that forms the basis for the solution of an inverse problem. As an example from one-dimensional elastodynamics, we consider in this section the problem of determining the unknown function or constant defining the elastic property of a layer. The configuration consists of two elastic domains defined by $x \leq 0$ and $x \geq l$ and the layer defined by $0 \leq x \leq l$; see Fig. 5.4.

The elastic domains $x \leq 0$ and $x \geq l$ are homogeneous with elastic constant C_0. The layer is inhomogeneous, with elastic property defined by $C(x)$.

Now let us consider an incident wave propagating in the positive x-direction towards the interface at $x = 0$. The wave is reflected with a reflection coefficient R, but it also enters the layer and is transmitted into the region $x \geq l$ with a transmission coefficient T. The following combination of waves is chosen as state A:

$$x \leq 0: \qquad u^A = U^A e^{ik_0 x} + RU^A e^{-ik_0 x}, \tag{5.9.1}$$

$$x \geq l: \qquad u^A = TU^A e^{ik_0 x}, \tag{5.9.2}$$

$$0 \leq x \leq l: \qquad u^A = U^A e^{ik_0 x} + u^{sc}(x, \hat{k}), \tag{5.9.3}$$

where $k_0 = \omega/c_0$, and $\hat{k} = \omega/\hat{c}$, $\hat{c}$ being the wave velocity in the layer. For state B, i.e., for the virtual wave, we choose

$$-\infty < x < \infty: \quad u^B = U^B e^{ik_0 x}. \tag{5.9.4}$$

It is noted that in the layer we have solutions u^A and u^B corresponding to different materials. Hence we should use the form of the reciprocity theorem given by Eq. (5.5.13) to account for the difference between C_0 and $C(x)$. When Eq. (5.5.13) is applied for $a < 0$ and $b > l$, we have

$$\int_0^l [C_0 - C(x)] \frac{\partial u^A}{\partial x} \frac{\partial u^B}{\partial x} \, dx = -\left[\tau^A u^B - \tau^B u^A\right]_{x=a}^{x=b}. \tag{5.9.5}$$

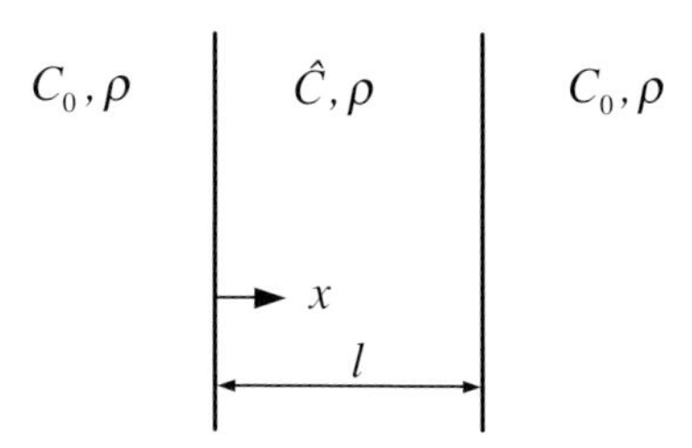

Figure 5.4 Inhomogeneous layer in between homogeneous domains.

As discussed earlier in this chapter, only the contributions from the counter-propagating waves enter the right-hand side of this equation. With this simplification Eq. (5.9.5) yields

$$R(k_0) = \frac{1}{2}\frac{1}{C_0}\frac{1}{U^A}\int_0^l \Delta C(x)\frac{\partial u^A}{\partial x}e^{ik_0x}dx, \tag{5.9.6}$$

where

$$\Delta C(x) = C(x) - C_0. \tag{5.9.7}$$

Here R would be the measured quantity. Unfortunately, in the integral not only $\Delta C(x)$ but also the term $\partial u^A/\partial x$ is unknown.

5.10 Comparison of Born approximation with exact solution

In the Born approximation it is assumed that the actual field in the inhomogeneity is not very different from the incident wave, $U^A \exp(ik_0x)$, which implies that $u^{sc}(x)$ can be neglected in Eq. (5.9.3). Substitution of the incident wave for $u^A(x)$ into Eq. (5.9.6) yields

$$R(k_0) = \frac{i}{2}\frac{k_0}{C_0}\int_0^l \Delta C(x)e^{2ik_0x}dx. \tag{5.10.1}$$

Since $\Delta C(x)$ vanishes outside the interval $(0, l)$, Eq. (5.10.1) may also be written as

$$R(k_0) = \frac{i}{2}\frac{k_0}{C_0}\int_{-\infty}^{\infty} e^{2ik_0x}\Delta C(x)\,dx. \tag{5.10.2}$$

Equation (5.10.2) has the formal appearance of an exponential Fourier transform. Hence $C(x)$ may be written in terms of the inverse transform as

$$\Delta C(x) = -4\frac{i}{\pi}C_0 \int_{-\infty}^{\infty} e^{-2ik_0x}\frac{1}{k_0}R(k_0)\,dk_0. \tag{5.10.3}$$

It is of interest to check some of the results for the Born approximation against an exact solution. The layer problem defined by Fig. 5.4 is exactly solvable when the elastic behavior of the layer is governed by a constant, i.e., when

$$C(x) = \hat{C}. \tag{5.10.4}$$

For this case the reflection coefficient given by Eq. (5.10.1) becomes

$$R(k_0) = \frac{1}{4}\frac{\hat{C} - C_0}{C_0}\left(e^{2ik_0l} - 1\right). \tag{5.10.5}$$

It can be verified that substitution of $R(k_0)$ as given by Eq. (5.10.5) into the formula for the inverse transform given by Eq. (5.10.3) yields

$$\Delta C(x) = (\hat{C} - C_0)H(x)H(l - x), \tag{5.10.6}$$

just as it should.

To obtain the exact solution for the problem of reflection by a layer of elastic constant $\hat{C}$ we recognize that the field inside the layer can be written as

$$0 \le x \le l: \quad u^A(x) = U^+e^{i\hat{k}x} + U^-e^{-i\hat{k}x}. \tag{5.10.7}$$

Here $\hat{k}$ is the wavenumber for propagation in the layer:

$$\hat{k} = \frac{\omega}{\hat{c}} = \frac{\omega}{(\hat{C}/\rho)^{1/2}}, \tag{5.10.8}$$

$$k_0 = \frac{\omega}{c_0} = \frac{\omega}{(C_0/\rho)^{1/2}}. \tag{5.10.9}$$

These expressions imply that ρ is taken as the same for the layer and the embedding. Equations (5.9.1), (5.9.2) and (5.10.7) now define the problem. The object is to determine the reflection coefficient as a function of the wavenumber k_0 that defines wave propagation in the embedding half-spaces. The reflection coefficient is one of four unknowns, namely, R, U^+, U^- and T. There are also four algebraic equations that define the continuities of displacement and traction at $x = 0$ and $x = l$. These equations are

$$\text{at} \quad x = 0: \qquad 1 + R = \overline{U}^+ + \overline{U}^-, \tag{5.10.10}$$

$$1 - R = \overline{C}\overline{U}^+ - \overline{C}\overline{U}^-; \tag{5.10.11}$$

$$\text{at} \quad x = l: \quad \overline{U}^+ e^{i\hat{k}l} + \overline{U}^- e^{-i\hat{k}l} = T e^{ik_0 l}, \tag{5.10.12}$$

$$\overline{C}\overline{U}^+ e^{i\hat{k}l} - \overline{C}\overline{U}^- e^{-i\hat{k}l} = T e^{ik_0 l}. \tag{5.10.13}$$

In these equations,

$$\overline{U}^+ = \frac{U^+}{U}, \quad \overline{U}^- = \frac{U^-}{U} \tag{5.10.14}$$

and

$$\overline{C} = \frac{i\hat{k}\hat{C}}{ik_0 C_0} = \frac{\hat{C}^{1/2}}{C_0^{1/2}}. \tag{5.10.15}$$

In Eq. (5.10.13) we have used the expressions for $\hat{k}$ and k_0 given by Eqs. (5.10.8) and (5.10.9).

The following solutions are easily obtained from Eqs. (5.10.10)–(5.10.13):

$$R = \frac{1}{2}[(1 - \overline{C})\overline{U}^+ + (1 + \overline{C})\overline{U}^-], \tag{5.10.16}$$

$$\overline{U}^- = -\frac{1 - \overline{C}}{1 + \overline{C}} e^{2i\hat{k}l}\, \overline{U}^+, \tag{5.10.17}$$

$$\overline{U}^+ = \frac{2(1 + \overline{C})}{(1 + \overline{C})^2 - (1 - \overline{C})^2 e^{2i\hat{k}l}}. \tag{5.10.18}$$

The explicit expression for R can then be obtained as

$$R = \frac{2[(1 + \overline{C}) - (1 - \overline{C})e^{2i\hat{k}l}]}{(1 + \overline{C})^2 - (1 - \overline{C})^2 e^{2i\hat{k}l}} - 1. \tag{5.10.19}$$

The general formula for the reflection coefficient that is given in Eq. (5.9.6) should of course produce the result stated by Eq. (5.10.19) if the appropriate expressions for $\Delta C(x)$ and $u^A(x)$ are substituted. By employing Eqs. (5.10.6) and (5.10.7) we obtain from (5.9.6)

$$R(k_0) = \frac{1}{2}\frac{\hat{C} - C_0}{C_0}\hat{k}\left[\frac{\overline{U}^+}{\hat{k} + k_0}\left(e^{i(\hat{k}+k_0)l} - 1\right) + \frac{\overline{U}^-}{\hat{k} - k_0}\left(e^{-i(\hat{k}-k_0)l} - 1\right)\right]. \tag{5.10.20}$$

It is not immediately apparent that this expression is the same as the one given by Eq. (5.10.16). A little manipulation using Eqs. (5.10.8) and (5.10.9) shows, however, that

$$\frac{\hat{C} - C_0}{C_0}\frac{\hat{k}}{\hat{k} + k_0} = 1 - \overline{C}, \tag{5.10.21}$$

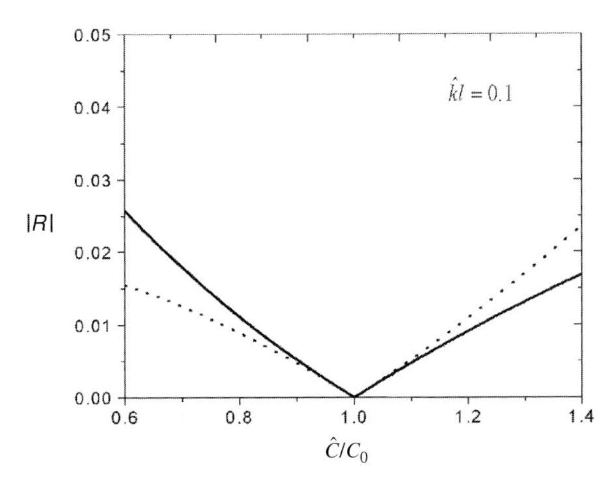

Figure 5.5 Exact reflection coefficient, Eq. (5.10.19) (solid line), and the Born approximation, Eq. (5.10.5) (dotted line), as functions of $\hat{C}/C_0$ for $\hat{k}l - 0.1$.

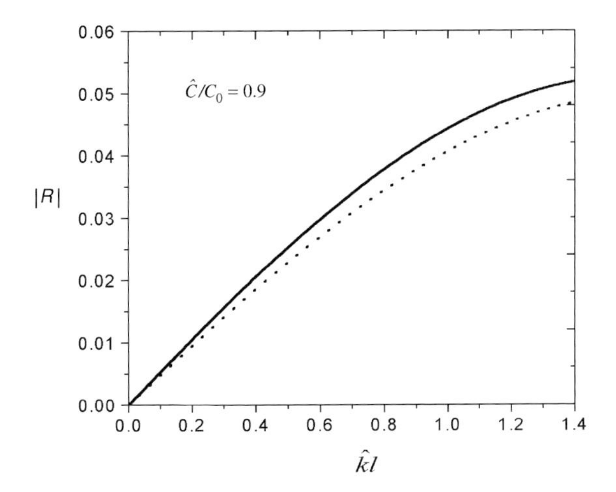

Figure 5.6 Exact reflection coefficient, Eq. (5.10.19) (solid line), and the Born approximation, Eq. (5.10.5), (dotted line), as functions of $\hat{k}l$ for $\hat{C}/C_0 = 0.9$.

$$\frac{\hat{C} - C_0}{C_0} \frac{\hat{k}}{\hat{k} - k_0} = 1 + \overline{C}, \tag{5.10.22}$$

where $\overline{C}$ is defined by Eq. (5.10.15). Subsequent replacement of $\overline{U}^{-}$ by the use of (5.10.17), but only in the term with $\exp[-i(\hat{k} - k_0)l]$, then reduces Eq. (5.10.20) to (5.10.16).

How good is the Born approximation? The question can be answered for the special case of normal incidence on a homogeneous layer of different elastic constants located in between two homogeneous domains. Figure 5.5 shows that for $\hat{k}l = 0.1$ the approximation is quite good for $0.8 < \hat{C}/C < 1.2$. Figure 5.6. shows that for $\hat{C}/C_0 = 0.9$ the accuracy decreases as $\hat{k}l$ increases, i.e., as the wavelength decreases and the frequency increases.

The Born approximation can be considered as the first term of a perturbation expansion for small values of $\Delta C(x)$. For the one-dimensional case, a complete perturbation expansion approach was presented by Cohen and Bleistein (1977).

6
Reciprocity in two- and three-dimensional elastodynamics

6.1 Introduction

In Chapter 1, a formal definition of a reciprocity theorem for elastodynamic states was stated as: "A reciprocity theorem relates, in a specific manner, two admissible elastodynamic states that can occur in the same time-invariant linearly elastic body. Each of the two states can be associated with its own set of time-invariant material parameters and its own set of loading conditions. The domain to which the reciprocity theorem applies may be bounded or unbounded."

Reciprocity theorems for elastodynamics in one-dimensional geometries were stated in Chapter 5. In the present chapter analogous theorems for three-dimensional elastodynamics are presented, as well as some applications. The most useful reciprocity theorems are for elastodynamic states in the frequency and Laplace transform domains. We also discuss reciprocity in a two-material body and reciprocity theorems for linearly viscoelastic solids.

For the time-harmonic case a number of applications of reciprocity in elastodynamics are considered. Some of the examples are concerned with the reciprocity of fields generated by point forces in bounded and unbounded elastic bodies. Other cases are concerned with the solution of the wave equation with polar symmetry and with reciprocity for plane waves reflected from a free surface.

Another purpose of the chapter is to provide insight on the applicability of reciprocity considerations, together with the use of a virtual wave, as a tool to obtain solutions for elastodynamic problems. Some examples are concerned with two-dimensional cases for anti-plane strain. These examples are very simple. In the literature their solutions have usually been obtained by the use of Fourier transform techniques. For example, the anti-plane elastodynamic field generated by a line load can be solved in an equally simple or simpler way by a number of other methods. An interesting case is provided by the displacement field in a half-space for the case where an anti-plane line load is applied in its interior, and the surface of the half-space is constrained by a boundary condition. For that case, the surface wave calculated by reciprocity considerations is verified by the result obtained by the conventional integral transform method. An example in the next to last section of this chapter deals with the establishment of domain integral equations for scattering by an inclusion. In the last section we briefly review some examples of the application of reciprocity from the technical literature.

In later chapters we discuss more complicated applications of the determination of elastodynamic fields by reciprocity considerations in conjunction with the use of a virtual wave.

6.2 Reciprocity theorems

The point of departure for the statement of reciprocity theorems for three-dimensional elastodynamics is Eq. (2.2.8). This equation is written for two elastodynamic states indicated by superscripts A and B:

$$\tau_{ij,i}^{A} + f_j^{A} = \rho \ddot{u}_j^{A}, \tag{6.2.1}$$

$$\tau_{ij,i}^{B} + f_j^{B} = \rho \ddot{u}_j^{B}. \tag{6.2.2}$$

The first and the second equation are multiplied by $u_j^B(\boldsymbol{x}, t)$ and $u_j^A(\boldsymbol{x}, t)$, respectively. Subtraction of the second equation so obtained from the first yields

$$\left(f_j^A - \rho \ddot{u}_j^A\right) u_j^B - \left(f_j^B - \rho \ddot{u}_j^B\right) u_j^A = -\left(\tau_{ij,i}^A u_j^B - \tau_{ij,i}^B u_j^A\right). \tag{6.2.3}$$

Now consider

$$\left(\tau_{ij,i}^A u_j^B - \tau_{ij,i}^B u_j^A\right) = \left(\tau_{ij}^A u_j^B - \tau_{ij}^B u_j^A\right)_{,i} - \left(\tau_{ij}^A u_{j,i}^B - \tau_{ij}^B u_{j,i}^A\right). \tag{6.2.4}$$

By using (2.2.3) and (2.2.4), the second pair of terms on the right-hand side can be simplified as follows:

$$\begin{aligned}
\tau_{ij}^A u_{j,i}^B - \tau_{ij}^B u_{j,i}^A &= \tau_{ij}^A \left(\varepsilon_{ij}^B - \omega_{ij}^B\right) - \tau_{ij}^B \left(\varepsilon_{ij}^A - \omega_{ij}^A\right) \\
&= \tau_{ij}^A \varepsilon_{ij}^B - \tau_{ij}^B \varepsilon_{ij}^A,
\end{aligned}$$

where the symmetry of τ_{ij} and ε_{ij} and the antisymmetry of ω_{ij} have been used. Next we use Hooke's law as given by Eq. (2.2.10) to obtain

$$\tau_{ij}^A u_{j,i}^B - \tau_{ij}^B u_{j,i}^A = C_{ijkl} \varepsilon_{kl}^A \varepsilon_{ij}^B - C_{ijkl} \varepsilon_{kl}^B \varepsilon_{ij}^A = 0.$$

By using this result in Eqs. (6.2.3) and (6.2.4) we find

$$\left(f_j^A - \rho \ddot{u}_j^A\right) u_j^B - \left(f_j^B - \rho \ddot{u}_j^B\right) u_j^A = \left(\tau_{ij}^B u_j^A - \tau_{ij}^A u_j^B\right)_{,i}. \qquad (6.2.5)$$

This is the local reciprocity theorem. The global theorem is obtained by integrating Eq. (6.2.5) over a region V with boundary S. The volume integration of the right-hand side is converted into a surface integral by the use of Gauss' theorem. The result is

$$\int_V \left[\left(f_j^A - \rho \ddot{u}_j^A\right) u_j^B - \left(f_j^B - \rho \ddot{u}_j^B\right) u_j^A\right] dV = \int_S \left(\tau_{ij}^B u_j^A - \tau_{ij}^A u_j^B\right) n_i \, dS, \qquad (6.2.6)$$

where $\boldsymbol{n}$ is the unit vector along the outward normal to S. It is noted that all terms under the integrals depend both on position $\boldsymbol{x}$ and time t. The time does not have to be the same for the two elastodynamic states defined by superscripts A and B; they can refer to different times t_1 and t_2.

For the steady-state time-harmonic case where

$$f_i(\boldsymbol{x}, t) = f_i(\boldsymbol{x}) e^{-i\omega t}, \quad u_i(\boldsymbol{x}, t) = u_i(\boldsymbol{x}) e^{-i\omega t}, \quad \tau_{ij}(\boldsymbol{x}, t) = \tau_{ij}(\boldsymbol{x}) e^{-i\omega t}, \qquad (6.2.7)$$

the terms corresponding to second-order time derivatives cancel each other, and Eq. (6.2.6) reduces to

$$\int_V \left(f_j^A u_j^B - f_j^B u_j^A\right) dV = \int_S \left(\tau_{ij}^B u_j^A - \tau_{ij}^A u_j^B\right) n_i \, dS. \qquad (6.2.8)$$

The terms in the integrals now depend only on position $\boldsymbol{x}$. It is evident that the equation is the same for the static and time-harmonic cases.

In the derivations of Eqs. (6.2.6) and (6.2.8) it has been assumed only that a linear relation exists between stresses and strains. The solid may be anisotropic, inhomogeneous or viscoelastic.

6.3 Reciprocity in the time domain

Equations (6.2.1) and (6.2.2) can be used as point of departure for a reciprocity theorem in the time domain. This theorem can be obtained in an easy way by application of the Laplace transform to the two equations. The Laplace transform was first used for this purpose by Graffi (1946).

After application of the Laplace transform, Eqs. (6.2.1) and (6.2.2) become

$$\overline{\tau}^A_{ij,i} + \overline{f}^A_j = \rho s^2 \overline{u}^A_j - \rho\left[s u^A_j(0^+) + \dot{u}^A(0^+)\right], \tag{6.3.1}$$

$$\overline{\tau}^B_{ij,i} + \overline{f}^B_j = \rho s^2 \overline{u}^B_j - \rho\left[s u^B_j(0^+) + \dot{u}^B(0^+)\right]. \tag{6.3.2}$$

It is convenient to introduce a new body-force term as

$$\overline{g}^A_j = \overline{f}^A_j + \rho\left[s u^A_j(0^+) + \dot{u}^A_j(0^+)\right],$$

with an analogous definition for $\overline{g}^B_j$. Next we observe that Eqs. (6.3.1) and (6.3.2) then become completely equivalent to the corresponding equations for the time-harmonic case, except that $-\rho\omega^2$ must be replaced by ρs^2. Manipulations analogous to the ones that follow Eqs. (6.2.1) and (6.2.2) then lead to the following equation, where all field quantities are Laplace transforms:

$$\int_V \left(\overline{g}^A_j \overline{u}^B_j - \overline{g}^B_j \overline{u}^A_j\right) dV = \int_S \left(\overline{\tau}^B_{ij} \overline{u}^A_j - \overline{\tau}^A_{ij} \overline{u}^B_j\right) n_i \, dS. \tag{6.3.3}$$

In terms of the original body forces, and using Eq. (2.2.6), we can then write

$$\begin{aligned} &\int_S \overline{t}^A_j \overline{u}^B_j dS + \int_V \{\overline{f}^A_j + \rho[s u^A_j(0^+) + \dot{u}^A_j(0^+)]\}\overline{u}^B_j \, dV \\ &\quad = \int_S \overline{t}^B_j \overline{u}^A_j dS + \int_V \{\overline{f}^B_j + \rho[s u^B_j(0^+) + \dot{u}^B_j(0^+)]\}\overline{u}^A_j \, dV, \end{aligned} \tag{6.3.4}$$

where the surface traction $\overline{t}_j = \overline{\tau}_{ij} n_i$, i.e., Eq. (2.2.6), has been used. By using the convolution theorem, the inverse Laplace transform of Eq. (6.3.4) is subsequently obtained as

$$\begin{aligned} &\int_S t^A_j(\mathbf{x}, t) * u^B_j(\mathbf{x}, t) \, dS + \int_V f^A_j(\mathbf{x}, t) * u^B_j(\mathbf{x}, t) \, dV \\ &+ \rho \int_V \left[u^A_j(\mathbf{x}, 0^+)\dot{u}^B_j(\mathbf{x}, t) + \dot{u}^A_j(\mathbf{x}, 0^+) u^B_j(\mathbf{x}, t)\right] dV \\ &\quad = \int_S t^B_j(\mathbf{x}, t) * u^A_j(\mathbf{x}, t) \, dS + \int_V f^B_j(\mathbf{x}, t) * u^A_j(\mathbf{x}, t) \, dV \\ &\quad + \rho \int_V \left[u^B_j(\mathbf{x}, 0^+)\dot{u}^A_j(\mathbf{x}, t) + \dot{u}^B_j(\mathbf{x}, 0^+) u^A_j(\mathbf{x}, t)\right] dV. \end{aligned} \tag{6.3.5}$$

Equation (6.3.5) is valid for $t > 0$.

Just as for acoustics, other reciprocity theorems, often referred to as power reciprocity theorems, can be formulated. In their application to elastodynamics these theorems are not further discussed in this book. They can be found in the book by de Hoop (1995).

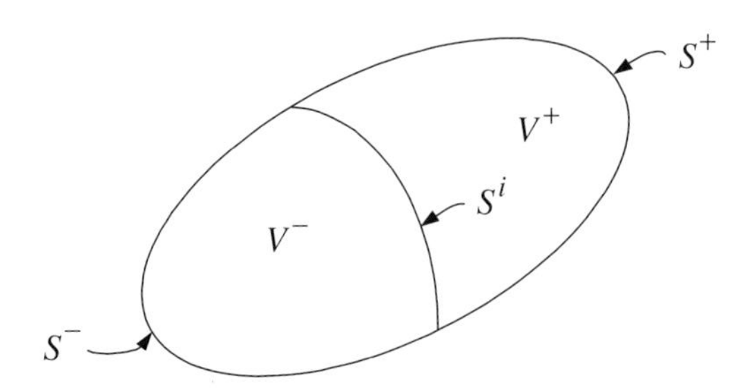

Figure 6.1 Body composed of two bodies of different materials with interface S^i.

6.4 Reciprocity in a two-material body

Let us now consider a body with an interface which separates two parts that are of different materials. The geometry is shown in Fig. 6.1. The field variables and the elastic constants on one side of the interface are indicated by plus signs, while on the other side they are indicated by minus signs. The interface is indicated by S^i. The external boundaries of the body are indicated by S^+ and S^-, while the corresponding internal regions are V^+ and V^-.

We will focus our attention on the time-harmonic case. For the volume V^+, Eq. (6.2.8) may then be rewritten as

$$\int_{V^+} \left(f_j^{A+}u_j^{B+} - f_j^{B+}u_j^{A+}\right)dV = \int_{S^+ + S^i} \left(\tau_{ij}^{B+}u_j^{A+} - \tau_{ij}^{A+}u_j^{B+}\right)n_i\, dS. \quad (6.4.1)$$

Similarly for the volume V^- we have

$$\int_{V^-} \left(f_j^{A-}u_j^{B-} - f_j^{B-}u_j^{A-}\right)dV = \int_{S^- + S^i} \left(\tau_{ij}^{B-}u_j^{A-} - \tau_{ij}^{A-}u_j^{B-}\right)n_i\, dS. \quad (6.4.2)$$

Across the interface the tractions and displacements are continuous. Since the normals along S^i are in opposite directions their contributions cancel when the two equations are added. We obtain

$$\int_{V^+} (\quad)dV + \int_{V^-} (\quad)dV = \int_{S^+} (\quad)n_i\, dS + \int_{S^-} (\quad)n_i\, dS. \quad (6.4.3)$$

The expressions in the parentheses follow from (6.4.1) and (6.4.2). The difference in the materials enters when the stress–strain relations are substituted.

6.5 Reciprocity theorems for linearly viscoelastic solids

In the frequency and Laplace transform domains, reciprocity theorems for linearly viscoelastic solids follow immediately from the corresponding reciprocity theorems for elastodynamics by simply replacing the stress–strain relations. For example, in the frequency domain we start with Eq. (6.2.8),

$$\int_V \left(f_j^A u^B - f_j^B u_j^A\right) dV = \int_S \left(\tau_{ij}^B u_j^A - \tau_{ij}^A u_j^B\right) n_i \, dS. \qquad (6.5.1)$$

However, we now use the stress–strain relation with complex moduli given by Eq. (2.5.13),

$$\tau_{ij} = \delta_{ij}\left[G_B(\omega) - \tfrac{2}{3}G_S(\omega)\right]\varepsilon_{kk} + 2G_S(\omega)\varepsilon_{ij}. \qquad (6.5.2)$$

In the Laplace transform domain the equation to start with is Eq. (6.3.3),

$$\int_V \left(\overline{g}_j^A \overline{u}_j^B - \overline{g}_j^B \overline{u}_j^A\right) dV = \int_S \left(\overline{\tau}_{ij}^B \overline{u}_j^A - \overline{\tau}_{ij}^A \overline{u}_j^B\right) n_i \, dS. \qquad (6.5.3)$$

The relevant stress–strain relation follows from the Laplace transform of Eq. (2.5.7) as

$$\overline{\tau}_{ij} = \delta_{ij}\left[\overline{G}_B(s) - \tfrac{2}{3}\overline{G}_S(s)\right]\overline{\varepsilon}_{kk} + 2\overline{G}_S(s)\overline{\varepsilon}_{ij}, \qquad (6.5.4)$$

where $\overline{G}_B(s)$ and $\overline{G}_S(s)$ are the Laplace transforms of the relaxation functions $G_B(t)$ and $G_S(t)$.

6.6 Anti-plane line load in an unbounded elastic solid

We now discuss some examples of the application of reciprocity considerations to obtain the solutions to elastodynamic problems directly. For example, in this section it is shown that the field radiated by a time-harmonic anti-plane line load can be determined by the use of reciprocity considerations. The governing equations for the anti-plane case are given by Eqs. (2.4.3), (2.4.4). The one displacement component is parallel to the applied line load, and the displacement field is axially symmetric with respect to the load. The line load is time harmonic. In polar coordinates the displacement is denoted by $w(r)$, where the term $\exp(-i\omega t)$ has been omitted. Writing the Laplacian in polar coordinates we have

$$\frac{d^2w}{dr^2} + \frac{1}{r}\frac{dw}{dr} + k^2 w + \frac{f(r)}{\mu} = 0, \qquad k = \omega/c_T, \qquad (6.6.1)$$

where $r^2 = x_1^2 + x_2^2$. The ratio ω/c_T is usually denoted by k_T, but for simplicity of notation we omit the subscript in this section. The relevant stress is τ_{rz}, given by

$$\tau_{rz} = \mu \frac{\partial w}{\partial r}. \tag{6.6.2}$$

The solution of the homogeneous form of Eq. (6.6.1), which represents an outgoing wave compatible with $\exp(-i\omega t)$, is

$$w(r) = H_0^{(1)}(kr). \tag{6.6.3}$$

We can equally well consider an incoming wave, i.e., a wave that converges on the origin:

$$\overline{w}(r) = H_0^{(2)}(kr). \tag{6.6.4}$$

Here it should be recalled that the Hankel functions of the first and second kind are defined as, see e.g. McLachlan (1961),

$$H_\nu^{(1)}(z) = J_\nu(z) + iY_\nu(z), \tag{6.6.5}$$

$$H_\nu^{(2)}(z) = J_\nu(z) - iY_\nu(z). \tag{6.6.6}$$

For large values of z, asymptotic approximations to these functions are defined as

$$H_\nu^{(1)}(z) \sim \left(\frac{2}{\pi z}\right)^{1/2} e^{i(z-\pi/4-\nu\pi/2)}, \tag{6.6.7}$$

$$H_\nu^{(2)}(z) \sim \left(\frac{2}{\pi z}\right)^{1/2} e^{-i(z-\pi/4-\nu\pi/2)}. \tag{6.6.8}$$

In conjunction with $\exp(-i\omega t)$ these asymptotic approximations clearly define an outgoing and an incoming wave, respectively.

For two distinct axially symmetric time-harmonic anti-plane states A and B, the reciprocity identity given by Eq. (6.2.8) is now applied over a circular area of radius a. It takes the form

$$\int_V \left(f_z^A w^B - f_z^B w^A\right) dV = 2\pi a \left(\tau_{rz}^B w^A - \tau_{rz}^A w^B\right)_{r=a},$$

where axial symmetry has been used. For state A we take the displacement and stress generated by the line load. Thus

$$f_z^A(x_1, x_2) = F\delta(x_1)\delta(x_2), \tag{6.6.9}$$

where F has the dimension of force/length, and

$$w^A(r) = RH_0^{(1)}(kr), \tag{6.6.10}$$

$$\tau_{rz}^A(r) = -\mu k R H_1^{(1)}(kr), \tag{6.6.11}$$

where R is a radiation constant that is to be determined. For state B we select a virtual wave consisting of the sum of an outgoing and a converging wave,

$$w^B(r) = \frac{1}{2}\left[H_0^{(1)}(kr) + H_0^{(2)}(kr)\right], \tag{6.6.12}$$

with corresponding stress

$$\tau_{rz}^B(r) = -\frac{1}{2}\mu k\left[H_1^{(1)}(kr) + H_1^{(2)}(kr)\right]. \tag{6.6.13}$$

The combination of waves represented by Eqs. (6.6.12) and (6.6.13) has the advantage that its fields are bounded at $r = 0$. Substitution of Eqs. (6.6.5)–(6.6.9) into the reciprocity relation yields

$$\begin{aligned}\frac{F}{2\pi a} = &-\frac{1}{2}\mu k R H_0^{(1)}(ka)\left[H_1^{(1)}(ka) + H_1^{(2)}(ka)\right] \\ &+ \frac{1}{2}\mu k R\left[H_0^{(1)}(ka) + H_0^{(2)}(ka)\right]H_1^{(1)}(ka).\end{aligned} \tag{6.6.14}$$

The only terms that do not cancel each other are those corresponding to counter-propagating waves, and Eq. (6.6.14) reduces to

$$\frac{F}{2\pi a} = -\frac{1}{2}\mu k R\left[H_0^{(1)}(ka)H_1^{(2)}(ka) - H_1^{(1)}(ka)H_0^{(2)}(ka)\right]. \tag{6.6.15}$$

This relation can be simplified by using the general formula

$$H_{\nu+1}^{(1)}(z)H_\nu^{(2)}(z) - H_\nu^{(1)}(z)H_{\nu+1}^{(2)}(z) = -\frac{4i}{\pi z}, \tag{6.6.16}$$

for $\nu = 0$; see Abramowitz and Stegun (1964), p. 360. By combining Eqs. (6.6.15) and (6.6.16) we then find

$$\frac{F}{2\pi a} = -\frac{1}{2}\mu k R \frac{4i}{\pi k a}, \tag{6.6.17}$$

or

$$R = \frac{i}{4}\frac{F}{\mu}. \tag{6.6.18}$$

The same problem can also be worked out in a somewhat more complicated manner by considering for state B a plane wave propagating in the x-direction:

$$w^B(r, \theta) = e^{ikr\cos\theta}, \tag{6.6.19}$$

$$\tau_{rz}^B(r, \theta) = i\mu k\cos\theta\, e^{ikr\cos\theta}. \tag{6.6.20}$$

Now we do have to integrate over the angle θ, for $r = a$. The reciprocity relation yields:

$$F = a\mu kR \int_0^{2\pi} \left[i H_0^{(1)}(ka) \cos\theta \, e^{ika\cos\theta} + e^{ika\cos\theta} H_1^{(1)}(ka) \right] d\theta, \quad (6.6.21)$$

or

$$F = 2\pi a\mu kRI(ka), \quad (6.6.22)$$

where

$$I(ka) = \frac{1}{2\pi} H_0^{(1)}(ka) \int_0^{2\pi} i \cos\theta \, e^{ika\cos\theta} \, d\theta + \frac{1}{2\pi} H_1^{(1)}(ka) \int_0^{2\pi} e^{ika\cos\theta} \, d\theta. \quad (6.6.23)$$

To evaluate the integrals we need some further results on Bessel functions. In particular we will use

$$J_n(z) = \frac{i^{-n}}{2\pi} \int_0^{2\pi} \cos(n\theta) \, e^{iz\cos\theta} \, d\theta; \quad (6.6.24)$$

see McLachlan (1961), p. 192, and a relation similar to Eq. (6.6.24), given by Abramowitz and Stegan (1964), p. 360, as

$$J_{\nu+1}(z)Y_\nu(z) - J_\nu(z)Y_{\nu+1}(z) = \frac{2}{\pi z}. \quad (6.6.25)$$

Now using the definitions of $H_0^{(1)}(ka)$ and $H_1^{(1)}(ka)$ given by (6.6.5), together with Eq. (6.6.25), the expression for I given by Eq. (6.6.23) can be rewritten as

$$I(ka) = -[J_0(ka) + iY_0(ka)]J_1(ka) + [J_1(ka) + iY_1(ka)]J_0(ka). \quad (6.6.26)$$

This expression reduces to

$$I(ka) = -iY_0(ka)J_1(ka) + iY_1(ka)J_0(ka). \quad (6.6.27)$$

By the use of Eq. (6.6.25) we subsequently obtain

$$I(ka) = -\frac{2i}{\pi ka}. \quad (6.6.28)$$

Substitution of this result into Eq. (6.6.22) yields finally

$$R = -\frac{F}{2\pi a\mu k} \frac{\pi ka}{2i} = \frac{i}{4} \frac{F}{\mu}, \quad (6.6.29)$$

just the same as Eq. (6.6.18).

It is of interest to note that if the time factor had been taken as $\exp(+i\omega t)$, Eq. (6.6.3) would have been $w(r) = H_0^{(2)}(kr)$ and Eq. (6.6.18) would have become $R = -(i/4)(F/\mu)$.

6.7 Anti-plane surface waves

Anti-plane surface waves are waves that propagate with the displacement parallel to a surface and with an amplitude that decays exponentially with distance away from the surface. It is easy to show that such surface waves do not exist when the surface is free of tractions. For the two-dimensional case it was discovered by Love (1967) that anti-plane surface waves do exist in a half-space that is covered by an elastic layer. The presence of the layer gives rise to a dispersion equation that relates the surface wave velocity to the angular frequency of the wave.

In a two-dimensional geometry, relative to an xyz-coordinate system, the anti-plane displacement, $w(\boldsymbol{x})\exp(-i\omega t)$, is governed by the two-dimensional Helmholtz equation

$$\nabla^2 w + k_T^2 w = 0, \tag{6.7.1}$$

where $k_T = \omega/c_T$ and $\nabla^2 = \partial^2/\partial x^2 + \partial^2/\partial y^2$. We consider a half-space defined by $y \geq 0$, as shown in Fig. 6.2. Anti-plane surface waves in the half-space are then of the general form

$$w(x, y) = A\exp(\pm ik_S x - \gamma_S y), \tag{6.7.2}$$

where $k_S = \omega/c_S$ is the wavenumber of the surface wave, c_S is its phase velocity and $\mathcal{R}(\gamma_S) > 0$. Since (6.7.2) must satisfy the wave equation (6.7.1), we have

$$\gamma_S = \left(k_S^2 - k_T^2\right)^{1/2}. \tag{6.7.3}$$

For simplicity we consider a boundary condition at $y = 0$ rather than a covering elastic layer. The condition states that

$$\frac{\partial w}{\partial y} = -k_T C w, \tag{6.7.4}$$

where C is a constant. This boundary condition applies when the half-space is covered by a layer of material of thickness h that has no internal cohesion but whose mass must be taken into account. By considering the motion of an element of the layer we can write

$$\tau_{yz} = \mu\frac{\partial w}{\partial y} = \rho' h(-\omega^2)w, \tag{6.7.5}$$

where ρ' is the mass density of the layer. It follows that the constant C in Eq. (6.7.4) is

$$C = k_T h\frac{\rho'}{\rho}.$$

Application of the boundary condition yields

$$\gamma_S = k_T C \qquad \text{or} \qquad k_S^2 = (1 + C^2)k_T^2. \tag{6.7.6}$$

For free surface waves this equation gives the velocity of the surface waves.

Let us now consider the anti-plane surface wave motion generated by an anti-plane line load at $x = 0$, $y = l$. In the reciprocity theorem Eq. (6.2.8), we then have

$$f_z^A = F\delta(x)\delta(y - l). \tag{6.7.7}$$

This force radiates surface waves of the following form

$$x > 0: \qquad w^A = Re^{ik_S x - \gamma_S y}, \tag{6.7.8}$$

$$x < 0: \qquad w^A = Re^{-ik_S x - \gamma_S y}, \tag{6.7.9}$$

where R is a radiation constant that will be determined by reciprocity considerations.

For the present problem the reciprocity relation given by Eq. (6.2.8) is of the form

$$\int_V \left(f_z^A w^B - f_z^B w^A\right) dV = \int_S \left(\tau_{\alpha z}^B w^A - \tau_{\alpha z}^A w^B\right) n_\alpha \, dS, \qquad \alpha = x, y. \tag{6.7.10}$$

The contour for integration along S is chosen as a rectangle with integrations at $x = b$, $\infty > y \geq 0$, and $x = a$, $0 \leq y < \infty$. The part of the contour at $y \to \infty$ is left out since the surface waves have exponentially decayed. The contribution along $y = 0$ vanishes because $\partial w^A/\partial y$ has the same relation to w^A as $\partial w^B/\partial y$ to w^B, both being given by Eq. (6.7.4). The line load also generates a cylindrical body wave but, at large enough values of a and b, these waves are negligible since they decay as $r^{-1/2}$. The geometry is shown in Fig. 6.2.

State A is represented by Eqs. (6.7.8), (6.7.9). For state B we choose $f_z^B \equiv 0$ and a virtual wave propagating in the positive x-direction,

$$w^B = e^{ik_S x - \gamma_S y}, \tag{6.7.11}$$

$$\tau_{xz}^B = ik_S \mu e^{ik_S x - \gamma_S y}. \tag{6.7.12}$$

Substitution of the expressions for states A and B into Eq. (6.7.10) yields

$$Fe^{-\gamma_S l} = ik_S \mu R \left\{ \int_\infty^0 \left[e^{2(ik_S b - \gamma_S y)} - e^{2(ik_S b - \gamma_S y)}\right] dy - \int_0^\infty \left[e^{-2\gamma_S y} + e^{-2\gamma_S y}\right] dy \right\}. \tag{6.7.13}$$

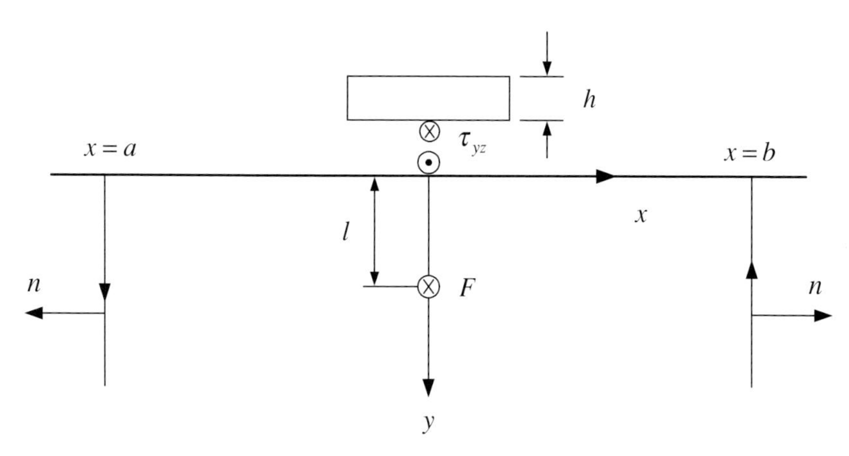

Figure 6.2 Anti-plane line load in a half-space.

Note that the waves propagating in the same direction in the first integral at $x = b$ cancel each other, while the counter-propagating waves in the second integral at $x = a$ produce a contribution that allows us to calculate R. The result is

$$R = -\frac{\gamma_S}{ik_S\mu} F e^{-\gamma_S l} = \frac{ik_T C}{k_S\mu} F e^{-\gamma_S l}. \tag{6.7.14}$$

It should be noted that the strength of the surface wave decays exponentially as the distance of the load from the surface increases.

Conventional calculation of surface waves

To verify the results of this section, the surface waves will now be calculated by the usual integral transform technique. After all, the concentrated force produces both cylindrical body waves and plane surface waves, but the cylindrical waves are totally ignored in the application of the reciprocity theorem. However, as we shall see, the reciprocity theorem gives the correct result.

The governing equation is

$$\frac{\partial^2 w}{\partial x^2} + \frac{\partial^2 w}{\partial y^2} + k_T^2 w = -\frac{F}{\mu}\delta(x)\delta(y - l). \tag{6.7.15}$$

The exponential Fourier transform of $w(x, y)$ is defined in the usual manner as

$$\tilde{w}(k, y) = \int_{-\infty}^{\infty} e^{-ikx} w(x, y)\, dx. \tag{6.7.16}$$

The inverse transform is

$$w(x, y) = \frac{1}{2\pi} \int_{-\infty}^{\infty} e^{ikx} \tilde{w}(k, y)\, dk. \tag{6.7.17}$$

Application of the exponential Fourier transform to Eq. (6.7.15) reduces that equation to an ordinary differential equation,

$$\frac{d^2\tilde{w}}{dy^2} - \gamma^2 \tilde{w} = -\frac{F}{\mu}\delta(y - l), \tag{6.7.18}$$

where

$$\gamma^2 = k^2 - k_T^2. \tag{6.7.19}$$

The solution of Eq. (6.7.18) suffers a discontinuity in its first-order derivative at $x = l$. At $x = l$ we have, therefore,

$$\tilde{w}(k, l^+) = \tilde{w}(k, l^-), \tag{6.7.20}$$

$$\frac{d\tilde{w}}{dy}(k, l^+) - \frac{d\tilde{w}}{dy}(k, l^-) = -\frac{F}{\mu}. \tag{6.7.21}$$

In addition we have

$$y = 0: \qquad \frac{d\tilde{w}}{dy} = -k_T C w, \tag{6.7.22}$$

$$y \to \infty: \qquad \tilde{w}(k, y) \sim e^{-\gamma y}. \tag{6.7.23}$$

It may be verified that the solution satisfying Eqs. (6.7.18)–(6.7.23) is

$$0 \le y \le l: \quad \tilde{w}(k, y) = \frac{F}{2\mu\gamma}\left[e^{-\gamma(l-y)} + Re^{-\gamma(y+l)}\right] \tag{6.7.24}$$

$$y \ge l: \quad \tilde{w}(k, y) = \frac{F}{2\mu\gamma}\left[e^{-\gamma(y-l)} + Re^{-\gamma(y+l)}\right], \tag{6.7.25}$$

where

$$R = \frac{\gamma + k_T C}{\gamma - k_T C}. \tag{6.7.26}$$

The expression for R can be rewritten as

$$R = \frac{(\gamma + k_T C)^2}{k^2 - k_T^2(1 + C^2)} = \frac{(\gamma + k_T C)^2}{(k + k_S)(k - k_S)}, \tag{6.7.27}$$

where k_S follows from Eq. (6.7.6).

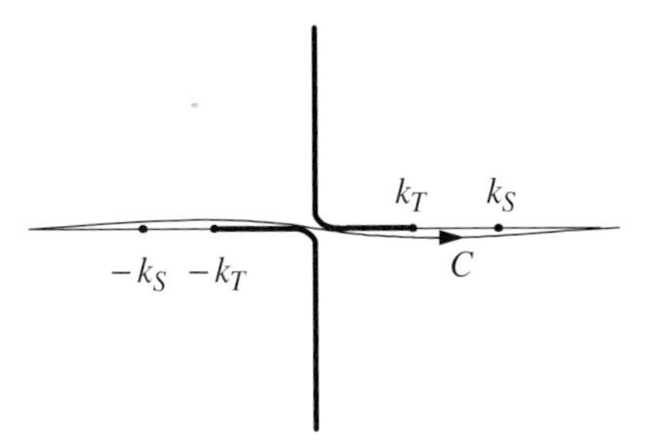

Figure 6.3 Branch cuts, poles and path of integration, C, in the k-plane.

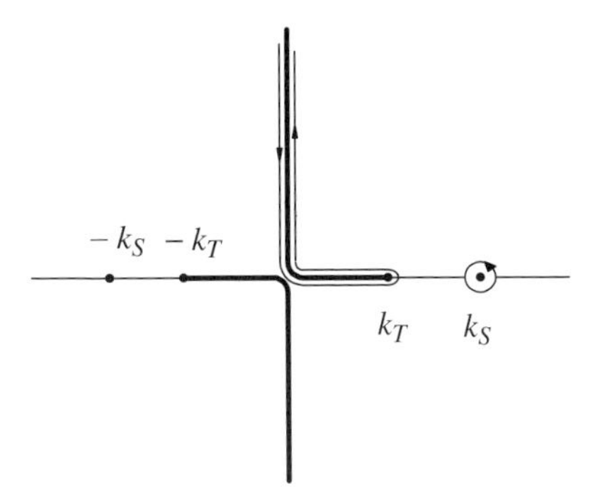

Figure 6.4 Integration around branch cut for cylindrical body waves.

In the domain $0 \leq y \leq l$, the inverse transform of $\tilde{w}(k, y)$ may now be written as

$$w(x, y) = \frac{F}{4\pi\mu} \int_C e^{ikx} \frac{1}{\gamma} \left[e^{-\gamma(l-y)} + Re^{-\gamma(y+l)} \right] dk, \qquad (6.7.28)$$

where C is a contour that is slightly deformed from the real axis in the k-plane, as shown in Fig. 6.3. It is noted that the presence of γ requires the introduction of branch cuts with branch points at $k = \pm k_T$ to render γ single valued, while the presence of R gives rise to poles at $k = \pm k_S$. An appropriate selection of branches such that $\mathcal{R}(\gamma) > 0$ in the whole k-plane and $\mathcal{I}(\gamma) < 0$ in the second and fourth quadrants, through which C passes, has been discussed in some detail by Harris (2001, p. 52). The branch cuts, the contour C and the poles are shown in Fig. 6.3. In the usual manner the contour is now closed in the upper half-plane, and the integration along C is replaced by the contribution from the pole at $k = k_S$ and the contribution from the integration around the branch cut in the upper half-plane. These contributions are indicated by Fig. 6.4. The branch cut integral corresponds to two cylindrical waves, one emanating from $(0, l)$ and the other from $(0, -l)$.

The contribution from the pole corresponds to the surface wave. Using Cauchy's residue theorem, the latter is calculated as

$$w_S(x, y) = \frac{F}{4\pi\mu} 2\pi i \frac{1}{k_T C} \frac{(2k_T C)^2}{2k_S} e^{ik_S x} e^{-k_T C(y+l)} \tag{6.7.29}$$

$$= \frac{ik_T C}{\mu k_S} F e^{ik_S x} e^{-k_T C(y+l)}. \tag{6.7.30}$$

It should be noted that the constant multiplying the exponentials agrees with the one calculated, in a much simpler manner, in Eq. (6.7.14) by the use of reciprocity considerations.

6.8 The wave equation with polar symmetry

In polar symmetry the inhomogeneous wave equation for a time-harmonic source reduces to

$$\frac{1}{R^2}\frac{\partial}{\partial R}\left(R^2 \frac{\partial \varphi}{\partial R}\right) - \frac{1}{c^2}\frac{\partial^2 \varphi}{\partial t^2} = \frac{\delta(R)}{4\pi R^2} e^{-i\omega t}. \tag{6.8.1}$$

For the steady-state case with polar symmetry we seek a solution for the potential in the form

$$\varphi(R, t) = \varphi(R) e^{-i\omega t}. \tag{6.8.2}$$

It then follows from the results of Section 3.3 that an outgoing wave is of the form

$$\varphi(R) = \frac{Q}{R} e^{ikR}, \tag{6.8.3}$$

where $k = \omega/c$. The amplitude constant Q is determined as another example of the application of a reciprocity relation.

For the case of polar symmetry the reciprocity relation follows from Eq. (4.5.4) as

$$4\pi \int_0^{R_0} \left[q^A(R)\varphi^B(R) - q^B(R)\varphi^A(R)\right] R^2\, dR$$

$$= 4\pi R_0^2 \left[\varphi^B(R)\frac{\partial \varphi^A}{\partial R} - \varphi^A(R)\frac{\partial \varphi^B}{\partial R}\right]_{R=R_0}, \tag{6.8.4}$$

where R_0 is the radius of the spherical domain V.

Now we take

$$q^A(R) = \frac{\delta(R)}{4\pi R^2}, \tag{6.8.5}$$

and we seek a solution for $\varphi^A(R)$ in the form given by Eq. (6.8.3). Similarly to earlier analysis, we select a virtual-wave solution for $\varphi^B(R)$. This solution must not have singularities in the domain V. A simple incoming or outgoing spherical wave will not do, because the expressions for these waves both have singular behavior at $R = 0$. The following combination will suit our purpose, however:

$$\varphi^B(R) = \frac{1}{2ikR}\left(e^{ikR} - e^{-ikR}\right) \qquad \text{and} \qquad f^B(R) = 0. \tag{6.8.6}$$

Substitution of (6.8.3), (6.8.5) and (6.8.6) into Eq. (6.8.4) yields

$$1 = 4\pi R_0^2 \frac{Q}{2ikR_0}\left(e^{ikR_0} - e^{-ikR_0}\right)\left(-\frac{1}{R_0^2}e^{ikR_0} + \frac{ik}{R_0}e^{ikR_0}\right)$$
$$- 4\pi R_0^2 \frac{Q}{R_0}e^{ikR_0}\left[-\frac{1}{2ikR_0^2}\left(e^{ikR_0} - e^{-ikR_0}\right) + \frac{1}{2R_0}\left(e^{ikR_0} + e^{-ikR_0}\right)\right].$$

The solution to this equation is

$$Q = -\frac{1}{4\pi}.$$

In summary, we have shown that the solution of

$$\nabla^2\varphi + k^2\varphi = \frac{\delta(R)}{4\pi R^2} \tag{6.8.7}$$

is given by

$$\varphi(R) = -\frac{e^{ikR}}{4\pi R}. \tag{6.8.8}$$

This result, which can be derived in a number of other ways, is well known.

As we saw in Section 3.8, the wave equation appears in elastodynamics as the equation governing displacement potentials.

6.9 Reciprocity for waves reflected from a free surface

In Section 3.4 it was shown that a longitudinal wave incident on a free surface generates reflected longitudinal and transverse waves. The reflection coefficients are $R_L^L(\theta_L)$ and $R_T^L(\theta_L)$. Analogously, an incident transverse wave generates reflected longitudinal and transverse waves with reflection coefficients $R_L^T(\theta_T)$ and $R_T^T(\theta_T)$. By reciprocity considerations it can be shown that a simple relation exists between $R_T^L(\theta_L)$ and $R_L^T(\theta_T)$.

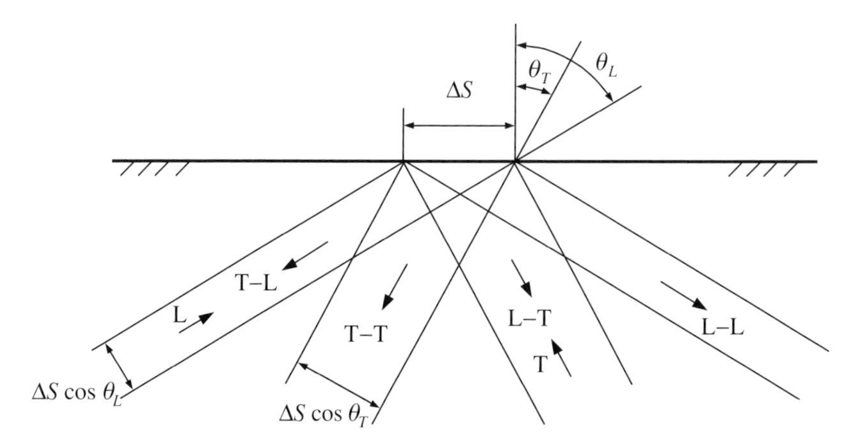

Figure 6.5 State *A* (L, L–L and L–T) and state *B* (T, T–T and T–L).

For state *A* we consider an incident longitudinal wave, indicated in Fig. 6.5 by L, while state *B* is an incident transverse wave indicated by T.

We now consider an element ΔS of the free surface, and we consider the beams that are incident on and reflected from this element. In the absence of body forces the reciprocal relation reduces to

$$\int_S \left(u_i^B \tau_{ij}^A - u_i^A \tau_{ij}^B\right) n_j \, dS = 0. \tag{6.9.1}$$

For the surface *S* we now take ΔS, the sides of the beams and the beam cross sections, indicated by $\Delta S \cos\theta_L$ and $\Delta S \cos\theta_T$ in Fig. 6.5. We note that waves for both state *A* and state *B* propagate in the beams labeled L, T–L and L–T, T in Fig. 6.5. The other two beams contain only waves due to either state *A* or state *B*. Since the waves are counter-propagating in the beams containing waves for both states *A* and *B*, Eq. (6.9.1) reduces to the following simple form:

$$ik_L(\lambda + 2\mu) R_L^T(\theta_T) \Delta S_L + ik_T \mu R_T^L(\theta_L) \Delta S_T = 0.$$

Using the relations

$$\Delta S_L = \Delta S \cos\theta_L \qquad \text{and} \qquad \Delta S_T = \Delta S \cos\theta_T,$$

we obtain

$$R_L^T(\theta_T) = -\frac{c_T}{c_L} \frac{\cos\theta_T}{\cos\theta_L} R_T^L(\theta_L).$$

This result can be checked from the expressions for R_L^T and R_T^L given in Section 3.4.

6.10 Reciprocity for fields generated by point loads in a bounded body

Let us again consider a body of volume V and surface S that is subjected to two point forces, $\boldsymbol{f}^A(t)\delta(\boldsymbol{x}-\boldsymbol{x}^A)$ at point A and $\boldsymbol{f}^B(t)\delta(\boldsymbol{x}-\boldsymbol{x}^B)$ at point B, both in the body's interior; $\delta(\quad)$ is the delta function in three-dimensional space. The surface of the body is free of surface tractions except at points of rigid support, where the displacement vanishes. For the time-harmonic case we write

$$f^A(t)\delta(\boldsymbol{x}-\boldsymbol{x}^A) = \boldsymbol{F}^A e^{-i\omega t}\delta(\boldsymbol{x}-\boldsymbol{x}^A), \tag{6.10.1}$$

$$f^B(t)\delta(\boldsymbol{x}-\boldsymbol{x}^B) = \boldsymbol{F}^B e^{-i\omega t}\delta(\boldsymbol{x}-\boldsymbol{x}^B). \tag{6.10.2}$$

Omitting the exponential terms, the corresponding displacements are $\boldsymbol{u}^A(\boldsymbol{x}-\boldsymbol{x}^A)$ and $u^B(\boldsymbol{x}-\boldsymbol{x}^B)$. Since the surface S is either free of tractions or free of displacements, the right-hand side of Eq. (6.2.8), which describes the role of the surface tractions and surface displacements, disappears. Using the sifting property of the delta function, the reciprocity relation then reduces to

$$F_j^A u_j^B(\boldsymbol{x}^A-\boldsymbol{x}^B) = F_j^B u_j^A(\boldsymbol{x}^B-\boldsymbol{x}^A). \tag{6.10.3}$$

In $u_j^B(\boldsymbol{x}^A-\boldsymbol{x}^B)$, $\boldsymbol{x}^B$ is the point of application of force F^B and $\boldsymbol{x}^A$ is the point of application of F^A. For equal point forces of unit magnitude, applied in the l-direction at $\boldsymbol{x}^A$ and the k-direction at $\boldsymbol{x}^B$, we may write

$$u_{l;k}^B(\boldsymbol{x}^A-\boldsymbol{x}^B) = u_{k;l}^A(\boldsymbol{x}^B-\boldsymbol{x}^A). \tag{6.10.4}$$

Here the first subscript in $u_{l;k}$ defines the direction of the displacement and the second the direction of the applied force. Using the notation of Eq. (1.2.2) for the static case, Eq. (6.10.4) states that

$$\delta_{AB} = \delta_{BA}. \tag{6.10.5}$$

In words, the time-harmonic elastodynamic displacement at point A in the direction of the point force at point A, but due to the unit point force at point B, equals the displacement at point B in the direction of the point force at point B, but due to the unit point force at point A.

An analogous result follows from Eq. (6.3.5) for loads of arbitrary time dependence. Now we consider a body that is either free of surface tractions or free of surface displacements and in addition has quiescent initial conditions. The point loads are now defined as

$$\boldsymbol{f}^A(\boldsymbol{x},t) = \boldsymbol{F}^A(t)\delta(\boldsymbol{x}-\boldsymbol{x}^A), \tag{6.10.6}$$

$$\boldsymbol{f}^B(\boldsymbol{x},t) = \boldsymbol{F}^B(t)\delta(\boldsymbol{x}-\boldsymbol{x}^B). \tag{6.10.7}$$

Substitution into Eq. (6.3.5) yields

$$F_j^A(t) * u_j^B(\boldsymbol{x}^A - \boldsymbol{x}^B,\ t) = F_j^B(t) * u_j^A(\boldsymbol{x}^B - \boldsymbol{x}^A,\ t), \qquad (6.10.8)$$

where the convolution integral is defined by Eq. (4.3.6). This result simplifies further when

$$F_l^A(t) = F_k^B(t) = \delta(t). \qquad (6.10.9)$$

The convolution integrals in Eq. (6.10.8) can then be evaluated and, equivalently to (6.10.4), we obtain

$$u_{l;k}^B(\boldsymbol{x}^A - \boldsymbol{x}^B,\ t) = u_{k;l}^A(\boldsymbol{x}^B - \boldsymbol{x}^A,\ t). \qquad (6.10.10)$$

It is left to the reader to derive results for other combinations of the boundary and initial conditions.

6.11 Reciprocity for point loads in an unbounded body

The simple applications of the elastodynamic reciprocity theorem given by Eq. (6.10.3) for a bounded body can easily be extended to an unbounded body subjected to two point loads. The forces, which are applied at $\boldsymbol{x} = \boldsymbol{x}^A$ and $\boldsymbol{x} = \boldsymbol{x}^B$, respectively, are represented by Eqs. (6.10.1) and (6.10.2). The outer surface of the body under consideration is taken as a sphere of radius $|\boldsymbol{x}|$, where $\boldsymbol{x} = \hat{\boldsymbol{x}}|\boldsymbol{x}|$ and $\hat{\boldsymbol{x}}$ is a unit vector. In the limit as $|\boldsymbol{x}| \to \infty$ the reciprocity theorem then becomes

$$F_j^A u_j^B(\boldsymbol{x}^A - \boldsymbol{x}^B) - F_j^B u_j^A(\boldsymbol{x}^B - \boldsymbol{x}^A) - I_{|\boldsymbol{x}|} = 0, \qquad (6.11.1)$$

where

$$I_{|\boldsymbol{x}|} = \lim_{|\boldsymbol{x}|\to\infty} \int_{S_x} \left[\tau_{ij}^B(\boldsymbol{x} - \boldsymbol{x}^B) u_j^A(\boldsymbol{x} - \boldsymbol{x}^A) - \tau_{ij}^A(\boldsymbol{x} - \boldsymbol{x}^A) u_j^B(\boldsymbol{x} - \boldsymbol{x}^B)\right] n_i \, dS_x. \qquad (6.11.2)$$

Here S_x is the surface of the sphere of radius $|\boldsymbol{x}|$.

Now we employ the far-field expressions given in Section 3.10, particularly Eqs. (3.10.1)–(3.10.5) and (3.10.6)–(3.10.10). The first term in the integrand in Eq. (6.11.2) becomes

$$\left[\tau_{ij;k}^{G;L}(\boldsymbol{x} - \boldsymbol{x}^B) + \tau_{ij;k}^{G;T}(\boldsymbol{x} - \boldsymbol{x}^B)\right]\left[u_{j;l}^{G;L}(\boldsymbol{x} - \boldsymbol{x}^A) + u_{j;l}^{G;T}(\boldsymbol{x} - \boldsymbol{x}^A)\right] n_i. \qquad (6.11.3)$$

The second term is of the same form, but all superscripts A must be replaced by superscripts B and vice versa. Note that (6.11.3) is written for the case where the loads are applied in the k- and l-directions. To simplify the notation for the present purpose we set

$$\frac{\exp(ik_\gamma|\boldsymbol{x}|)}{4\pi|\boldsymbol{x}|}\exp(-ik_\gamma\hat{\boldsymbol{x}}\cdot\boldsymbol{x}^A) = K_\gamma^A, \tag{6.11.4}$$

$$\frac{\exp(ik_\gamma|\boldsymbol{x}|)}{4\pi|\boldsymbol{x}|}\exp(-ik_\gamma\hat{\boldsymbol{x}}\cdot\boldsymbol{x}^B) = K_\gamma^B, \tag{6.11.5}$$

where $\gamma =$ L or T. The total integrand then becomes

$$\begin{aligned}&\left(ik_L B_{ij;k}^{G;L} K_L^B + ik_T B_{ij;k}^{G;T} K_T^B\right)\left(A_{j;l}^{G;L} K_L^A + A_{j;l}^{G;T} K_T^A\right) n_i \\ &- \left(ik_L B_{ij;l}^{G;L} K_L^A + ik_T B_{ij;l}^{G;T} K_T^A\right)\left(A_{j;k}^{G;L} K_L^B + A_{j;k}^{G;T} K_T^B\right) n_i.\end{aligned} \tag{6.11.6}$$

It now clearly follows from the definitions of $B_{ij;k}^{G;L}$, $A_{i;l}^{G;L}$ etc., see Eqs. (3.9.11), (3.9.12), that

$$I_{|\boldsymbol{x}|} = 0. \tag{6.11.7}$$

Hence the reciprocity relation given by Eq. (6.10.4) also holds for an unbounded body.

The question may be raised whether the higher-order terms in the far-field expressions, which are not given in Section 3.10, may make a contribution to $I_{|\boldsymbol{x}|}$. However, these terms are of order $O(|\boldsymbol{x}|^{-2})$, and thus the contributions of their products vanish as the radius of the sphere increases beyond bounds.

6.12 Distribution of body forces

The elastic wave motion generated by a time-harmonic distribution of body forces $f_j^A(\boldsymbol{x})\exp(-i\omega t)$, applied over the region $\hat{V}$, is governed by

$$\tau_{ij,i}^A + f_j^A(\boldsymbol{x}) = -\rho\omega^2 u_j^A. \tag{6.12.1}$$

For state B we take the field generated by a point load applied in the x_k-direction at $\boldsymbol{x} = \boldsymbol{x}^B$:

$$f_k^B(\boldsymbol{x}, t) = \delta(\boldsymbol{x} - \boldsymbol{x}^B)e^{-i\omega t}\boldsymbol{e}_k. \tag{6.12.2}$$

The resulting displacement field for state B is denoted by $u_{i;k}^G(\boldsymbol{x} - \boldsymbol{x}^B)$. The complete expression for $u_{i;k}^G$ is given by Eq. (3.9.21). As noted before, $u_{i;k}^G$ is the displacement in the x_i-direction at point $\boldsymbol{x}$ due to a unit point load applied in the x_k-direction at point $\boldsymbol{x}^B$.

Next we apply the reciprocity theorem given by Eq. (6.2.8). For an unbounded domain the surface integral vanishes, and we have

$$\int_V f_j^A u_j^B \, dV_x = \int_V f_j^B u_j^A \, dV_x. \tag{6.12.3}$$

For states A and B we now select the fields defined by Eqs. (6.12.1) and (6.12.2), respectively. The reciprocity relation (6.12.3) then yields

$$\int_{\hat{V}} f_j^A(\boldsymbol{x}) u_{j;k}(\boldsymbol{x} - \boldsymbol{x}^B) \, dV_x = u_k^A(\boldsymbol{x}^B). \tag{6.12.4}$$

By virtue of Eq. (6.10.4) we have

$$u_{j;k}(\boldsymbol{x} - \boldsymbol{x}^B) = u_{k;j}(\boldsymbol{x}^B - \boldsymbol{x}), \tag{6.12.5}$$

and Eq. (6.12.4) may then also be written as

$$u_k(\boldsymbol{x}) = \int_{\hat{V}} f_j(\boldsymbol{y}) u_{k;j}(\boldsymbol{x} - \boldsymbol{y}) \, dV_y. \tag{6.12.6}$$

Here, in addition to the introduction of (6.12.5) into (6.12.4), the symbols $\boldsymbol{x}$ and $\boldsymbol{y}$ have been interchanged and $\boldsymbol{x}^B$ has subsequently been replaced by $\boldsymbol{y}$.

6.13 Domain integral equation for scattering by an inclusion

An inclusion that occupies a volume $\hat{V}$ with boundary $\hat{S}$ is embedded in a homogeneous linearly elastic solid of infinite extent. The embedding solid has uniform mass density ρ and elastic properties defined by elastic constants C_{ijkl}. The corresponding quantities for the inhomogeneous inclusion are $\hat{\rho}(\boldsymbol{x})$ and $\hat{C}_{ijkl}(\boldsymbol{x})$.

A steady-state displacement field defined by $u_i^{\text{in}}(\boldsymbol{x}) \exp(-i\omega t)$ is incident on the inclusion and gives rise to a scattered field. The resulting total field is written as

$$u_i = u_i^{\text{in}} + u_i^{\text{sc}}. \tag{6.13.1}$$

The incident field is defined everywhere, i.e., also within the confines of the inclusion. It satisfies

$$\tau_{ij,j}^{\text{in}} + \rho\omega^2 u_i^{\text{in}} = 0. \tag{6.13.2}$$

For simplicity we use the general form of Hooke's law,

$$\tau_{ij}^{\text{in}} = C_{ijkl} u_{k,l}^{\text{in}}. \tag{6.13.3}$$

Inside the inclusion the total field must satisfy

$$(\hat{C}_{ijkl}u_{k,l}),_j + \hat{\rho}\omega^2 u_i = 0. \tag{6.13.4}$$

Let

$$\hat{C}_{ijkl} = C_{ijkl} + \Delta C_{ijkl}(\boldsymbol{x}), \tag{6.13.5}$$

$$\hat{\rho} = \rho + \Delta\rho(\boldsymbol{x}). \tag{6.13.6}$$

Then, inside $\hat{V}$, Eq. (6.13.4) can be rewritten as

$$C_{ijkl}u_{k,lj} + \rho\omega^2 u_i = -(\Delta C_{ijkl}u_{k,l}),_j - \Delta\rho\omega^2 u_i. \tag{6.13.7}$$

By substituting τ_{ij}^{in}, Eq. (6.13.3), into Eq. (6.13.2) and subtracting the result from (6.13.7), we obtain

$$C_{ijkl}u_{k,lj}^{\text{sc}} + \rho\omega^2 u_i^{\text{sc}} = \begin{cases} -(\Delta C_{ijkl}u_{k,l}),_j - \Delta\rho\omega^2 u_i & \text{in } \hat{V}, \\ 0 & \text{outside } \hat{V}. \end{cases} \tag{6.13.8}$$

It is now noted that the right-hand side of the last equation can be considered as a distribution of body forces in $\hat{V}$, applied in an unbounded solid. By using the results of the previous section, specifically Eq. (6.12.6), the solution to Eq. (6.13.8) may then be written formally as

$$u_m^{\text{sc}}(\boldsymbol{x}) = \int_{\hat{V}} \frac{\partial}{\partial y_j}[\Delta C_{ijkl}(\boldsymbol{y})u_{k,l}(\boldsymbol{y})]u_{m;i}^G(\boldsymbol{x}-\boldsymbol{y})\,dV_y$$
$$+ \omega^2 \int_{\hat{V}} \Delta\rho(\boldsymbol{y})u_i(\boldsymbol{y})u_{m;i}^G(\boldsymbol{x}-\boldsymbol{y})\,dV_y. \tag{6.13.9}$$

The equation can be simplified somewhat by using

$$u_{m;i}^G \frac{\partial}{\partial y_j}(\Delta C_{ijkl}u_{k,l}) = \frac{\partial}{\partial y_j}\left(u_{m;i}^G \Delta C_{ijkl}u_{k,l}\right) - \frac{\partial}{\partial y_j}\left(u_{m;i}^G\right)\Delta C_{ijkl}u_{k,l}.$$

The volume integral of the first term on the right-hand side can be converted into an integral over the surface $\hat{S}$ by the use of Gauss' theorem. That integral will vanish, however, if $\Delta C_{ijkl}(\boldsymbol{x})$ vanishes on $\hat{S}$. Equation (6.13.9) then becomes

$$u_m^{\text{sc}}(\boldsymbol{x}) = -\int_{\hat{V}} \Delta C_{ijkl}(\boldsymbol{y})\left[\frac{\partial}{\partial y_j}u_{m;i}^G(\boldsymbol{x}-\boldsymbol{y})\right]u_{k,l}(\boldsymbol{y})\,dV_y$$
$$+ \omega^2 \int_{\hat{V}} \Delta\rho(\boldsymbol{y})u_i(\boldsymbol{y})u_{m;i}^G(\boldsymbol{x}-\boldsymbol{y})\,dV_y. \tag{6.13.10}$$

This equation was discussed by Mal and Knopoff (1967). For the special case where ΔC_{ijkl} and $\Delta\rho$ are constants, Eq. (6.13.10) yields

$$u_m^{sc}(\boldsymbol{x}) = \Delta C_{ijkl} \int_{\hat{V}} \left[\frac{\partial}{\partial y_j} u_{m;i}^G(\boldsymbol{x}-\boldsymbol{y}) \right] u_{k,l}(\boldsymbol{y})\, dV_y + \omega^2 \Delta\rho \int_{\hat{V}} u_i(\boldsymbol{y}) u_{m;i}^G(\boldsymbol{x}-\boldsymbol{y})\, dV_y. \qquad (6.13.11)$$

When considered as equations for ΔC_{ijkl} and $\Delta\rho$, Eqs. (6.13.10) and (6.13.11) can generally only be solved numerically, since $u_{k,l}(\boldsymbol{y})$ and $u_i(\boldsymbol{y})$ are not known a priori inside the inclusion. An approximate solution can be obtained, however, by the Born approximation. In this approximation, which was also discussed in the two previous chapters, the total displacement field inside the inclusion is approximated by the incident field.

For further details on applications of the Born approximation to elastodynamic scattering problems we refer to the work of Gubernatis *et al.* (1977). To obtain numerically exact solutions it is generally more convenient to cast the formulation of the scattering problem in terms of boundary integral equations over the surface $\hat{S}$ of the inclusion. Such equations and the boundary element method for their numerical solution are discussed in Chapter 11.

6.14 Examples from the technical literature

From time to time, authors have recognized the utility of elastodynamic reciprocity relations for the study of wave motion in elastic bodies. In this section we briefly review some interesting examples from the technical literature.

Knopoff and Gangi (1959) drew attention to reciprocity for applications in seismology. They noted that the usual derivation of vector reciprocity depends upon the demonstration that the differential equations describing the field are self-adjoint and that the derivation of reciprocity relations can be carried out for a solid of arbitrary inhomogeneity and anisotropy. Their result is the same as that given by Eqs. (6.10.3) and (6.10.4). Knopoff and Gangi (1959) also presented an experimental demonstration of elastodynamic reciprocity for the case where two transducers, one transmitting, the other receiving, are placed on a large granite block. The signal at the receiving transducer is noted to contain a longitudinal signal (P) and a surface wave (R). A cylindrical brass block is subsequently placed on the granite block between the transducers. The interaction of the incident signal with the brass block converts some of the longitudinal wave into a surface wave (PR) and part of the surface wave into a longitudinal wave (RP). For transducers measuring normal displacements, the experiment shows that when the brass block is placed an equal distance at different sides of the mid-distance between the transducers, the wave (PR) for the first position equals the wave (RP) for the second position. Essentially the results confirm experimentally that there is reciprocity between the conversion of (P) into (R) and (R) into (P).

DiMaggio and Bleich (1959) used the sub-surface vertical displacement due to a normal surface force in conjunction with the elastodynamic reciprocity theorem to determine the vertical surface displacement due to a sub-surface vertical load. Interesting applications to the determination of the transient wave field of an elastic body acted upon by a moving point force were considered by Payton (1964). The case of a moving point load in an unbounded solid is quite simple. The solution is based on the use of Eq. (6.3.5) for the case where the body is unbounded and originally at rest. Equation (6.3.5) then reduces to

$$\int_V f_j^A(\mathbf{x}, t) * u_j^B(\mathbf{x}, t)\, dV = \int_V f_j^B(\mathbf{x}, t) * u_j^A(\mathbf{x}, t)\, dV. \quad (6.14.1)$$

Payton (1964) considered a moving point force acting in the direction of its line of motion. In our notation this would be state A, with

$$f_z^A(\mathbf{x}, t) = F_0 \delta(x)\delta(y)\delta(z - vt)H(t),$$

where v is the constant velocity of the force. To determine the displacement in the x-direction, state B is defined by

$$f_x^B(\mathbf{x}, t) = \delta(x - x_0)\delta(y - y_0)\delta(z - z_0)\delta(t).$$

The corresponding displacement field for state B is known. Substitution of the available information into Eq. (6.14.1) yields an expression for $u_x^A(\mathbf{x}_0, y_0, z_0, t)$. The same approach can be followed to determine the surface motion generated by a normal point force moving with velocity v over the surface of a half-space. The results were presented in Payton (1964).

Keller and Karal (1964) pointed out an interesting application of reciprocity in the geometrical theory of elastic surface-wave excitation and propagation. The theory involves complex rays that travel from a sub-surface point of excitation to the surface, then along the surface and finally from the surface to observation points in the solid. The reciprocity principle is stated as

$$u_x(P, Qy) = u_y(Q, Px). \quad (6.14.2)$$

Here $u_x(P, Qy)$ denotes the x-component of displacement at P due to a unit force in the y-direction applied at Q. Similarly, $u_y(Q, Px)$ is the y-component of displacement at Q due to a unit force in the x-direction at P. Equation (6.14.2) implies a useful relationship between the excitation and radiation coefficients. These coefficients control the conversion of wave amplitudes from body rays to surface rays and from surface rays to body rays, respectively. The specifics of the relationship are given in Keller and Karal (1964). As a consequence, only one coefficient, say the radiation coefficient, needs to be determined from a canonical problem in order to apply the ray theory.

We conclude this section with a brief summary of an interesting paper by Burridge and Knopoff (1964), in which the authors show that appropriate distributions of body forces generate radiation equivalent to that from seismic dislocations. Their results are based on still another form of the elastodynamic reciprocity relation. In a body of volume V and surface S, state A is taken to satisfy

$$\left[C_{ijpq}(\mathbf{x})u^A_{p,q}(\mathbf{x},t)\right]_{,j} - \rho(\mathbf{x})\ddot{u}^A_i(\mathbf{x},t) = -f^A_i(\mathbf{x},t), \qquad (6.14.3)$$

where the elastic solid is taken as inhomogeneous and anisotropic. Similarly for state B they write

$$\left[C_{ijpq}(\mathbf{x})u^B_{p,q}(\mathbf{x},t)\right]_{,j} - \rho(\mathbf{x})\ddot{u}^B_i(\mathbf{x},t) = -f^B_i(\mathbf{x},t). \qquad (6.14.4)$$

It is assumed that f^A_i and f^B_i vanish for $t < -T$, T being a constant, and that u^A_i and u^B_i also vanish for $t < -T$ by virtue of causality.

In the next step t is replaced by $-t$ in Eq. (6.14.4). Equation (6.14.4) then becomes the governing equation for state $\overline{B}$ with displacement components v^B_i :

$$\left[C_{ijpq}(\mathbf{x})v^B_{p,q}(\mathbf{x},t)\right]_{,j} - \rho(\mathbf{x})\ddot{v}^B_i(\mathbf{x},t) = -\overline{f}^B(\mathbf{x},t),$$

where

$$v^B_p(\mathbf{x},t) = u^B_p(\mathbf{x},-t), \qquad \overline{f}^B(\mathbf{x},t) = f^B(\mathbf{x},-t)$$

and

$$v^B_p(\mathbf{x},t) = 0 \qquad \text{for} \quad t > T.$$

Following the usual steps of combining the equations of states A and $\overline{B}$, but including an integration over t, the following global reciprocity relation is obtained:

$$\int_{-\infty}^{\infty} dt \int_V \left(u^A_i \overline{f}^B_i - v^B_i f^A_i\right) dV = \int_{-\infty}^{\infty} dt \int_S \left(v^B_i C_{ijpq} u^A_{p,q} - u^A_i C_{ijpq} v^B_{p,q}\right) n_j \, dS,$$

where the n_i are the components of the outward normal on S. The integrals over t are in fact finite since the integrands vanish outside the interval $(-T, T)$.

Further manipulations discussed by Burridge and Knopoff (1964) then lead to the following representation theorem,

$$
\begin{aligned}
u_n^A(\boldsymbol{y}, s) = &\int_{-\infty}^{\infty} dt \int_V G_{ni}(\boldsymbol{y}, s; \boldsymbol{x}, t) f_i^A(\boldsymbol{x}, t)\, dV_x \\
&+ \int_{-\infty}^{\infty} dt \int_S \big[G_{ni}(\boldsymbol{y}, s; \boldsymbol{x}, t) C_{ijpq}(\boldsymbol{x}) u_{p,q}^A(\boldsymbol{x}, t) \\
&\qquad - u_i^A(\boldsymbol{x}, t) C_{ijpq}(\boldsymbol{x}) G_{np,q'}(\boldsymbol{y}, s; \boldsymbol{x}, t)\big] n_j\, dS_{\boldsymbol{x}}. \quad (6.14.5)
\end{aligned}
$$

Here $G_{in}(\boldsymbol{x}, t; y, s)$ is the displacement in the i-direction at $(\boldsymbol{x}, t)$ due to an instantaneous point force of unit impulse in the n-direction at $(\boldsymbol{y}, s)$. It is also assumed that u_i and G_{in} satisfy the same homogeneous boundary conditions on S. Also,

$$
G_{np,q'}(\boldsymbol{y}, s; \boldsymbol{x}, t) = \frac{\partial}{\partial \boldsymbol{x}_q} G_{np}(\boldsymbol{y}, s; \boldsymbol{x}, t),
$$

where q' indicates that the subscript refers to the second set of arguments of G_{np}, and the summation convention still applies.

Burridge and Knopoff (1964) used the representation theorem given by Eq. (6.14.5) to find the radiation from prescribed discontinuities in the displacement and its derivatives across a surface Σ imbedded in V. They then proceeded to show that, with regard to radiation, the effect of prescribed discontinuities across Σ is the same as the effect of introducing specified body forces into an unfaulted medium.

7
Wave motion guided by a carrier wave

7.1 Introduction

In this chapter we seek solutions to the elastodynamic equations that represent a combination of a carrier wave propagating on a preferred plane and motions that are carried along by that wave. The carrier wave supports standing-wave motions in the direction normal to the plane of the carrier wave. Such combined wave motions include Rayleigh surface waves propagating along the free surface of an elastic half-space and Lamb waves in an elastic layer.

The usual way to construct solutions to the elastodynamic equations of motion for homogeneous, isotropic, linearly elastic solids is to express the components of the displacement vector as the sum of the gradient of a scalar potential and the curl of a vector potential, where the potentials must be solutions of classical wave equations whose propagation velocities are the velocities of longitudinal and transverse waves, respectively. The decomposition of the displacement vector was discussed in Section 3.8.

The approach using displacement potentials has generally been used also for surface waves propagating along a free surface or an interface and Lamb waves propagating along a layer. The particular nature of these guided wave motions suggests, however, an alternative formulation in terms of a membrane-like wave over the guiding plane that acts as a carrier wave of superimposed motions away from the plane. For time-harmonic waves the corresponding formulation, presented in this chapter, shows that the carrier wave satisfies a reduced wave equation, also known as the Helmholtz or membrane equation, in coordinates in the plane of the carrier wave. The functions representing the superimposed motions, which show the variation with the coordinate normal to the preferred plane, are solutions of ordinary differential equations. The formulation has the advantage that the carrier wave can be determined separately in a general form.

In section 7.2 we present forms for the displacement components in terms of a single potential that represents the carrier wave, and we show that these forms satisfy the displacement equations of motion. The connection with Rayleigh surface waves is discussed in Section 7.3. In the usual manner the condition of vanishing surface traction on the free surface yields the well-known equation for the velocity of Rayleigh surface waves. A few examples of solutions of the membrane equation, which governs the propagation of the carrier wave along the free surface are discussed in Section 7.4.

Based on the results of Section 7.2, classical Lamb waves in a homogeneous, isotropic, linearly elastic layer are considered in Section 7.6. For a layer, the displacement components are obtained in terms of thickness motions superimposed on a carrier wave that describes propagation along the mid-plane of the layer. Again, the carrier wave can be any solution of a Helmholtz equation in the mid-plane. The analysis of the thickness motions results in the usual Rayleigh–Lamb frequency equation relating the circular frequency, ω, and a wavenumber-like quantity, k. The nature of the selected solution of the reduced wave equation affects the interpretation of k but not the general form of the thickness motions.

We start the discussion of waves in an elastic layer with the much simpler case of equivoluminal waves. The results of Section 7.5 are needed for reference purposes in later chapters.

7.2 Guided wave motion

The wave motion is referred to an orthogonal coordinate system, (s_1, s_2, z), with two of its coordinates, s_1 and s_2, spanning the plane of carrier-wave propagation, the z-axis being normal to that plane. The coordinates in the plane of the carrier wave might be Cartesian, (x_1, x_2), polar, (r, θ), or any other suitable set of coordinates. In terms of these coordinates we now define a two-dimensional del operator $\overline{\nabla}$ and a Laplacian. For example, in Cartesian coordinates we have

$$\overline{\nabla} = \boldsymbol{i}_1 \frac{\partial}{\partial x_1} + \boldsymbol{i}_2 \frac{\partial}{\partial x_2},$$

$$\overline{\nabla}^2 = \frac{\partial^2}{\partial x_1^2} + \frac{\partial^2}{\partial x_2^2},$$

while the corresponding operators in polar coordinates are

$$\overline{\nabla} = \boldsymbol{i}_r \frac{\partial}{\partial r} + \boldsymbol{i}_\theta \frac{1}{r} \frac{\partial}{\partial \theta},$$

$$\overline{\nabla}^2 = \frac{\partial^2}{\partial r^2} + \frac{1}{r} \frac{\partial}{\partial r} + \frac{1}{r^2} \frac{\partial^2}{\partial \theta^2}.$$

Pursuing a slight extension of the formulation given by Achenbach (1998), we seek solutions for the displacement components in the forms

$$\bar{\boldsymbol{u}} = \frac{1}{k} V(z) e^{-i\omega t}\, \overline{\nabla}\varphi(s_1, s_2), \tag{7.2.1}$$

$$u_z = W(z) e^{-i\omega t} \varphi(s_1, s_2). \tag{7.2.2}$$

Here $\bar{\boldsymbol{u}}$ is the displacement vector in the $s_1 s_2$-plane and k is a wavenumber-like quantity, i.e., its dimension is reciprocal length. The function $\varphi(s_1, s_2)$ is dimensionless. In what follows we seek steady-state solutions, and we omit the factor $\exp(-i\omega t)$.

Equations (7.2.1) and (7.2.2) yield for the divergence of the displacement vector $\boldsymbol{u}$

$$\nabla \cdot \boldsymbol{u} = \frac{1}{k} V(z) \overline{\nabla}^2 \varphi + \frac{dW}{dz} \varphi, \tag{7.2.3}$$

where ∇ is the full three-dimensional operator, i.e.,

$$\nabla = \overline{\nabla} + \boldsymbol{i}_3 \frac{\partial}{\partial z}.$$

Substitution of (7.2.1)–(7.2.3) into the homogeneous vector form of the equations of motion, Eq. (2.2.20) without the body-force term, yields for the displacement components in the plane $z = 0$

$$(\lambda + \mu)\left[\frac{1}{k} V(z) \overline{\nabla}\, \overline{\nabla}^2 \varphi + \frac{dW}{dz} \overline{\nabla}\varphi\right] + \mu \left[\frac{1}{k} V(z) \overline{\nabla}^2\, \overline{\nabla}\varphi + \frac{1}{k}\frac{d^2 V}{dz^2} \overline{\nabla}\varphi\right]$$

$$= -\frac{\rho\omega^2}{k} V(z) \overline{\nabla}\varphi,$$

while for u_z we have

$$(\lambda + \mu)\left[\frac{1}{k}\frac{dV}{dz} \overline{\nabla}^2 \varphi + \frac{d^2 W}{dz^2} \varphi\right] + \mu \left[W(z) \overline{\nabla}^2 \varphi + \frac{d^2 W}{dz^2} \varphi\right]$$

$$= -\rho\omega^2 W(z)\varphi.$$

These equations can be rewritten as

$$\overline{\nabla}\left\{\frac{\lambda + 2\mu}{k} V(z) \overline{\nabla}^2 \varphi + \left[(\lambda + \mu)\frac{dW}{dz} + \frac{\mu}{k}\frac{d^2 V}{dz^2} + \frac{\rho\omega^2}{k} V(z)\right]\varphi\right\} = 0 \tag{7.2.4}$$

and

$$\left[\frac{\lambda + \mu}{k}\frac{dV}{dz} + \mu W(z)\right]\overline{\nabla}^2 \varphi + \left[(\lambda + 2\mu)\frac{d^2 W}{dz^2} + \rho\omega^2 W(z)\right]\varphi = 0. \tag{7.2.5}$$

We now seek solutions for $V(z)$ and $W(z)$ from (7.2.4) and (7.2.5) for the case where φ satisfies the following Helmholtz equation in the plane $z = 0$:

$$\overline{\nabla}^2 \varphi + k^2 \varphi = 0. \tag{7.2.6}$$

Eliminating $\overline{\nabla}^2 \varphi$ from (7.2.4) and (7.2.5) by the use of (7.2.6), we find that $V(z)$ and $W(z)$ must satisfy

$$(\lambda + \mu)\frac{dW}{dz} + \frac{\mu}{k}\frac{d^2 V}{dz^2} + \frac{\rho\omega^2}{k} V(z) = k^2 \frac{\lambda + 2\mu}{k} V(z), \tag{7.2.7}$$

$$(\lambda + 2\mu)\frac{d^2 W}{dz^2} + \rho\omega^2 W(z) = k^2 \left[\frac{\lambda + \mu}{k}\frac{dV}{dz} + \mu W(z) \right]. \tag{7.2.8}$$

Equations (7.2.7) and (7.2.8) are a set of coupled ordinary differential equations for $V(z)$ and $W(z)$. These equations must be solved for specific cases, subject to appropriate boundary conditions. In what follows we consider the cases of surface waves in a half-space and Lamb waves in a layer.

In addition to the wave motions described by Eqs. (7.2.1) and (7.2.2), we can also consider wave motions that are equivoluminal, with displacements that are parallel to the plane $z = 0$. The corresponding displacements are normal to the direction of wave propagation. We consider

$$\boldsymbol{u}(x) = \frac{1}{l} f(z) \nabla \wedge (\psi\, \boldsymbol{i}_z), \tag{7.2.9}$$

where

$$\overline{\nabla}^2 \psi + l^2 \psi = 0. \tag{7.2.10}$$

In Cartesian coordinates, the displacement components follow from Eq. (7.2.9) as

$$u_1 = \frac{1}{l} f(z) \frac{\partial \psi}{\partial x_2}, \qquad u_2 = -\frac{1}{l} f(z) \frac{\partial \psi}{\partial x_1}, \qquad u_3 \equiv 0. \tag{7.2.11}$$

Substitution into the displacement equations of motion shows that the displacements given by Eq. (7.2.11) are elastodynamic displacement solutions, provided that $f(z)$ satisfies

$$\frac{d^2 f}{dz^2} + \left(\frac{\omega^2}{c_T^2} - l^2 \right) f = 0. \tag{7.2.12}$$

7.3 Rayleigh surface waves

Waves traveling along the free surface of an elastic half-space in such a way that the displacements decay exponentially with distance from the free surface were first investigated by Lord Rayleigh (1887). Within the formulation presented in the previous section, surface waves on the half-space $z \geq 0$ can be investigated by seeking solutions of (7.2.7) and (7.2.8) of the general forms

$$V(z) = \overline{V}e^{-pz} \qquad \text{and} \qquad W(z) = \overline{W}e^{-pz}, \tag{7.3.1}$$

where $\overline{V}$ and $\overline{W}$ are constants. Substitution of (7.3.1) into Eqs. (7.2.7) and (7.2.8) yields a set of homogeneous algebraic equations for the constants $\overline{V}$ and $\overline{W}$. The condition that the determinant of the coefficients must vanish yields the following equation:

$$p^4 + \left(\frac{\omega^2}{c_L^2} + \frac{\omega^2}{c_T^2} - 2k^2\right)p^2 + \left(\frac{\omega^2}{c_L^2} - k^2\right)\left(\frac{\omega^2}{c_T^2} - k^2\right) = 0,$$

where c_L and c_T are the velocities of longitudinal and transverse waves, respectively. The two solutions of the above equation are

$$p^2 = k^2 - \frac{\omega^2}{c_L^2} \qquad \text{and} \qquad q^2 = k^2 - \frac{\omega^2}{c_T^2}. \tag{7.3.2}$$

The corresponding solutions of (7.2.7) and (7.2.8) may then be written as

$$V(z) = Ae^{-pz} + Be^{-qz}, \tag{7.3.3}$$

$$W(z) = -\frac{p}{k}Ae^{-pz} - \frac{k}{q}Be^{-qz}. \tag{7.3.4}$$

In summary, it has been shown that Eqs. (7.2.1) and (7.2.2) are solutions of the elastodynamic equations, provided that $\varphi(s_1, s_2)$ is a solution of Eq. (7.2.6) and $V(z)$ and $W(z)$ are given by Eqs. (7.3.3) and (7.3.4).

We still have to satisfy the boundary conditions at $z = 0$. For the case of Cartesian coordinates, these conditions are

$$\tau_{z\beta}(x_1, x_2, z)|_{z=0} = 0 \qquad \text{and} \qquad \tau_{zz}(x_1, x_2, z)|_{z=0} = 0,$$

where $\beta = 1, 2$. To obtain expressions for $\tau_{z\beta}$ and τ_{zz} we use Hooke's law, Eq. (2.2.11), in the form

$$\tau_{ij} = \rho\left(c_L^2 - 2c_T^2\right)u_{k,k}\delta_{ij} + \rho c_T^2(u_{i,j} + u_{j,i}). \tag{7.3.5}$$

By using Eqs. (7.3.3) and (7.3.4), the relevant stress components may be expressed as

$$\tau_{z\beta} = \rho c_T^2\left[W(z) + \frac{1}{k}\frac{dV}{dz}\right]\varphi,_\beta\,,$$

$$\tau_{zz} = \rho\left(c_L^2 - 2c_T^2\right)\left[\frac{1}{k}V(z)\varphi,_{\beta\beta} + \frac{dW}{dz}\varphi\right] + 2\rho c_T^2\frac{dW}{dz}\varphi,$$

where $\beta = 1, 2$. To satisfy the boundary conditions at $z = 0$, we must then have

$$\frac{1}{k}\frac{dV}{dz}(z)\Big|_{z=0} + W(z)\Big|_{z=0} = 0,$$
$$-k(c_L^2 - 2c_T^2)V(z)\Big|_{z=0} + c_L^2\frac{dW}{dz}(z)\Big|_{z=0} = 0,$$

where $\varphi_{,\beta\beta}$ has been eliminated by the use of Eq. (7.2.6). Substitution of $V(z)$ and $W(z)$ from Eqs. (7.3.3) and (7.3.4) yields a system of homogeneous algebraic equations for the constants A and B:

$$2\sqrt{1-\frac{c^2}{c_L^2}}\sqrt{1-\frac{c^2}{c_T^2}}A + \left(2-\frac{c^2}{c_T^2}\right)B = 0, \tag{7.3.6}$$

$$\left(2-\frac{c^2}{c_T^2}\right)A + 2B = 0, \tag{7.3.7}$$

where we have introduced $k = \omega/c$. The condition that the determinant of the coefficients must vanish yields the well-known equation for the phase velocity, $c = c_R$, of Rayleigh surface waves:

$$\left(2-\frac{c^2}{c_T^2}\right)^2 - 4\sqrt{1-\frac{c^2}{c_L^2}}\sqrt{1-\frac{c^2}{c_T^2}} = 0. \tag{7.3.8}$$

For future reference the surface-wave results in Cartesian coordinates, obtained in this chapter, are summarized as follows:

$$u_\alpha = A\frac{1}{k_R}V^R(z)\frac{\partial\varphi}{\partial x_\alpha}(x_1, x_2), \tag{7.3.9}$$

$$u_z = AW^R(z)\varphi(x_1, x_2), \tag{7.3.10}$$

where $\alpha = 1, 2$ and

$$\frac{\partial^2\varphi}{\partial x_1^2} + \frac{\partial^2\varphi}{\partial x_2^2} + k_R^2\varphi = 0, \qquad k_R = \omega/c_R. \tag{7.3.11}$$

Also,

$$V^R(z) = d_1e^{-pz} + d_2e^{-qz}, \tag{7.3.12}$$

$$W^R(z) = d_3e^{-pz} - e^{-qz}, \tag{7.3.13}$$

in Eqs. (7.3.12), (7.3.13) we have

$$d_1 = -\frac{1}{2}\frac{k_R^2+q^2}{k_Rp}, \qquad d_2 = \frac{q}{k_R} \tag{7.3.14}$$

$$d_3 = \frac{1}{2}\frac{k_R^2+q^2}{k_R^2}. \tag{7.3.15}$$

The quantities p and q are defined by Eq. (7.3.2) but with k replaced by k_R:

$$p^2 = k_R^2 - \frac{\omega^2}{c_L^2} \quad \text{and} \quad q^2 = k_R^2 - \frac{\omega^2}{c_T^2}. \tag{7.3.16}$$

The single constant A in Eqs. (7.3.9), (7.3.10) has been obtained by solving for B from Eq. (7.3.7) and then redefining A.

The relevant stresses are

$$\tau_{11} = AT_{11}^R(z)\varphi(x_1, x_2) - A\frac{\mu}{k_R}V^R(z)\frac{\partial^2\varphi}{\partial x_2^2}, \tag{7.3.17}$$

$$\tau_{z\beta} = AT_{z\beta}^R(z)\frac{1}{k_R}\frac{\partial\varphi}{\partial x_\beta}, \tag{7.3.18}$$

where

$$T_{11}^R(z) = \mu(d_4e^{-pz} + d_5e^{-qz}), \tag{7.3.19}$$

$$T_{z\beta}^R(z) = \mu(d_6e^{-pz} + d_7e^{-qz}). \tag{7.3.20}$$

Also,

$$d_4 = \frac{1}{2}\left(k_R^2 + q^2\right)\frac{2p^2 + k_R^2 - q^2}{pk_R^2}, \qquad d_5 = -2q \tag{7.3.21}$$

$$d_6 = \frac{k_R^2 + q^2}{k_R}, \qquad d_7 = -\frac{k_R^2 + q^2}{k_R} \tag{7.3.22}$$

An interesting result of this section is that the specific dependence on z displayed by Eqs. (7.3.9) and (7.3.10), which is well-known for the plane-strain case, applies to a general class of surface wave motions in the half-space, where $\varphi(x_1, x_2)$ is a solution of the reduced wave equation (7.3.11) on the free surface of the half-space.

7.4 Carrier waves

The carrier wave is a solution to the reduced membrane wave equation in the x_1x_2-plane given by Eq. (7.2.6). The multiplying factors in Eqs. (7.2.1), (7.2.2) are independent of the choice of carrier wave. In this section we present examples of the carrier wave.

First we consider plane-strain waves in the layer. This case corresponds to a solution of the membrane equation of the form

$$\varphi(x_2) = e^{ikx_2}. \tag{7.4.1}$$

Here k is the wavenumber, where $k = \omega/c$, c being the phase velocity. Equation (7.2.1) shows that the out-of-plane displacement, $u_1(x_2, x_3, t)$, vanishes.

As a generalization of Eq. (7.4.1) let us consider a wave standing in the x_1-direction but propagating in the x_2-direction, i.e.,

$$\varphi(x_1, x_2) = \sin(\xi x_1)e^{i\eta x_2},$$

where Eq. (7.2.6) yields

$$k^2 = \xi^2 + \eta^2. \tag{7.4.2}$$

Now ξ and η must satisfy (7.4.2). Thus if a wave of a specific frequency and wavenumber η in the propagation direction x_2 is being considered then the variation with x_1 is fixed and follows by obtaining ξ from (7.4.2).

Next let us consider axisymmetric motion. Equation (7.2.6) then becomes

$$\frac{d^2\varphi}{dr^2} + \frac{1}{r}\frac{d\varphi}{dr} + k^2\varphi = 0. \tag{7.4.3}$$

The solution of (7.4.3) for an outgoing wave compatible with $\exp(-i\omega t)$ is

$$\varphi(r) = H_0^{(1)}(kr).$$

Equation (7.2.1) yields

$$u_r = -V(z)e^{-iwt}H_1^{(1)}(kr). \tag{7.4.3}$$

When the motion is not axially symmetric, Eq. (7.2.6) becomes in polar coordinates

$$\frac{\partial^2\varphi}{\partial r^2} + \frac{1}{r}\frac{\partial\varphi}{\partial r} + \frac{1}{r^2}\frac{\partial^2\varphi}{\partial\theta^2} + k^2\varphi = 0. \tag{7.4.5}$$

A solution of Eq. (7.4.5) is

$$\varphi(r, \theta) = H_\upsilon^{(1)}(kr)e^{\pm i\upsilon\theta}. \tag{7.4.6}$$

For this case the radial and circumferential displacements follow from (7.2.1) as

$$u_r = \frac{1}{k}V(z)e^{-i\omega t}\frac{\partial\varphi}{\partial r}, \tag{7.4.7}$$

$$u_\theta = \frac{1}{k}V(z)e^{-i\omega t}\frac{1}{r}\frac{\partial\varphi}{\partial\theta}, \tag{7.4.8}$$

while u_z is still given by

$$u_z = W(z)e^{-i\omega t}\varphi(r, \theta). \tag{7.4.9}$$

The approach presented here also lends itself to investigation by an asymptotic approach of the propagation of more general waves. As an example, we consider a wave that propagates out radially over a sector spanned by an angle $2\hat{\theta}$ in such a way that the distribution along the wavefronts is Gaussian. For large values of kr we seek a solution of Eq. (7.4.5) of the form

$$\varphi(r,\theta) = A\frac{e^{+i(kr-\pi/4)}}{(kr)^{1/2}}e^{-\theta^2/(2\hat{\theta}^2)}\left[1+\frac{b_0+b_2\theta^2}{kr}+\frac{c_0+c_2\theta^2}{(kr)^2}+\cdots\right]. \tag{7.4.10}$$

This expression will satisfy Eq. (7.4.5) with greater accuracy, as more terms are included in the brackets. Substitution of Eq. (7.4.10) into (7.4.5) yields a sequence of terms of order $(kr)^{-(n+1)/2}$, where $n = 0, 1, \ldots$ Here we will keep terms up to order $(kr)^{-7/2}$. It can be checked that the terms of orders $(kr)^{-1/2}$ and $(kr)^{-3/2}$ vanish identically. The condition that the sum of the terms of order $(kr)^{-5/2}$ should be zero leads to the equation

$$\left(2ib_0-\frac{1}{4}+\frac{1}{\hat{\theta}^2}\right)-\theta^2\left(2ib_2-\frac{1}{\hat{\theta}^4}\right)=0.$$

The requirement that the two terms should vanish separately yields expressions for b_0 and b_2. It follows that $\varphi(r,\theta)$ may be written as

$$\varphi(r,\theta) = A\frac{e^{+i(kr-\pi/4)}}{(kr)^{1/2}}e^{-\theta^2/2\hat{\theta}^2}\left[1-\frac{i}{kr}\left(\frac{1}{8}-\frac{1}{2\hat{\theta}^2}+\frac{\theta^2}{2\hat{\theta}^4}\right)+O(kr)^{-2}\right]. \tag{7.4.11}$$

The actual displacements for this case then follow from Eqs. (7.4.7)–(7.4.9), with $\varphi(r,\theta)$ defined by Eq. (7.4.11).

It is of interest to note that for $\hat{\theta}\to\infty$, i.e., for a uniform distribution along the wavefront, Eq. (7.4.11) reduces to the first two terms of the asymptotic expression of $H_0^{(1)}(kr)$, in accordance with Eq. (7.4.3), provided A is taken as $(2/\pi)^{1/2}$.

7.5 Equivoluminal waves in an elastic layer

In Cartesian coordinates, equivoluminal waves in an elastic layer can be described by Eqs. (7.2.9), (7.2.10), where the carrier wave $\psi(x_1, x_2)$ propagates in the mid-plane of the layer. The displacements are parallel to the x_1x_2-plane. The corresponding wave modes are called horizontally polarized wave modes. The geometry is shown in Fig. 7.1.

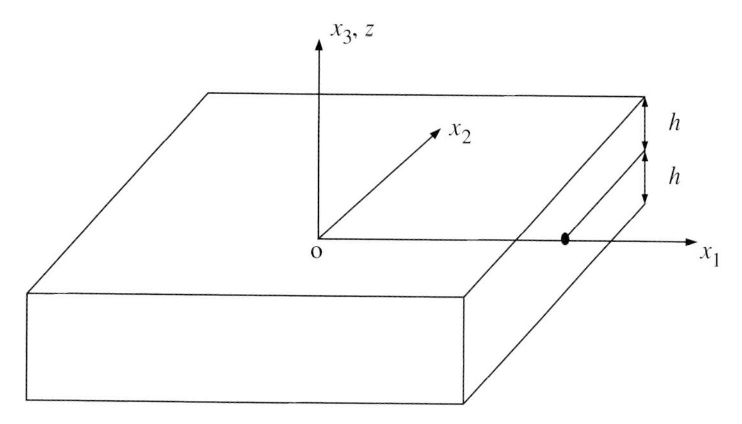

Figure 7.1 Configuration of the elastic layer.

Just as in Eq. (7.2.11) the horizontally polarized wave modes may be represented by

$$u_1 = \frac{1}{l_n} U_n(z) \frac{\partial \psi}{\partial x_2}, \tag{7.5.1}$$

$$u_2 = -\frac{1}{l_n} U_n(z) \frac{\partial \psi}{\partial x_1}, \tag{7.5.2}$$

and $u_z \equiv 0$. Since we expect n wave modes for this type, we have introduced the index n. Substitution into the displacement equation of motion shows that the displacements given by Eqs. (7.5.1) and (7.5.2) are elastodynamic displacement solutions provided that $\psi(x_1, x_2)$ is a solution of

$$\frac{\partial^2 \psi}{\partial x_1^2} + \frac{\partial^2 \psi}{\partial x_2^2} + l_n^2 \psi = 0 \tag{7.5.3}$$

and that

$$\frac{d^2 U_n}{dz^2} + \bar{q}_n^2 U_n = 0; \tag{7.5.4}$$

here

$$\bar{q}_n^2 = \left(\frac{\omega}{c_T}\right)^2 - l_n^2. \tag{7.5.5}$$

Let us consider the case where the faces of the layer are free of surface tractions, i.e.,

$$\text{at } z = \pm h: \qquad \tau_{zz} = 0 \quad \text{and} \quad \tau_{z1} = \tau_{z2} = 0. \tag{7.5.6}$$

It is not difficult to show that $\tau_{zz} \equiv 0$ through the thickness of the layer, while $\tau_{z1} = \tau_{z2} = 0$ implies that

$$\frac{dU_n}{dz} = 0 \qquad \text{at} \quad z = 0. \tag{7.5.7}$$

The general solution of Eq. (7.5.4) is

$$U_n(z) = B_1 \sin(\overline{q}_n z) + B_2 \cos(\overline{q}_n z), \tag{7.5.8}$$

where $U_n^S(z) = B_2 \cos(\overline{q}_n z)$ and $U_n^A(z) = B_1 \sin(\overline{q}_n z)$ represent symmetric and antisymmetric modes, respectively. The boundary conditions (7.5.7) yield

$$B_1 \cos(\overline{q}_n h) \pm B_2 \sin(\overline{q}_n h) = 0.$$

This equation can be satisfied in two ways:
either

$$B_1 = 0 \qquad \text{and} \qquad \sin(\overline{q}_n h) = 0 \tag{7.5.9}$$

or

$$B_2 = 0 \qquad \text{and} \qquad \cos(\overline{q}_n h) = 0. \tag{7.5.10}$$

For an arbitrarily specified value of the frequency ω, these equations yield an infinite number of solutions for the wavenumber l_n. A specific wave motion of the layer, called a mode of wave propagation, corresponds to each frequency–wavenumber combination satisfying (7.5.9) or (7.5.10). If $B_1 = 0$, the expression for $U^n(z)$ shows that the displacement is symmetric with respect to the mid-plane of the layer. The displacement is antisymmetric if $B_2 = 0$. In both cases the wavenumbers follow from

$$\overline{q}_n h = \frac{n\pi}{2}, \tag{7.5.11}$$

where, however, $n = 0, 2, 4, \ldots$ for symmetric modes, and $n = 1, 3, 5, \ldots$ for antisymmetric modes.

Equation (7.5.11) can be put in a more elegant form by introducing the dimensionless frequency

$$\Omega = \frac{2h\omega}{\pi c_T} \tag{7.5.12}$$

and the dimensionless wavenumber

$$\xi = \frac{2l_n h}{\pi}. \tag{7.5.13}$$

Equation (7.5.11) is then equivalent to

$$\Omega^2 = n^2 + \xi^2. \tag{7.5.14}$$

In the $\Omega\xi$-plane this frequency equation yields an infinite number of continuous curves, called *branches*, each corresponding to an integer value of n. A branch displays the relationship between the dimensionless frequency Ω and the dimensionless wavenumber ξ for a particular mode of propagation. The collection of branches constitutes the *frequency spectrum*.

For free motions the frequency should be taken as real-valued. It is now seen from the frequency equation (7.5.14) that for $\Omega < n$ this equation can be satisfied only if ξ is purely imaginary. Displacements associated with positive imaginary wavenumbers decay exponentially. Such displacements do not represent progressive waves but, rather, localized standing-wave motions. For a particular mode the frequency at which the wavenumber changes from real to imaginary values is called the *cut-off frequency*. It is noted that for horizontally polarized shear waves the cut-off frequencies are given by the frequencies at vanishing wavenumbers.

The frequency spectrum is well known. If ξ is real, the branches are hyperbolas with asymptotes $\Omega = \xi$ as $\xi \to \infty$. For imaginary wavenumbers, Eq. (7.5.14) represents circles with radii n. Since the frequency spectrum is symmetric in ξ, the second quadrant can be used to plot Ω versus positive imaginary values of ξ. Negative values of Ω need not be considered. Since the two relevant material constants, μ and ρ, and the one relevant geometrical parameter, h, appear in Ω and ξ only, the frequency spectrum can be employed for any homogeneous isotropic linearly elastic layer.

By using Eq. (7.2.9), the displacement components for the nth symmetric mode can be derived in polar coordinates as

$$u_r^n = \frac{1}{l_n} \cos\left(\frac{n\pi z}{2h}\right) \frac{1}{r}\frac{\partial \psi}{\partial \theta}, \tag{7.5.15}$$

$$u_\theta^n = -\frac{1}{l_n} \cos\left(\frac{n\pi z}{2h}\right) \frac{\partial \psi}{\partial r}. \tag{7.5.16}$$

7.6 Lamb waves

For plane-strain conditions, time-harmonic waves in an elastic layer were first investigated by Lamb (1917). It is well known that there is an infinite number of modes; they may be split into antisymmetric and symmetric modes relative to the mid-plane of the layer, corresponding to thickness shear and thickness stretch motions, respectively. To investigate Lamb waves in an elastic layer we return to Eqs. (7.2.1), (7.2.2), where $\varphi(x_1, x_2)$ is governed by Eq. (7.2.6) and $V(z)$ and $W(z)$ are the solutions of Eqs. (7.2.7) and (7.2.8). These equations are

$$(\lambda+\mu)\frac{dW}{dz} + \frac{\mu}{k}\frac{d^2V}{dz^2} + \frac{\rho\omega^2}{k}V(z) = k^2\frac{\lambda+2\mu}{k}V(z), \tag{7.6.1}$$

$$(\lambda+2\mu)\frac{d^2W}{dz^2} + \rho\omega^2 W(z) = k^2\left[\frac{\lambda+\mu}{k}\frac{dV}{dz} + \mu W(z)\right]. \tag{7.6.2}$$

For the layer shown in Fig. 7.1 we seek solutions of (7.6.1) and (7.6.2) of the general forms

$$V(z) = \overline{V}e^{ipz} \qquad \text{and} \qquad W(z) = \overline{W}e^{ipz},$$

where $\overline{V}$ and $\overline{W}$ are constants. Substitution of these expressions into Eqs. (7.6.1) and (7.6.2) yields a set of homogeneous algebraic equations for the constants $\overline{V}$ and $\overline{W}$. The condition that the determinant of the coefficients must vanish yields the following equation:

$$p^4 + \left(2k^2 - \frac{\omega^2}{c_L^2} - \frac{\omega^2}{c_T^2}\right)p^2 + \left(\frac{\omega^2}{c_L^2} - k^2\right)\left(\frac{\omega^2}{c_T^2} - k^2\right) = 0 \tag{7.6.3}$$

where c_L and c_T are the velocities of longitudinal and transverse waves, respectively. The two solutions of Eq. (7.6.3) are

$$p^2 = \frac{\omega^2}{c_L^2} - k^2 \tag{7.6.4}$$

and

$$q^2 = \frac{\omega^2}{c_T^2} - k^2. \tag{7.6.5}$$

The corresponding solutions of (7.5.1) and (7.5.2) may then be written as

$$V(z) = \overline{V}e^{ipz} + \overline{W}e^{iqz}, \tag{7.6.6}$$

$$W(z) = \frac{ip}{k}\overline{V}e^{ipz} + \frac{k}{iq}\overline{W}e^{iqz}. \tag{7.6.7}$$

Let us first consider displacements represented by the real parts of Eqs. (7.6.6) and (7.6.7), i.e.,

$$V(z) = \overline{V}\cos(pz) + \overline{W}\cos(qz), \tag{7.6.8}$$

$$W(z) = -\frac{p}{k}\overline{V}\sin(pz) + \frac{k}{q}\overline{W}\sin(qz). \tag{7.6.9}$$

These expressions represent motions that are symmetric relative to the mid-plane of the layer. They define the so-called thickness-stretch modes of the layer.

The displacement components given by Eqs. (7.2.1) and (7.2.2) with $V(z)$ and $W(z)$ defined by Eqs. (7.6.8) and (7.6.9) do not yet satisfy the conditions on the faces of the layer. These conditions are

$$\text{at } z = \pm h: \quad \tau_{z\beta} = 0 \quad \text{and} \quad \tau_{zz} = 0, \tag{7.6.10}$$

where $\beta = 1, 2$. To obtain expressions for $\tau_{z\beta}$ and τ_{zz} we use Hooke's law in the form given by Eq. (7.3.5). By using Eqs. (7.2.1) and (7.2.2) in Cartesian coordinates, the relevant stress components may be expressed as

$$\tau_{z\beta} = \rho c_T^2 \left[W(z) + \frac{1}{k}\frac{dV}{dz} \right] \varphi,_{\beta},$$

$$\tau_{zz} = \rho\left(c_L^2 - 2c_T^2\right)\left[\frac{1}{k}V(z)\varphi,_{\beta\beta} + \frac{dW}{dz}\varphi\right] + 2\rho c_T^2 \frac{dW}{dz}\varphi.$$

To satisfy the boundary conditions, Eq. (7.6.10), at $z = \pm h$, we must then have

$$W(h) + \left.\frac{1}{k}\frac{dV}{dz}\right|_{z=h} = 0, \tag{7.6.11}$$

$$-\rho k\left(c_L^2 - 2c_T^2\right)V(h) + \rho c_L^2 \left.\frac{dW}{dz}\right|_{z=h} = 0, \tag{7.6.12}$$

where $\varphi,_{\beta\beta}$ has been eliminated by the use of Eq. (7.2.6). Substitution of $V(z)$ and $W(z)$ from Eqs. (7.6.8) and (7.6.9) yields, after some manipulation, the following system of homogeneous algebraic equations for the constants $\overline{V}$ and $\overline{W}$:

$$-\frac{2p}{k}\sin(ph)\,\overline{V} + \frac{k^2 - q^2}{qk}\sin(qh)\,\overline{W} = 0 \tag{7.6.13}$$

$$\frac{k^2 - q^2}{k}\cos(ph)\,\overline{V} + 2k\cos(qh)\,\overline{W} = 0. \tag{7.6.14}$$

The condition that the determinant of the coefficients must vanish yields

$$\frac{\tan(qh)}{\tan(ph)} = -\frac{4pqk^2}{(q^2 - k^2)^2}. \tag{7.6.15}$$

Equation (7.6.15) is recognized as the well-known Rayleigh–Lamb frequency equation for symmetric modes of the layer; see e.g., Mindlin (1960) and Achenbach (1973).

In a similar manner we can consider the imaginary parts of Eqs. (7.6.6) and (7.6.7), i.e.,

$$V(z) = \overline{V}\ \sin(pz) + \overline{W}\ \sin(qz), \tag{7.6.16}$$

$$W(z) = \frac{p}{k}\overline{V}\ \cos(pz) - \frac{k}{q}\overline{W}\cos(qz). \tag{7.6.17}$$

These expressions represent motions that are antisymmetric relative to the mid-plane of the layer. They are the so-called thickness-shear modes. Substitution of (7.6.16) and (7.6.17) into (7.6.11) and (7.6.12) yields the Rayleigh–Lamb frequency equation for antisymmetric modes of the layer:

$$\frac{\tan(qh)}{\tan(ph)} = -\frac{(q^2 - k^2)^2}{4pqk^2}. \tag{7.6.18}$$

The Rayleigh–Lamb frequency equations (7.6.15) and (7.6.18) relate the wavenumber, k, to the frequency, ω. To obtain detailed and precise numerical information the roots of these equations have to be obtained numerically by choosing a real-valued frequency ω and calculating the infinite number of corresponding values of k. Each value of k corresponds to a mode of wave propagation in the layer. The calculated values of k may be real, imaginary or complex. Real-valued wavenumbers correspond to propagating wave modes. Imaginary and complex wavenumbers correspond to standing waves with decaying amplitudes. At small frequencies, Eqs. (7.6.15) and (7.6.18) each have one real-valued solution for k. All the other solutions for k are then either imaginary or complex. The two modes with real-valued k's correspond to the lowest symmetric and antisymmetric modes.

The result that interests us in this section is that the thickness motions governed by Eqs. (7.6.15) and (7.6.18), which are well known for the plane-strain case, apply to a general class of guided waves in the layer as long as $\varphi(x_1, x_2)$ is a solution of the reduced wave equation (7.2.6).

For the symmetric modes, Eq. (7.6.14) gives

$$\overline{V} = -\frac{2k^2\cos(qh)}{(k^2 - q^2)\cos(ph)}\overline{W} \tag{7.6.19}$$

Substitution of (7.6.19) into (7.6.8) and (7.6.9) yields, after a slight redefinition and the introduction of the arbitrary constant A_n^S,

$$u_\alpha^n = A_n^S \frac{1}{k_n} V_S^n(z)\varphi_{,\alpha}(x_1, x_2,), \tag{7.6.20}$$

$$u_z = A_n^S W_S^n(z)\varphi(x_1, x_2), \tag{7.6.21}$$

where $\alpha = 1, 2$, and

$$V_S^n(z) = s_1 \cos(pz) + s_2 \cos(qz), \tag{7.6.22}$$

$$W_S^n(z) = s_3 \sin(pz) + s_4 \sin(qz), \tag{7.6.23}$$

while

$$s_1 = 2\cos(qh), \qquad s_2 = -\left[\left(k_n^2 - q^2\right)/k_n^2\right]\cos(ph), \tag{7.6.24}$$

$$s_3 = -2(p/k_n)\cos(qh), \qquad s_4 = -\left[\left(k_n^2 - q^2\right)/(qk_n)\right]\cos(ph). \tag{7.6.25}$$

Here the index n indicates that for given ω the wavenumber k_n is the nth solution of Eq. (7.6.15), i.e., (7.6.22) and (7.6.23) represent the nth symmetric mode of the frequency spectrum for Lamb-wave modes. The quantities p and q are defined by Eqs. (7.6.4) and (7.6.5), except that k must be replaced by k_n.

The relevant displacement expressions for the nth antisymmetric mode can be obtained similarly to Eqs. (7.6.22) and (7.6.23). The mode shapes are

$$V_A^n(z) = a_1 \sin(pz) + a_2 \sin(qz), \tag{7.6.26}$$

$$W_A^n(z) = a_3 \cos(pz) + a_4 \cos(qz), \tag{7.6.27}$$

where

$$a_1 = 2\sin(qh), \qquad a_2 = -\left[\left(k_n^2 - q^2\right)/k_n^2\right]\sin(ph), \tag{7.6.28}$$

$$a_3 = 2(p/k_n)\sin(qh), \qquad a_4 = \left[\left(k_n^2 - q^2\right)/(qk_n)\right]\sin(ph). \tag{7.6.29}$$

As noted before, in these expressions p and q have been redefined as

$$p^2 = \frac{\omega^2}{c_L^2} - k_n^2 \qquad \text{and} \qquad q^2 = \frac{\omega^2}{c_T^2} - k_n^2. \tag{7.6.30}$$

The results of this chapter are used in the next three chapters in conjunction with reciprocity considerations.

8

Computation of surface waves by reciprocity considerations

8.1 Introduction

It is shown in this chapter that the reciprocity theorem can be used to calculate in a convenient manner, that is, without the use of integral transform techniques, the surface-wave motion generated by a time-harmonic line load or a time-harmonic point load applied in an arbitrary direction in the interior of a half-space. The virtual wave motion that is used in the reciprocity relation is also a surface wave. Hence the calculation does not include the body waves generated by the loads. For a point load applied normally to the surface of a half-space, it is shown in Section 8.6 that the surface-wave motion is the same as obtained in the conventional manner by the integral transform approach.

It is well known that the dynamic response to a time-harmonic point load normal to the surface of the half-space was solved by Lamb (1904), who also gave explicit expressions for the generated surface-wave motion. The surface-wave motion can be obtained as the contribution from the pole in inverse integral transform representations of the displacement components. The analogous transient time-domain problem for a point load normal to the surface of the half-space was solved by Pekeris (1955). The displacements generated by a transient tangential point load applied to the half-space surface were worked out by Chao (1960).

8.2 Surface waves generated by a line load

This is a two-dimensional problem. Figure 8.1 shows a half-space of a homogeneous, isotropic, linearly elastic solid, referred to a Cartesian coordinate system such that the plane $z = 0$ coincides with the surface of the half-space. The half-space is subjected to a time-harmonic line load at $z = z_0$. The load

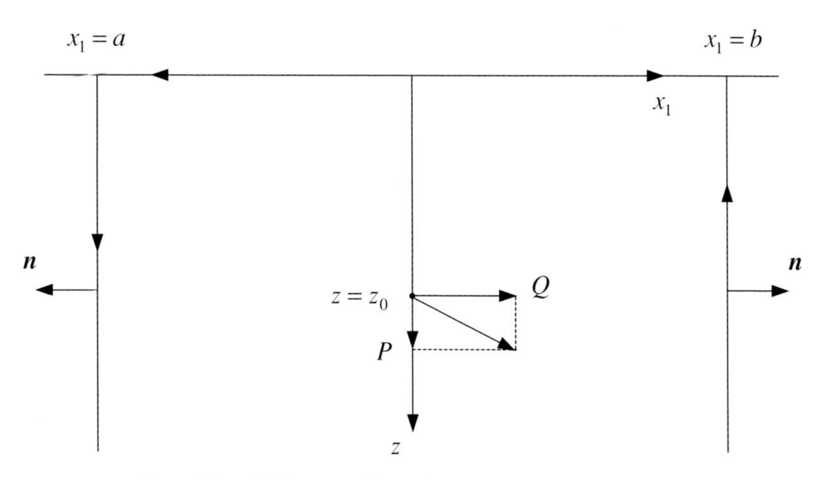

Figure 8.1 Half-space subjected to a time-harmonic line load.

has components in the x_1- and z-directions only. The surface-wave response of the half-space is sought as the superposition of the responses due to the vertical component, P, and the horizontal component, Q.

With reference to Eqs. (7.3.9), (7.3.10), and in conjunction with the time factor $\exp(-i\omega t)$, the relevant solutions for $\varphi(x_1, x_2)$ are

$$\varphi(x_1) = e^{ik_R x_1} \tag{8.2.1}$$

for surface waves propagating in the positive x_1-direction, and

$$\varphi(x_1) = e^{-ik_R x_1} \tag{8.2.2}$$

for propagation in the negative x_1-direction. Here $k_R = \omega/c_R$, where c_R is the solution of Eq. (7.3.8), i.e., the phase velocity of Rayleigh surface waves.

As shown in Section 7.3, Eqs. (7.3.9) and (7.3.10), the displacements may be written as

$$u_1(x_1, z) = \pm i A V^R(z) e^{\pm i k_R x_1}, \tag{8.2.3}$$

$$u_z(x_1, z) = A W^R(z) e^{\pm i k_R x_1}. \tag{8.2.4}$$

The relevant stresses follow from (7.3.17) and (7.3.18) as

$$\tau_{11}(x_1, z) = A T_{11}^R(z) e^{\pm i k_R x_1}, \tag{8.2.5}$$

$$\tau_{1z}(x_1, z) = \pm i A T_{1z}^R(z) e^{\pm i k_R x_1}. \tag{8.2.6}$$

In these expressions $\exp(-i\omega t)$ has been omitted. The second term in Eq. (7.3.17) disappears because φ does not depend on x_2.

The reciprocity relation will now be applied to the problem of a vertical time-harmonic line load applied inside a half-space. The line load generates surface waves, and we will determine the amplitude of these surface waves by using reciprocity with a virtual surface wave. For V we take the region defined by $a \leq x_1 \leq b,\ z \geq 0$. For state A the body force is the vertical force P shown in Fig. 8.1, i.e.,

$$f_z^A(x, z) = P\delta(x_1)\delta(z - z_0).$$

For the surface waves of state A, the displacements, which are symmetric with respect to $x_1 = 0$, and the stresses are given by Eqs. (8.2.3)–(8.2.6), where $\varphi = \exp(ik_R x_1)$ applies for $x_1 > 0$ and $\varphi = \exp(-ik_R x_1)$ for $x_1 < 0$. For state B, the virtual wave, we select a surface wave propagating in the positive x_1-direction:

$$u_1^B(x_1, z) = iBV^R(z)e^{ik_R x_1}, \tag{8.2.7}$$

$$u_z^B(x_1, z) = BW^R(z)e^{ik_R x_1}, \tag{8.2.8}$$

$$\tau_{11}^B(x_1, z) = BT_{11}^R(z)e^{ik_R x_1}, \tag{8.2.9}$$

$$\tau_{1z}^B(x_1, z) = iBT_{1z}^R(z)e^{ik_R x_1}. \tag{8.2.10}$$

For the problem at hand, and considering the contour shown in Fig. 8.1, the reciprocity relation Eq. (6.2.8) becomes

$$\int_V f_z^A u_z^B dV = \int_\infty^0 F_{AB}\big|_{x_1=b} dz - \int_0^\infty F_{AB}\big|_{x_1=a} dz. \tag{8.2.11}$$

The contribution from the integration along the line at constant z is not included since it vanishes as $z \to \infty$. In Eq. (8.2.11),

$$F_{AB}(x_1, z) = u_1^A \tau_{11}^B + u_z^A \tau_{1z}^B - u_1^B \tau_{11}^A - u_z^B \tau_{1z}^A. \tag{8.2.12}$$

As has been noted in earlier calculations the integrations along $x_1 = a,\ 0 \leq z < \infty$, and $x_1 = b,\ 0 \leq z < \infty$, only yield contributions from counter-propagating waves. By the present choice of virtual wave, state A and state B are counter-propagating at $x_1 = a,\ 0 \leq z < \infty$. Substitution of the expressions for states A and B into Eq. (8.2.11) then yields

$$PW^R(z_0) = 2iAI, \tag{8.2.13}$$

where

$$I = \int_0^\infty \left[T_{11}^R(z)V^R(z) - T_{1z}^R(z)W^R(z)\right] dz. \tag{8.2.14}$$

Hence the amplitude of the surface waves generated by force P is

$$A = \frac{PW^R(z_0)}{2iI}. \tag{8.2.15}$$

An analogous result can be obtained when a horizontal force Q is applied. For this case the surface waves are antisymmetric with respect to $x_1 = 0$.

8.3 Surface waves generated by a sub-surface point load

Figure 8.2 shows a half-space of a homogeneous, isotropic, linearly elastic solid referred to a Cartesian coordinate system such that the x_1x_2-plane coincides with the surface of the half-space. The half-space is subjected to a time-harmonic point load at $z = z_0$ pointing in an arbitrary direction. Without loss of generality the coordinate system can be chosen such that the load acts in the x_1z-plane. The surface-wave response of the half-space is sought as the superposition of the responses due to the vertical component, P, and the horizontal component, Q; see Achenbach (2000).

Let us first consider the wave motion generated by the time-harmonic point load, Q, applied in the direction of the x_1-axis. In cylindrical coordinates defined by $x_1 = r\cos\theta$, $x_2 = r\sin\theta$, z, Eqs. (7.2.1), (7.2.2) can be rewritten as

$$u_r = A\frac{1}{k}V(z)\frac{\partial\varphi}{\partial r}(r,\theta), \tag{8.3.1}$$

$$u_\theta = A\frac{1}{k}V(z)\frac{1}{r}\frac{\partial\varphi}{\partial\theta}(r,\theta), \tag{8.3.2}$$

$$u_z = AW(z)\varphi(r,\theta), \tag{8.3.3}$$

and Eq. (7.2.6) becomes

$$\frac{\partial^2\varphi}{\partial r^2} + \frac{1}{r}\frac{\partial\varphi}{\partial r} + \frac{1}{r^2}\frac{\partial^2\varphi}{\partial\theta^2} + k^2\varphi = 0. \tag{8.3.4}$$

The first point of consideration now is to select the correct solution of Eq. (8.3.4) for $\varphi(r,\theta)$. For this we take guidance from the displacement in an unbounded solid when a time-harmonic point force in the x_1-direction of magnitude F is applied at the origin. As shown in Section 3.9, in a Cartesian coordinate system this solution can be written as

$$u_i = \frac{1}{k_T^2}\frac{F}{\mu}\frac{\partial}{\partial x_i}\frac{\partial}{\partial x_1}[-G(k_LR) + G(k_TR)] + \frac{F}{\mu}G(k_TR)\delta_{i1}, \tag{8.3.5}$$

where δ_{i1} is the Kronecker delta and

$$G(k_\gamma R) = e^{ik_\gamma R}/(4\pi R). \tag{8.3.6}$$

Here $\gamma = L$ or T, $k_L^2 = \omega^2/c_L^2$, $k_T^2 = \omega^2/c_T^2$ and $R^2 = x_1^2 + x_2^2 + z^2$.

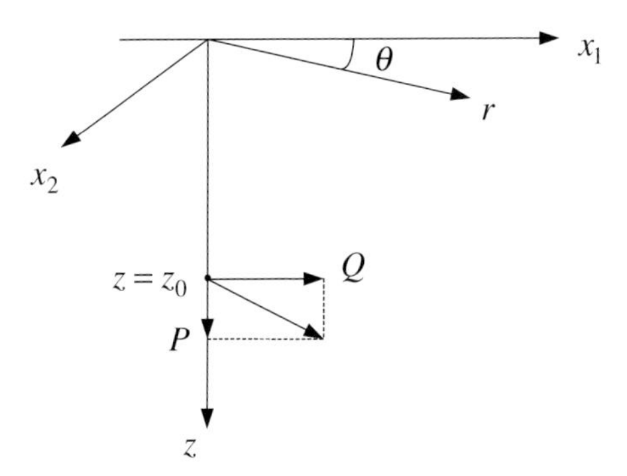

Figure 8.2 Homogeneous, isotropic, linearly elastic half-space subjected to a sub-surface time-harmonic point force.

The displacements in cylindrical coordinates follow from Eq. (8.3.5) as

$$u_r = \frac{F}{\mu}\left\{\frac{1}{k_T^2}\frac{\partial^2}{\partial r^2}[-G(k_L R) + G(k_T R)] + G(k_T R)\right\}\cos\theta, \quad (8.3.7)$$

$$u_\theta = \frac{F}{\mu}\left\{\frac{1}{k_T^2}\frac{1}{r}\frac{\partial}{\partial r}[G(k_L R) - G(k_T R)] - G(k_T R)\right\}\sin\theta, \quad (8.3.8)$$

$$u_z = \frac{F}{\mu}\frac{1}{k_T^2}\frac{\partial^2}{\partial z \partial r}[-G(k_L R) + G(k_T R)]\cos\theta, \quad (8.3.9)$$

where

$$R^2 = r^2 + z^2. \quad (8.3.10)$$

These solutions show a simple dependence on θ. It is to be expected that the presence of a traction-free surface parallel to the plane of r and θ does not change this dependence. This has been shown to be true by Chao (1960) for a point load applied tangentially to the surface of a half-space. By comparing Eqs. (8.3.1)–(8.3.3) with Eqs. (8.3.7)–(8.3.9) it immediately follows that we should take

$$\varphi(r, \theta) = \Phi(kr)\cos\theta, \quad (8.3.11)$$

where, according to Eq. (8.3.4), $\Phi(kr)$ must be the solution of

$$\frac{d^2\Phi}{dr^2} + \frac{1}{r}\frac{d\Phi}{dr} + \left(k^2 - \frac{1}{r^2}\right)\Phi = 0. \quad (8.3.12)$$

The solution to Eq. (8.3.12) for an outgoing wave compatible with the time factor $\exp(-i\omega t)$ is

$$\Phi(kr) = H_1^{(1)}(kr). \quad (8.3.13)$$

We can equally well consider an incoming wave, i.e., a wave that converges on the origin:

$$\overline{\Phi}(kr) = H_1^{(2)}(kr). \tag{8.3.14}$$

For simplicity of notation we will use $\Phi(kr)$ and $\overline{\Phi}(kr)$ in subsequent expressions rather than the Hankel functions. The notation $\Phi'(kr)$ is again used for the derivative with respect to the argument: $\Phi'(\xi) = d\Phi/d\xi$.

In polar coordinates the displacements for surface waves are given by Eqs. (8.3.1)–(8.3.3), provided that $V(z)$ and $W(z)$ are chosen as $V^R(z)$ and $W^R(z)$, which are given by Eqs. (7.3.12) and (7.3.13). For outgoing surface waves the displacements are then

$$u_r = AV^R(z)\Phi'(kr)\cos\theta, \tag{8.3.15}$$

$$u_z = AW^R(z)\Phi(kr)\cos\theta, \tag{8.3.16}$$

$$u_\theta = AV^R(z)\left(\frac{-1}{kr}\right)\Phi(kr)\sin\theta. \tag{8.3.17}$$

In these expressions $k = k_R$, where $k_R = \omega/c_R$, but to simplify the notation we omit the subscript R. The relevant stress components are

$$\tau_{rz} = AT_{rz}^R(z)\Phi'(kr)\cos\theta, \tag{8.3.18}$$

$$\tau_{rr} = AT_{rr}^R(z)\Phi(kr)\cos\theta - A\overline{T}_{rr}^R(z)\left[\frac{1}{kr}\Phi'(kr) - \frac{1}{(kr)^2}\Phi(kr)\right]\cos\theta, \tag{8.3.19}$$

$$\tau_{r\theta} = AT_{r\theta}^R(z)\left[\frac{1}{kr}\Phi'(kr) - \frac{1}{(kr)^2}\Phi(kr)\right]\sin\theta, \tag{8.3.20}$$

where

$$T_{rz}^R(z) = \mu k\left(\frac{1}{k}\frac{dV^R}{dz} + W^R\right) = \mu(d_6 e^{-pz} + d_7 e^{-qz}) = T_{1z}^R(z),$$

$$T_{rr}^R(z) = -\mu\left[\frac{c_L^2}{c_T^2}\left(kV^R - \frac{dW^R}{dz}\right) + 2\frac{dW^R}{dz}\right] = T_{11}^R(z),$$

$$\overline{T}_{rr}^R(z) = 2\mu kV^R = 2\mu k(d_1 e^{-pz} + d_2 e^{-qz}),$$

$$T_{r\theta}^R(z) = -2\mu kV^R = -2\mu k(d_1 e^{-pz} + d_2 e^{-qz}).$$

The expression for $T_{11}^R(z)$ is given by Eq. (7.3.19), and the quantities d_1, d_2, d_6 and d_7 are defined in Section 7.3. In addition to the displacements defined by Eqs. (7.2.1) and (7.2.2), we may also consider wave motions that are equivoluminal, with displacements that are parallel to the x_1x_2-plane. The corresponding wave motions are horizontally polarized. They may be represented by Eqs. (7.2.9)–(7.2.12). There are no solutions of Eqs. (7.2.10)–(7.2.12) that represent surface waves; hence horizontally polarized waves are not further considered in this chapter.

8.4 Application of elastodynamic reciprocity in cylindrical coordinates

In this section we use the reciprocity theorem to determine the amplitude A of the surface wave generated by the horizontal component of the sub-surface point force shown in Fig. 8.2. The force is represented by

$$f_1 = Q\delta(x_1)\delta(x_2)\delta(z - z_0)e^{-i\omega t}. \tag{8.4.1}$$

For V we take the region defined by $0 \leq r \leq b, 0 \leq z < \infty, 0 \leq \theta \leq 2\pi$. The displacements for state A are taken as Eqs. (8.3.15)–(8.3.17), i.e.,

$$u_r^A = AV^R(z)\Phi'(k_R r)\cos\theta, \tag{8.4.2}$$

$$u_z^A = AW^R(z)\Phi(k_R r)\cos\theta, \tag{8.4.3}$$

$$u_\theta^A = AV^R(z)\left(\frac{-1}{k_R r}\right)\Phi(k_R r)\sin\theta, \tag{8.4.4}$$

where A is the unknown amplitude factor, $k_R = \omega/c_R$ and $V^R(z)$ and $W^R(z)$ are defined by Eqs. (7.3.12) and (7.3.13), while $\Phi(k_R r)$ is given by Eq. (8.3.13). For state B we take a virtual wave consisting of the sum of an outgoing and a converging wave:

$$u_r^B = \tfrac{1}{2}BV^R(z)[\Phi'(k_R r) + \overline{\Phi}'(k_R r)]\cos\theta, \tag{8.4.5}$$

$$u_z^B = \tfrac{1}{2}BW^R(z)[\Phi(k_R r) + \overline{\Phi}(k_R r)]\cos\theta, \tag{8.4.6}$$

$$u_\theta^B = \tfrac{1}{2}BV^R(z)\left(\frac{-1}{k_R r}\right)[\Phi(k_R r) + \overline{\Phi}(k_R r)]\sin\theta, \tag{8.4.7}$$

where $\overline{\Phi}(k_R r)$ is defined by Eq. (8.3.14).

The displacements in Eqs. (8.4.5)–(8.4.7) are bounded at $r = 0$. It can be verified that the left-hand side of the reciprocity relation (6.2.8) becomes

$$Qu_r^B(0, 0, z_0) = \tfrac{1}{2}QBV^R(z_0). \tag{8.4.8}$$

Hence we obtain

$$\frac{1}{2}QBV^R(z_0) = b\int_0^{2\pi}\int_0^{\infty}\left[\left(u_r^A\tau_{rr}^B - u_r^B\tau_{rr}^A\right) + \left(u_z^A\tau_{rz}^B - u_z^B\tau_{rz}^A\right)\right.$$
$$\left. + \left(u_\theta^A\tau_{r\theta}^B - u_\theta^B\tau_{r\theta}^A\right)\right] d\theta\, dz. \tag{8.4.9}$$

In Eq. (8.4.9) all field variables are evaluated at $r = b$.

It turns out that the right-hand side of Eq. (8.4.9) becomes a quite simple expression. To show this we write out the terms in detail, except that we consider it understood that V^R, W^R, T_{rr}^R, $\overline{T}_{rr}^R$, T_{rz}^R and $T_{r\theta}^R$ are functions of z and that Φ and $\overline{\Phi}$ are evaluated at $k_R b$. We have

$$\begin{aligned}
u_r^A\tau_{rr}^B &= \tfrac{1}{2}ABV^R\Phi' T_{rr}^R(\Phi + \overline{\Phi})\cos^2\theta \\
&\quad - \tfrac{1}{2}ABV^R\Phi'\overline{T}_{rr}^R\left[\frac{1}{k_R r}(\Phi' + \overline{\Phi}') - \frac{1}{(k_R r)^2}(\Phi + \overline{\Phi})\right]\cos^2\theta, \\
u_r^B\tau_{rr}^A &= \tfrac{1}{2}ABV^R(\Phi' + \overline{\Phi}')T_{rr}^R\Phi\cos^2\theta \\
&\quad - \tfrac{1}{2}ABV^R(\Phi' + \overline{\Phi}')\overline{T}_{rr}^R\left(\frac{1}{k_R r}\Phi' - \frac{1}{(k_R r)^2}\Phi\right)\cos^2\theta, \\
u_z^A\tau_{rz}^B &= \tfrac{1}{2}ABW^R\Phi T_{rz}^R(\Phi' + \overline{\Phi}')\cos^2\theta, \\
u_z^B\tau_{rz}^A &= \tfrac{1}{2}ABW^R(\Phi + \overline{\Phi})T_{rz}^R\Phi'\cos^2\theta, \\
u_\theta^A\tau_{r\theta}^B &= \tfrac{1}{2}ABV^R\left(-\frac{1}{k_R r}\right)\Phi T_{r\theta}^R\left[\frac{1}{k_R r}(\Phi' + \overline{\Phi}') - \frac{1}{(k_R r)^2}(\Phi + \overline{\Phi})\right]\sin^2\theta, \\
u_\theta^B\tau_{r\theta}^A &= \tfrac{1}{2}ABV^R\left(-\frac{1}{k_R r}\right)(\Phi + \overline{\Phi})T_{r\theta}^R\left[\frac{1}{k_R r}\Phi' - \frac{1}{(k_R r)^2}\Phi\right]\sin^2\theta.
\end{aligned}$$

Upon substitution of these expressions into Eq. (8.4.9) we find that several terms cancel. The result is

$$QV^R(z_0) = AIH, \tag{8.4.10}$$

where I is defined by

$$I = \int_0^{\infty}\left[T_{rr}^R(z)V^R(z) - T_{rz}^R(z)W^R(z)\right]dz. \tag{8.4.11}$$

Since $T_{rr}^R(z) = T_{11}^R(z)$ and $T_{rz}^R(z) = T_{1z}^R(z)$, this is the same integral as in Eq. (8.2.14). Also,

$$H = \pi b\left[\Phi'(k_R b)\overline{\Phi}(k_R b) - \overline{\Phi}'(k_R b)\Phi(k_R b)\right]. \tag{8.4.12}$$

In evaluating (8.4.9) we have used that

$$\int_0^{2\pi} \cos^2\theta \, d\theta = \int_0^{2\pi} \sin^2\theta \, d\theta = \pi \qquad \text{and} \qquad T_{r\theta}(z) = -\overline{T}_{rr}(z).$$

The expression for H can be further simplified by using the following identity for Hankel functions, see McLachlan (1961, p. 198):

$$\left[\frac{d}{d\zeta} H_v^{(1)}(\zeta)\right] H_v^{(2)}(\zeta) - H_v^{(1)}(\zeta) \left[\frac{d}{d\zeta} H_v^{(2)}(\zeta)\right] = \frac{4i}{\pi\zeta}. \tag{8.4.13}$$

We obtain

$$H = \frac{4i}{k_R} \tag{8.4.14}$$

and Eq. (8.4.10) yields

$$A = \frac{k_R}{4i} \frac{QV^R(z_0)}{I}. \tag{8.4.15}$$

Substitution of the expressions for $T_{rr}^R(z)$, $T_{rz}^R(z)$, $V^R(z)$ and $W^R(z)$, but with $k = k_R$, into Eq. (8.4.11) yields a relatively simple integral over z, which can be evaluated easily. The parameter k_R appearing in the integral is the wavenumber for Rayleigh waves.

To simplify further manipulation we introduce the dimensionless Rayleigh wave velocity by

$$\xi = \frac{\omega}{c_T} \frac{1}{k_R}. \tag{8.4.16}$$

We also introduce

$$q_R^2 = 1 - \xi^2 \tag{8.4.17}$$

and

$$p_R^2 = 1 - \frac{\xi^2}{\kappa^2}, \tag{8.4.18}$$

where κ^2 is found from Eq. (3.2.6). Carrying out the integration in Eq. (8.4.11) yields after some further manipulation

$$I = \mu J, \tag{8.4.19}$$

where

$$J = -\frac{1 + 3q_R^2}{2q_R} + \left(1 + q_R^2\right)\left[\frac{1 + q_R^2}{2p_R} + \frac{q_R}{p_R}(p_R - q_R) + \frac{1}{p_R}\right] - \frac{\left(1 + q_R^2\right)^2}{8p_R^3}\left(1 + 4p_R^2 - q_R^2\right). \tag{8.4.20}$$

By using (8.4.3), (8.4.15) and (8.4.19) the vertical displacement at position (r, θ, z_1) may then be written as

$$\frac{\mu}{Q}\frac{u_z^R}{k_R} = \frac{1}{4i}\frac{1}{J}V^R(z_0)W^R(z_1)H_1^{(1)}(k_R r)\cos\theta, \tag{8.4.21}$$

where $V^R(z_0)$ and $W^R(z_1)$ follow from Eqs. (7.3.12) and (7.3.13).

8.5 Surface wave motion generated by a vertical force *P*

For the axially symmetric case the relevant solution $\varphi(k_R r)$ of Eq. (8.3.4) is

$$\varphi(k_R r) = H_0^{(1)}(k_R r). \tag{8.5.1}$$

The displacements for state A, which are the surface-wave displacements generated by P, are of the general form

$$u_r^A = -AV^R(z)H_1^{(1)}(k_R r), \tag{8.5.2}$$

$$u_z^A = AW^R(z)H_0^{(1)}(k_R r), \tag{8.5.3}$$

where $V^R(z)$ and $W^R(z)$ are given by Eqs. (7.3.12) and (7.3.13) and A is an unknown amplitude factor. The relevant corresponding stresses may be written as

$$\tau_{rz}^A = -AT_{rz}(z)H_1^{(1)}(k_R r), \tag{8.5.4}$$

$$\tau_{rr}^A = A\left[T_{rr}(z)H_0^{(1)}(k_R r) + \overline{T}_{rr}(z)\frac{1}{k_R r}H_1^{(1)}(k_R r)\right], \tag{8.5.5}$$

where $T_{rz}(z)$, $T_{rr}(z)$ and $\overline{T}_{rr}(z)$ are defined in Section 8.3. For state B we select the virtual-wave solution

$$u_r^B = -\tfrac{1}{2}BV^R(z)\left[H_1^{(1)}(k_R r) + H_1^{(2)}(k_R r)\right],$$

$$u_z^B = \tfrac{1}{2}BW^R(z)\left[H_0^{(1)}(k_R r) + H_0^{(2)}(k_R r)\right].$$

The reciprocity relation now yields

$$PW^R(z_0) = -\frac{4i}{k}AI \qquad \text{or} \qquad A = -\frac{k_R}{4i}\frac{PW^R(z_0)}{I}, \tag{8.5.6}$$

where Eq. (8.4.13) has been used, and I is defined by Eq. (8.4.11).

By using Eqs. (8.5.2), (8.5.3), (8.5.6) and (8.4.19) the displacements at position (r, θ, z_1) may then be written as

$$\frac{\mu}{P}\frac{u_r^R}{k_R} = \frac{1}{4i}\frac{1}{J}W^R(z_0)V^R(z_1)H_1^{(1)}(k_R r), \tag{8.5.7}$$

$$\frac{\mu}{P}\frac{u_z^R}{k_R} = -\frac{1}{4i}\frac{1}{J}W^R(z_0)W^R(z_1)H_0^{(1)}(k_R r). \tag{8.5.8}$$

Here $V^R(z_1)$, $W^R(z_0)$ and $W^R(z_1)$ follow from Eqs. (7.3.12) and (7.3.13).

8.6 Check of calculated surface-wave amplitude

As has been shown in this chapter, the surface-wave displacements due to a sub-surface time-harmonic point load are relatively easy to obtain by using the reciprocity relation. It will be of interest, however, to verify the results for internal consistency, and by comparison with analogous results obtained by the use of Hankel transform techniques. First we will check the reciprocity of the solutions given by Eqs. (8.4.21) and (8.5.7). The simplest check of reciprocity is for the case where a vertical unit load is applied at $r = 0$, $z = z_0$, and a horizontal unit load is applied at $r = r$, $z = z_1$. In this case the displacement u_r at $r = r$, $z = z_1$, $\theta = 0$ due to the vertical load should be just the same as the displacement u_z at $r = r$, $z = z_0$, $\theta = \pi$ due to the horizontal load. It is easily verified that this equality is indeed satisfied.

For the special case of a vertical point load applied at $r = 0$, $z = 0$ the vertical surface-wave displacement at $(r, 0)$ is available from a number of sources. Achenbach (1973, p. 310) used the Laplace and Hankel transforms to determine the response of a half-space to a normal point load of arbitrary time dependence. For a time-harmonic point load the Hankel transform of the vertical displacement at $z = 0$ follows from Achenbach (1973, p. 313, Eq. (7.233)) by replacing the Laplace transform parameter p by $i\omega$:

$$u_z^H = -\frac{P}{2\pi}\frac{1}{\mu}\frac{\omega^2}{c_T^2}\frac{p(k)}{F(k)}, \tag{8.6.1}$$

where $p(k)$ is defined by Eq. (7.3.2) and

$$F(k) = (k^2 + q^2)^2 - 4k^2 pq.$$

By application of the inverse Hankel transform we then find

$$u_z = -\frac{P}{2\pi}\frac{1}{\mu}\frac{\omega^2}{c_T^2}\int_0^\infty \frac{p(k)}{F(k)} k J_0(kr)\, dk. \tag{8.6.2}$$

To obtain the Rayleigh-wave contribution from this integral we use

$$J_0(kr) = \tfrac{1}{2}H_0^{(1)}(kr) + \tfrac{1}{2}H_0^{(2)}(kr) = \tfrac{1}{2}H_0^{(1)}(kr) - \tfrac{1}{2}H_0^{(1)}(-kr). \quad (8.6.3)$$

Substitution of this result into Eq. (8.6.2) allows the integral to be rewritten as

$$u_z = -\frac{P}{4\pi}\frac{1}{\mu}\frac{\omega^2}{c_T^2}\int_{-\infty}^{\infty}\frac{p(k)}{F(k)}kH_0^{(1)}(kr)\,dk. \quad (8.6.4)$$

The Rayleigh wave is the contribution from the pole at the point $k = k_R$, where k_R is the solution of $F(k) = 0$. We find

$$u_z^R = \frac{P}{2}\frac{i}{\mu}\frac{\omega^2}{c_T^2}\frac{\left(k_R^2 - \omega^2/c_L^2\right)^{1/2}}{F'(k_R)}k_R H_0^{(1)}(k_R r), \quad (8.6.5)$$

where

$$F'(k_R) = \left.\frac{dF(k)}{dk}\right|_{k=k_R}. \quad (8.6.6)$$

It is convenient to recast Eq. (8.6.5) in the form

$$\frac{\mu}{P}\frac{u_z^R}{k_R} = U_z(\xi)H_0^{(1)}(k_R r); \quad (8.6.7)$$

here

$$U_z(\xi) = \frac{i}{8}\frac{\xi^2(1-\xi^2/\kappa^2)^{1/2}}{D(\xi)}, \quad (8.6.8)$$

where ξ is defined by Eq. (8.4.16) and

$$D(\xi) = 2(2-\xi^2) - 2p_R q_R - \frac{q_R^2 + p_R^2}{p_R q_R}. \quad (8.6.9)$$

Casting (8.5.8) in the form (8.6.7) for $z_0 = 0$ and $z_1 = 0$ yields

$$U_z(\xi) = \frac{i}{16}\frac{\left(1-q_R^2\right)^2}{J}. \quad (8.6.10)$$

Equations (8.6.8) and (8.6.10) both represent the surface wave generated by a time-harmonic point load applied normally to the surface of a homogeneous,isotropic, linearly elastic half-space. The question of interest is whether they do in fact give the same numerical result. This can easily be checked by a simple calculation. Let us consider the case where Poisson's ratio $\upsilon = 0.25$. For this case, Eq. (3.2.6) yields $\kappa^2 = 3$, and the solution of $D(\xi) = 0$ is

$$\xi = 0.919402.$$

Substitution of this value in Eqs. (8.6.8) and (8.6.10) yields for both cases exactly the same value, namely,

$$U_z = -0.917429i.$$

The method presented here applies also to layered half-spaces,to a transversely isotropic half-space with the symmetry axis normal to the free surface and to certain cases of continuous inhomogeneity in the z-direction.

9

Reciprocity considerations for an elastic layer

9.1 Introduction

In Chapter 7 it was shown that for time-harmonic wave motion there is an infinite number of wave modes that can propagate in a linearly elastic layer. These modes correspond to standing waves across the thickness of the layer and propagation along the layer. It was shown that for the isotropic case the standing waves consist of thickness-stretch and thickness-shear motions carried by the wave propagating along the layer, the carrier wave. The carrier wave acts like a membrane wave in the mid-plane of the layer in that it is governed by a reduced wave equation in that plane. As was discussed earlier, the carrier wave may be a plane, cylindrical or any other kind of wave as long as it is a solution of the membrane equation. The important point is that the thickness motions remain the same; they are independent of the form of the carrier wave. For a specific real-valued frequency a wavenumber-like quantity is the solution to the Rayleigh–Lamb frequency equations. The curves that represent the solution for the real, imaginary or complex-valued wavenumber versus the frequency define the frequency spectrum. Each line of frequency versus wavenumber is called a branch and defines a mode of wave propagation in the layer. The modes can be separated into symmetric and antisymmetric modes. For a detailed discussion of the frequency spectrum of Lamb waves we refer to Mindlin (1960) or Achenbach (1973).

The modes are independent from each other in that they satisfy orthogonality relations. These relations are simple and quite evident for the equivoluminal modes, but they are more complicated for the more general modes. An elegant way of deriving the orthogonality condition is by the use of reciprocity relations. We first show this for two-dimensional anti-plane shear modes. We then proceed to more general equivoluminal modes, both in rectangular and polar coordinates. Next we consider Lamb waves in plane strain but, to show the more general applicability of the approach and the results, we also consider Lamb waves in polar coordinates. It is shown that appropriate orthogonality relations can be derived in a straightforward manner.

9.2 Two-dimensional horizontally polarized transverse waves

The simplest solution to Eq. (7.5.3) for the carrier wave is a plane wave. Consistent with the time factor $\exp(-i\omega t)$,

$$\psi = e^{il_n x_1}$$

represents a plane wave propagating in the positive x_1-direction. The corresponding displacements are

$$u_1 \equiv 0, \qquad u_2 = U_n(z)e^{il_n x_1}, \tag{9.2.1}$$

where the factor $-i$ has been included in $U_n(z)$. This expression represents a two-dimensional horizontally polarized transverse wave mode, usually referred to as an anti-plane shear mode.

As in Chapter 7, the x_1x_2-plane coincides with the mid-plane of the layer, and the layer is defined by $|z| \leq h$; see Fig. 7.1. Following the results of Section 7.5, we consider as state A a symmetric wave mode of the form

$$u_2^A = U_n^A \cos(\overline{q}_n z)\, e^{il_n x_1}, \tag{9.2.2}$$

where U_n^A is a constant. This mode propagates in the positive x_1-direction and $\overline{q}$ follows from Eq. (7.5.5) as

$$\overline{q}_n^2 = \frac{\omega^2}{c_T^2} - l_n^2. \tag{9.2.3}$$

The time-harmonic term $\exp(-i\omega t)$ has been omitted, as it will be in the sequel. For a given circular frequency ω, the wavenumber l_n follows from the frequency equation (7.5.11). Equation (9.2.2) represents a symmetric mode of horizontally polarized transverse wave motion in the layer. The stress component corresponding to Eq. (9.2.2) is

$$\tau_{12}^A = il_n\mu U_n^A \cos\left(\overline{q}_n z\right) e^{il_n x_1}. \tag{9.2.4}$$

For this simple two-dimensional case of anti-plane shear, the appropriate domain V for the reciprocity theorem is defined by

$$a \leq x_1 \leq b, \qquad -h \leq z \leq h. \tag{9.2.5}$$

Only the tractions on the cross sections $x_1 = a$ and $x_1 = b$ enter the reciprocity relation. Let us first consider the case where there are no body forces in the domain defined by (9.2.5). We then have from Eq. (6.2.8)

$$\int_{-h}^{h} \left(\tau_{12}^B u_2^A - \tau_{12}^A u_2^B\right)_{x_1=b} dz - \int_{-h}^{h} \left(\tau_{12}^B u_2^A - \tau_{12}^A u_2^B\right)_{x_1=a} ds = 0. \tag{9.2.6}$$

For state A we now take the displacement and stress defined by Eqs. (9.2.2) and (9.2.4), while for state B we take another symmetric mode, also propagating in the positive x_1-direction:

$$u_2^B = U_m^B \cos(\overline{q}_m z)\, e^{il_m x_1},$$
$$\tau_{12}^B = il_m \mu U_m^B \cos(\overline{q}_m z)\, e^{il_m x_1}.$$

We now define for the symmetric modes

$$I_{mn}^S = \int_{-h}^{h} \cos(\overline{q}_m z) \cos(\overline{q}_n z)\, dz. \tag{9.2.7}$$

Equation (9.2.6) then becomes

$$I_{mn}^S (l_m - l_n) \left[e^{i(l_m+l_n)b} - e^{i(l_m+l_n)a}\right] = 0.$$

This equation is satisfied for $m = n$, i.e., $l_m = l_n$, but it can be satisfied only if

$$I_{mn}^S = 0 \qquad \text{for} \qquad m \neq n. \tag{9.2.8}$$

Equation (9.2.8) is the orthogonality condition for the symmetric modes. We can easily obtain

$$I_{mn}^S = h\delta_{mn}, \tag{9.2.9}$$

except for $m = n = 0$ when $\overline{q} = 0$, and

$$I_{00}^S = 2h. \tag{9.2.10}$$

In a similar manner we can derive for the antisymmetric modes

$$I_{mn}^A = \int_{-h}^{h} \sin(\overline{q}_m z) \sin(\overline{q}_n z)\, dz = h\delta_{mn}, \tag{9.2.11}$$

except for $m = n = 0$, when $I_{00}^A = 2h$.

9.3 Equivoluminal wave modes

A more interesting case of equivoluminal wave modes is provided by

$$\psi = \cos(\xi x_2)\, e^{i\eta x_1}, \tag{9.3.1}$$

where

$$\xi^2 + \eta^2 = k^2. \tag{9.3.2}$$

Now the displacements follow from (7.5.1) and (7.5.2) as

$$u_1 = -\frac{\xi}{l_n} U_n(z) \sin(\xi x_2)\, e^{i\eta x_1}, \tag{9.3.3}$$

$$u_2 = -i\frac{\eta}{l_n} U_n(z) \cos(\xi x_2)\, e^{i\eta x_1}. \tag{9.3.4}$$

These displacements satisfy

$$\frac{\partial u_1}{\partial x_1} + \frac{\partial u_2}{\partial x_2} = 0,$$

as they should for an equivoluminal wave mode. For $\xi = 0$ the mode reduces to a mode of anti-plane shear, as discussed in Section 9.2. Since ψ as given by Eq. (9.3.1) is just another solution of the membrane equation (7.5.3), the function $U(z)$ is the same as for the simple anti-plane shear case.

In polar coordinates the function $\psi(r, \theta)$ must be the solution of the membrane equation (7.5.3) in polar coordinates. Let us consider a solution of the form

$$\psi = \Psi_j(lr) \cos(j\theta), \tag{9.3.5}$$

where j is an integer. We then find that

$$\Psi_j(lr) = H_j^{(1)}(lr). \tag{9.3.6}$$

Equivalently to (7.5.1), (7.5.2), the displacements for the nth mode become in polar coordinates

$$u_r^n = \frac{1}{l_n} U_n(z) \frac{1}{r} \frac{\partial \psi}{\partial \theta}, \tag{9.3.7}$$

$$u_\theta^n = -\frac{1}{l_n} U_n(z) \frac{\partial \psi}{\partial r}. \tag{9.3.8}$$

Equations (9.3.7) and (9.3.8) satisfy the displacement equations of motion. Because $U_n(z)$ applies for any ψ, the traction-free conditions on the faces of the elastic layer are also satisfied. Let us consider for $U(z)$ a symmetric mode,

$$u_r^{nj} = U_S^n \cos(\overline{q}_n z) \frac{1}{l_n r} \frac{\partial}{\partial \theta} \left[H_j^{(1)}(l_n r) \cos(j\theta) \right], \tag{9.3.9}$$

$$u_\theta^{nj} = -U_S^n \cos(\overline{q}_n z) \frac{1}{l_n} \frac{\partial}{\partial r} \left[H_j^{(1)}(l_n r) \cos(j\theta) \right], \tag{9.3.10}$$

$$u_z^{nj} \equiv 0, \tag{9.3.11}$$

where U_S^n is a constant. Equations (9.3.9), (9.3.10) can be further simplified to

$$u_r^{nj} = -U_S^n \cos(\overline{q}_n r) \frac{j}{l_n r} H_j^{(1)}(l_n r) \sin(j\theta), \tag{9.3.12}$$

$$u_\theta^{nj} = -U_S^n \cos(\overline{q}_n z) \frac{1}{l_n} \frac{d}{dr} \left[H_j^{(1)}(l_n r) \right] \cos(j\theta). \tag{9.3.13}$$

In polar coordinates the change in volume of an element is

$$\Delta = \frac{\partial u_r}{\partial r} + \frac{u_r}{r} + \frac{1}{r} \frac{\partial u_\theta}{\partial \theta} + \frac{\partial u_z}{\partial z}. \tag{9.3.14}$$

Substitution of Eqs. (9.3.12), (9.3.13) and (9.3.11) into Eq. (9.3.14) yields

$$\Delta = \left\{ -\frac{d}{dr} \left[\frac{1}{r} H_j^{(1)}(l_n r) \right] - \frac{1}{r^2} H_j^{(1)}(l_n r) \right.$$
$$\left. + \frac{1}{r} \frac{d}{dr} H_j^{(1)}(l_n r) \right\} \frac{j}{l_n} \left(U_S^n \right)^2 \cos(\overline{q}_n z) \sin(j\theta).$$

We note that $\Delta = 0$, as it must be for an equivoluminal wave.

For $j = 0$, i.e., for axially symmetric wave motion, we have $u_r^{nj} \equiv 0$. This case corresponds to rotary or torsional shear motions, which involve circumferential displacements only.

9.4 Lamb waves in plane strain

In Section 7.6 it was shown that the thickness motions of an elastic layer, governed by Eqs. (7.6.15) and (7.6.18), apply to a general class of time-harmonic wave motions, called Lamb waves, as long as $\varphi(x_1, x_2)$ is a solution of the reduced wave equation (7.2.6). To investigate Lamb waves in plane strain we refer to Fig. 7.1. Thus, the $x_1 x_2$-plane coincides with the mid-plane of the layer, and the layer is defined by $|z| \leq h$.

For motions in plane strain the following solution of Eq. (7.2.6) applies:

$$\varphi(x_1) = e^{\pm i k_n x_1}. \tag{9.4.1}$$

For *symmetric* modes the relevant displacements and stresses corresponding to Eq. (9.4.1) follow from Eqs. (7.2.1), (7.2.2) and Hooke's law:

$$u_1^n = \pm i A_n^S V_S^n(z) e^{\pm i k_n x_1}, \tag{9.4.2}$$

$$u_z^n = A_n^S W_S^n(z) e^{\pm i k_n x_1}, \tag{9.4.3}$$

$$\tau_{1z}^n = \pm i A_n^S T_{1z}^{Sn}(z) e^{\pm i k_n x_1}, \tag{9.4.4}$$

$$\tau_{zz}^n = A_n^S T_{zz}^{Sn}(z) e^{\pm i k_n x_1}, \tag{9.4.5}$$

$$\tau_{11}^n = A_n^S T_{11}^{Sn}(z) e^{\pm i k_n x_1}. \tag{9.4.6}$$

Here the plus or minus sign applies for propagation in the positive or negative x_1-direction, respectively. The time-harmonic factor $\exp(-i\omega t)$ has been omitted. For a given circular frequency ω, the wavenumber k_n follows from the frequency equation (7.6.15). The functions $V_S^n(z)$ and $W_S^n(z)$ are defined by Eqs. (7.6.22) and (7.6.23), and $T_{1z}^{Sn}(z)$, T_{zz}^{Sn} and $T_{11}^{Sn}(z)$ follow from Hooke's law, Eq. (7.3.5):

$$T_{1z}^{Sn}(z) = \mu[s_5 \sin(pz) + s_6 \sin(qz)], \tag{9.4.7}$$

$$T_{zz}^{Sn}(z) = \mu[s_7 \cos(pz) + s_8 \cos(qz)], \tag{9.4.8}$$

$$T_{11}^{Sn}(z) = \mu[s_9 \cos(pz) + s_{10} \cos(qz)]. \tag{9.4.9}$$

In these expressions, we have

$$s_5 = 4p\cos(qh), \qquad s_6 = \left[\left(k_n^2 - q^2\right)^2 / \left(qk_n^2\right)\right]\cos(ph),$$

$$s_7 = \left[2\left(k_n^2 - q^2\right)/k_n\right]\cos(qh), \qquad s_8 = -\left[2\left(k_n^2 - q^2\right)/k_n\right]\cos(ph),$$

$$s_9 = \left[2\left(2p^2 - k_n^2 - q^2\right)/k_n\right]\cos(qh), \qquad s_{10} = \left[2\left(k_n^2 - q^2\right)/k_n\right]\cos(ph).$$

Similarly, for the *antisymmetric* Lamb-wave modes we have

$$u_1^n = \pm i A_n^A V_A^n(z) e^{\pm i k_n x_1}, \tag{9.4.10}$$

$$u_z^n = A_n^A W_A^n(z) e^{\pm i k_n x_1}, \tag{9.4.11}$$

$$\tau_{1z}^n = \pm i A_n^A T_{1z}^{An}(z) e^{\pm i k_n x_1}, \tag{9.4.12}$$

$$\tau_{zz}^n = A_n^A T_{zz}^{An}(z) e^{\pm i k_n x_1}, \tag{9.4.13}$$

$$\tau_{11}^n = A_n^A T_{11}^{An}(z) e^{\pm i k_n x_1}, \tag{9.4.14}$$

where $V_A^n(z)$ and $W_A^n(z)$ are defined by Eqs. (7.6.26) and (7.6.27) and

$$T_{1z}^{An}(z) = \mu[a_5 \cos(pz) + a_6 \cos(qz)], \tag{9.4.15}$$

$$T_{zz}^{An}(z) = \mu[a_7 \sin(pz) + a_8 \sin(qz)], \tag{9.4.16}$$

$$T_{11}^{An}(z) = \mu[a_9 \sin(pz) + a_{10} \sin(qz)]. \tag{9.4.17}$$

In these expressions, we have

$$a_5 = -4p\sin(qh), \qquad a_6 = -\left[\left(k_n^2 - q^2\right)^2/\left(qk_n^2\right)\right]\sin(ph),$$
$$a_7 = \left[2\left(k_n^2 - q^2\right)/k_n\right]\sin(qh), \qquad a_8 = -\left[2\left(k_n^2 - q^2\right)/k_n\right]\sin(ph),$$
$$a_9 = \left[2\left(2p^2 - k_n^2 - q^2\right)/k_n\right]\sin(qh), \qquad a_{10} = \left[2\left(k_n^2 - q^2\right)/k_n\right]\sin(ph).$$

We will now apply the reciprocity relation to two free Lamb-wave modes in plane strain. An appropriate domain for the reciprocity relation is the same as for anti-plane strain, i.e., $a \leq x_1 \leq b$, $-h \leq z \leq h$. Since we are considering free wave modes, the left-hand side of Eq. (6.2.8) vanishes. For states A and B we choose two symmetric Lamb-wave modes, mode m with wavenumber k_m and mode n with wavenumber k_n. The reciprocity theorem, Eq. (6.2.8,) then yields

$$0 = \int_{-h}^{h} F_{mn}|_{x_1=b}\, dz - \int_{-h}^{h} F_{mn}|_{x_1=a}\, ds, \tag{9.4.18}$$

where

$$F_{mn} = \tau_{11}^{n} u_1^{m} + \tau_{1z}^{n} u_z^{m} - \tau_{11}^{m} u_1^{n} - \tau_{1z}^{m} u_z^{n}. \tag{9.4.19}$$

In Eq. (9.4.18) the field variables are evaluated at $x_1 = a$ or $x_1 = b$, as indicated.

Now let us consider two cases: a set of two counter-propagating antisymmetric modes and a set of two antisymmetic modes that propagate in the same direction. First we consider the counter-propagating modes. For mode m, which propagates in the positive x_1-direction, the relevant expressions are given by Eqs. (9.4.10)–(9.4.14) with positive signs. For mode n, which propagates in the negative x_1-direction, the negative sign applies in the exponentials and in u_1^n and τ_{1z}^n. Substitution of the results into Eq. (9.4.19) and subsequently into (9.4.18) yields

$$\left(I_{mn}^{A} + I_{nm}^{A}\right)\left[e^{i(k_m - k_n)b} - e^{i(k_m - k_n)a}\right] = 0, \tag{9.4.20}$$

where

$$I_{mn}^{A} = \int_{-h}^{h} \left[T_{11}^{Am}(z) V_A^n(z) - T_{1z}^{An}(z) W_A^m(z)\right] dz, \tag{9.4.21}$$

$$I_{nm}^{A} = \int_{-h}^{h} \left[T_{11}^{An}(z) V_A^m(z) - T_{1z}^{Am}(z) W_A^n(z)\right] dz. \tag{9.4.22}$$

Equation (9.4.20) must be satisfied for arbitrary values of a and b. For $m = n$ the equation is satisfied. For $m \neq n$, Eq. (9.4.20) can be satisfied only if

$I^A_{mn} = I^A_{nm} \equiv 0$. Thus

$$I^A_{mn} = 0 \qquad \text{for} \quad m \neq n. \tag{9.4.23}$$

Equation (9.4.23) is an orthogonality condition.

For the two modes propagating in the same direction the relevant expressions to be substituted in Eq. (9.4.19) are given by Eqs. (9.4.10)–(9.4.14) with all signs taken as positive. Application of Eq. (9.4.18) yields

$$\left(I^A_{mn} - I^A_{nm}\right)\left[e^{i(k_m+k_n)b} - e^{i(k_m+k_n)a}\right] = 0, \tag{9.4.24}$$

where I^A_{mn} and I^A_{nm} are defined by Eqs. (9.4.21) and (9.4.22), respectively.

It is noted that Eq. (9.4.24) implies that the contributions from the two cross sections at $x_1 = a$ and $x_1 = b$ vanish independently. The same is not the case for Eq. (9.4.20), where both cross sections produce a contribution for $m = n$. This distinction is important for forced motions, where the amplitudes can be calculated at the side of the counter-propagating actual and virtual wave modes. For the plane strain case an example is given in Section 10.3.

By using the displacements and stresses for the symmetric modes we obtain in the same manner

$$I^S_{mn} = 0 \qquad \text{for} \quad m \neq n, \tag{9.4.25}$$

where

$$I^S_{mn} = \int_{-h}^{h} \left[T^{Sm}_{11}(z)V^n_S(z) - T^{Sn}_{1z}(z)W^m_S(z)\right] dz. \tag{9.4.26}$$

9.5 Lamb waves in polar coordinates

Even though it is expected that the results for plane strain will simply carry over, it is of interest to consider separately and in some detail Lamb waves in polar coordinates. Anticipating that we will need to know the response of an elastic layer to a concentrated load parallel to the faces of the layer, we now consider an expression for $\varphi(r, \theta)$ of the form

$$\varphi(r, \theta) = \Phi(k_n r)\cos\theta, \tag{9.5.1}$$

where the index n indicates that (9.5.1) is associated with the nth mode. According to Eq. (7.2.6), $\Phi(k_n r)$ must be a solution of

$$\frac{d^2\Phi}{dr^2} + \frac{1}{r}\frac{d\Phi}{dr} + \left(k_n^2 - \frac{1}{r^2}\right)\Phi = 0. \tag{9.5.2}$$

The solution to Eq. (9.5.2) for an outgoing wave compatible with the time factor $\exp(-i\omega t)$ is

$$\Phi(k_n r) = H_1^{(1)}(k_n r), \tag{9.5.3}$$

while that for an incoming wave, i.e., a wave that converges on the origin, is

$$\overline{\Phi}(k_n r) = H_1^{(2)}(k_n r). \tag{9.5.4}$$

For simplicity of notation we will use $\Phi(k_n r)$ in subsequent expressions rather than the Hankel functions. Again, the notation $\Phi'(k_n r)$ is used for the derivative with respect to the argument: $\Phi'(\xi) = d\Phi/d\xi$.

For the outgoing *symmetric* Lamb-wave modes, the displacements and stresses corresponding to Eq. (9.5.1) are

$$u_r^n = A_n^S V_S^n(z)\Phi'(k_n r)\cos\theta, \tag{9.5.5}$$

$$u_z^n = A_n^S W_S^n(z)\Phi(k_n r)\cos\theta, \tag{9.5.6}$$

$$u_\theta^n = A_n^S V_S^n(z)\left(\frac{-1}{k_n r}\right)\Phi(k_n r)\sin\theta, \tag{9.5.7}$$

$$\tau_{rz}^n = A_n^S T_{rz}^{Sn}(z)\Phi'(k_n r)\cos\theta, \tag{9.5.8}$$

$$\tau_{zz}^n = A_n^S T_{zz}^{Sn}(z)\Phi(k_n r)\cos\theta, \tag{9.5.9}$$

$$\tau_{rr}^n = A_n^S T_{rr}^{Sn}(z)\Phi(k_n r)\cos\theta - A_n^S \overline{T}_{rr}^{Sn}(z)\left[\frac{1}{k_n r}\Phi'(k_n r) - \frac{1}{(k_n r)^2}\Phi(k_n r)\right]\cos\theta, \tag{9.5.10}$$

$$\tau_{\theta z}^n = A_n^S T_{\theta z}^{Sn}(z)\left(\frac{-1}{k_n r}\right)\Phi(k_n r)\sin\theta, \tag{9.5.11}$$

$$\tau_{r\theta}^n = A_n^S T_{r\theta}^{Sn}(z)\left[\frac{1}{k_n r}\Phi'(k_n r) - \frac{1}{(k_n r)^2}\Phi(k_n r)\right]\sin\theta, \tag{9.5.12}$$

where $V_S^n(z)$ and $W_S^n(z)$ are defined by Eqs. (7.6.22) and (7.6.23), $T_{zz}^{Sn}(z)$ by (9.4.8), and

$$T_{rz}^{Sn}(z) = T_{1z}^{Sn}(z),$$

$$T_{rr}^{Sn}(z) = T_{11}^{Sn}(z),$$

$$\overline{T}_{rr}^{Sn}(z) = \mu[s_{11}\cos(pz) + s_{12}\cos(qz)],$$

$$T_{r\theta}^{Sn}(z) = -\overline{T}_{rr}^{Sn}(z), \qquad T_{\theta z}^{Sn}(z) = -T_{rz}^{Sn}(z).$$

In these expressions $T_{1z}^{Sn}(z)$ and $T_{11}^{Sn}(z)$ are given by Eqs. (9.4.7) and (9.4.9) and

$$s_{11} = 4k_n \cos(qh), \qquad s_{12} = -\left[2\left(k_n^2 - q^2\right)/k_n\right]\cos(ph).$$

Similarly, for the *antisymmetric* Lamb-wave modes we have

$$u_r^n = A_n^A V_A^n(z)\Phi'(k_n r)\cos\theta, \tag{9.5.13}$$

$$u_z^n = A_n^A W_A^n(z)\Phi(k_n r)\cos\theta, \tag{9.5.14}$$

$$u_\theta^n = A_n^A V_A^n(z)\left(\frac{-1}{k_n r}\right)\Phi(k_n r)\sin\theta, \tag{9.5.15}$$

$$\tau_{rz}^n = A_n^A T_{rz}^{An}(z)\Phi'(k_n r)\cos\theta, \tag{9.5.16}$$

$$\tau_{zz}^n = A_n^A T_{zz}^{An}(z)\Phi(k_n r)\cos\theta, \tag{9.5.17}$$

$$\begin{aligned}\tau_{rr}^n &= A_n^A T_{rr}^{An}(z)\Phi(k_n r)\cos\theta \\ &\quad - A_n^A \overline{T}_{rr}^{An}(z)\left[\frac{1}{k_n r}\Phi'(k_n r) - \frac{1}{(k_n r)^2}\Phi(k_n r)\right]\cos\theta,\end{aligned} \tag{9.5.18}$$

$$\tau_{\theta z}^n = A_n^A T_{\theta z}^{An}(z)\left(\frac{-1}{k_n r}\right)\Phi(k_n r)\sin\theta, \tag{9.5.19}$$

$$\tau_{r\theta}^n = A_n^A T_{r\theta}^{An}(z)\left[\frac{1}{k_n r}\Phi'(k_n r) - \frac{1}{(k_n r^2)}\Phi(k_n r)\right]\sin\theta, \tag{9.5.20}$$

where $V_A^n(z)$ and $W_A^n(z)$ are defined by Eqs. (7.6.26) and (7.6.27), $T_{zz}^{An}(z)$ by (9.4.16), and

$$\begin{aligned}
T_{rz}^{An}(z) &= T_{1z}^{An}(z),\\
T_{rr}^{An}(z) &= T_{11}^{An}(z),\\
\overline{T}_{rr}^{An}(z) &= \mu[a_{11}\sin(pz) + a_{12}\sin(qz)],\\
T_{r\theta}^{An}(z) &= -\overline{T}_{rr}^{An}(z), \qquad T_{\theta z}^{An}(z) = -T_{rz}^{An}(z).
\end{aligned}$$

In these expressions, $T_{1z}^{An}(z)$ and $T_{11}^{An}(z)$ are given by (9.4.15) and (9.4.17), and

$$a_{11} = 4k_n \sin(qh), \qquad a_{12} = -\left[2\left(k_n^2 - q^2\right)/k_n\right]\sin(ph).$$

It is of interest to check the conditions at $z = \pm h$. On the faces of the layer the surface tractions should vanish, which implies that

$$\tau_{zz}^n(h) = \tau_{zr}^n(h) = \tau_{z\theta}^n(h) \equiv 0.$$

It may be checked that $T_{rz}^{Sn}(h) = 0$ and $T_{z\theta}^{Sn} = 0$ yield the Raleigh–Lamb frequency equation for symmetric modes, given by Eq. (7.6.15), while $T_{zz}^{Sn}(h) = 0$

is identically satisfied. Similarly $T_{zz}^{An}(h) = 0$ is identically satisfied, while $T_{rz}^{An} = 0$ and $T_{z\theta}^{An}(h) = 0$ yield the Rayleigh–Lamb frequency equation for antisymmetric modes, given by Eq. (7.6.18). For a specified value of the frequency ω, Eqs. (7.6.15) and (7.6.18) are equations for k_n. For each solution k_n, Eqs. (9.5.5)–(9.5.12) and (9.5.13) – (9.5.20) define a specific mode, i.e., a set of displacements and stresses defining symmetric or antisymmetric Lamb-wave motions of the layer.

Let us now consider the reciprocal identity in the annular domain $|z| \leq h$, $a \leq r \leq b$. For the two solutions we choose two symmetric Lamb-wave modes, mode m with wavenumber k_m and mode n with wavenumber k_n. Since the singularities of these modes are at $r = 0$, the reciprocity relation becomes

$$Q_{mn}(b) - Q_{mn}(a) = 0 \qquad (9.5.21)$$

where

$$Q_{mn}(b) = b \int_0^{2\pi} \int_{-h}^{h} \left[\left(u_r^m \tau_{rr}^n - u_r^n \tau_{rr}^m \right) + \left(u_z^m \tau_{rz}^n - u_z^n \tau_{rz}^m \right) \right.$$
$$\left. + \left(u_\theta^m \tau_{r\theta}^n - u_\theta^n \tau_{r\theta}^m \right) \right] d\theta \, dz. \qquad (9.5.22)$$

In Eq. (9.5.22) all field variables are evaluated at $r = b$. By substitution of the displacements and stresses for the symmetric modes we find that, just as in Eq. (8.4.9), several terms cancel, and we obtain

$$Q_{mn}^S(b) = \pi b A_n^S A_m^S \left[I_{nm}^S \Phi'(k_m b) \Phi(k_n b) - I_{mn}^S \Phi'(k_n b) \Phi(k_m b) \right], \qquad (9.5.23)$$

where we have also used that $\overline{T}_{rr}^{Sn} = -T_{r\theta}^{Sn} = 2\mu k_n V_S^n(z)$, while I_{mn}^S is defined as

$$I_{mn}^S = \int_{-h}^{h} \left[T_{rr}^{Sm}(z) V_S^n(z) - T_{rz}^{Sn}(z) W_S^m(z) \right] dz. \qquad (9.5.24)$$

Since $T_{rr}^{Sm}(z) = T_{11}^{Sm}(z)$ and $T_{rz}^{Sn}(z) = T_{1z}^{Sn}(z)$, as stated earlier in this section, the definitions of I_{mn}^S given by (9.4.26) and (9.5.24) are the same.

The analogous expression for $Q_{mn}(a)$ is obtained by replacing b by a in Eq. (9.5.23). Equation (9.5.21) then becomes

$$I_{nm}^S [\Phi'(k_m b) \Phi(k_n b) - \Phi'(k_m a) \Phi(k_n a)]$$
$$- I_{mn}^S [\Phi'(k_n b) \Phi(k_m b) - \Phi'(k_n a) \Phi(k_m a)] = 0. \qquad (9.5.25)$$

This equation must be satisfied for arbitrary values of a and b. For $m = n$, the two terms in brackets cancel each other. For $m \neq n$, Eq. (9.5.25) can be satisfied only if $I_{mn}^S = I_{nm}^S \equiv 0$, in agreement with Eq. (9.4.25). These results do in fact imply that the two groups of terms in Eq. (9.5.25) vanish independently. By using the displacements and stresses for the antisymmetric modes we obtain in the same manner the orthogonality relation already given by Eq. (9.4.23).

When mode n is a counter-propagating mode, we have $\overline{\Phi}(k_n\mathrm{r})$ instead of $\Phi(k_n r)$. Equation (9.5.23) then becomes

$$\overline{Q}^{S}_{nm}(b) = \pi b A^{S}_{n} A^{S}_{m}\left[I^{S}_{nm}\Phi'(k_m b)\overline{\Phi}(k_n b) - I^{S}_{mn}\overline{\Phi}'(k_n b)\Phi(k_m b)\right]. \quad (9.5.26)$$

The analogous expression $\overline{Q}^{S}_{mn}(a)$ is obtained by replacing b with a. For $m \neq n$ both $\overline{Q}^{S}_{mn}(b)$ and $\overline{Q}^{S}_{mn}(a)$ vanish, and Eq. (9.5.21) is satisfied. For $m = n$, Eq. (9.5.26) becomes

$$\overline{Q}^{S}_{nn}(b) = \pi b A^{S}_{n} A^{S}_{n} I^{S}_{nn}[\Phi'(k_n b)\overline{\Phi}(k_n b) - \overline{\Phi}'(k_n b)\Phi(k_n b)]. \quad (9.5.27)$$

By using the Wronskian given by Eq. (8.4.13), the expression in brackets simplifies to $4i/(\pi k_n b)$, and Eq. (9.5.27) becomes

$$\overline{Q}^{S}_{nn}(b) = \frac{4i}{k_n} A^{S}_{n} A^{S}_{n} I^{S}_{nn}. \quad (9.5.28)$$

Since the expression for $\overline{Q}^{S}_{nn}(a)$ is the same, Eq. (9.5.21) is again satisfied. The expressiongiven by Eq. (9.5.28) does not depend on b. As will be shown in Chapter 10, this makes it possible to use a counter-propagating virtual wave to determine the amplitudes of modes generated by the forced motion of an elastic layer.

10

Forced motion of an elastic layer

10.1 Introduction

As discussed in Chapter 9, the modes of wave propagation in an elastic layer are well known from Lamb's (1917) classical work. The Rayleigh–Lamb frequency equations, as well as the corresponding equations for horizontally polarized wave modes, have been analyzed in considerable detail; see Achenbach (1973) and Mindlin (1960). It appears, however, that a simple direct way of expressing wave fields due to the time-harmonic loading of a layer in terms of mode expansions, and a suitable method to obtain the coefficients in the expansions by reciprocity considerations, has so far not been recognized. Of course, wave modes have entered the solutions to problems of the forced wave motion of an elastic layer, at least in the case of surface forces applied normally to the faces of the layer, but via the more cumbersome method of integral transform techniques and the subsequent evaluation of Fourier integrals by contour integration and residue calculus. For examples, we refer to the work of Lyon (1955) for the plane-strain case, and that of Vasudevan and Mal (1985) for axial symmetry.

In this chapter the displacements excited by a time-harmonic point load of arbitrary direction, either applied internally or to one of the surfaces of the layer, are obtained directly as summations over symmetric and/or antisymmetric modes of wave propagation along the layer. This is possible by virtue of an application of the reciprocity relation between time-harmonic elastodynamic states. In Chapter 9 the reciprocity relation was used to derive an orthogonality relation for modes of wave propagation in the layer. In the present chapter the summation coefficients in the modal expansion of the displacements are determined by the use of reciprocity considerations in conjunction with virtual wave modes. For the wave response to a point load the virtual modes must be carefully selected, as discussed in this chapter.

The general formulation used in the present analysis is based on the formulation of Chapter 7, where the displacement fields were expressed in terms of thickness motions superposed on a carrier wave that defines the propagation along the layer. The carrier wave can be any solution of a reduced wave equation in the mid-plane of the layer. For the principal problem considered in this chapter, Hankel functions represent the appropriate carrier waves. Analysis of the thickness motions results in the usual Rayleigh–Lamb frequency equations.

Numerous treatments of the point and line-source excitation of elastic layers can be found in the technical literature. The review article on guided waves by Chimenti (1997) has a section on this topic, in which are listed a quite large number of papers that are based on the application of integral transforms and/or numerical techniques, particularly papers of a more recent origin and papers dealing with anisotropic plates. In addition to the work of Lyon (1955) and Vasudevan and Mal (1985), we also mention the papers by Miklowitz (1962), Weaver and Pao (1982) and Santosa and Pao (1989). These five papers are on transient loads, and they use integral transforms, but the role of wave modes is discussed in some detail in them.

Figure 10.1 shows a layer of homogeneous, isotropic, linearly elastic material of thickness $2h$, referred to a Cartesian coordinate system such that the x_1x_2-plane coincides with the mid-plane of the layer. The layer is subjected to a time-harmonic point load in an arbitrary direction. Without loss of generality the coordinate system is chosen such that the load acts in the x_1z-plane. The dynamic response of the layer is sought as the superposition of the responses due to the vertical component, P, and the horizontal component, Q. These components are indicated in Fig. 10.1(*a*). It is convenient to split each of these two problems further, into two other problems whose solutions are symmetric and antisymmetric, respectively, relative to the mid-plane of the layer. These divisions are illustrated in Fig. 10.1(*b*) and 10.1(*c*). In this chapter we express the wave motion generated by the four loading configurations in terms of symmetric and antisymmetric modes of wave propagation in the layer. The displacements due to the horizontal load Q, have been derived in detail; see also Achenbach and Xu (1999a). The displacements due to the vertical load, P, can be obtained similarly, and the expressions are also given in this chapter.

To introduce this approach, the time-harmonic elastodynamic response of an elastic layer to an anti-plane line load is discussed, in Section 10.2. To illustrate a further use of this displacement solution, mode superposition is subsequently used to construct the displacement response to a pulsed line load. Next, the response to an in-plane line load normal to the layer is presented, in Section 10.3.

In section 10.4 we discuss the wave modes generated by a time-harmonic point load. A summary of relevant expressions is given in Section 10.5. It is

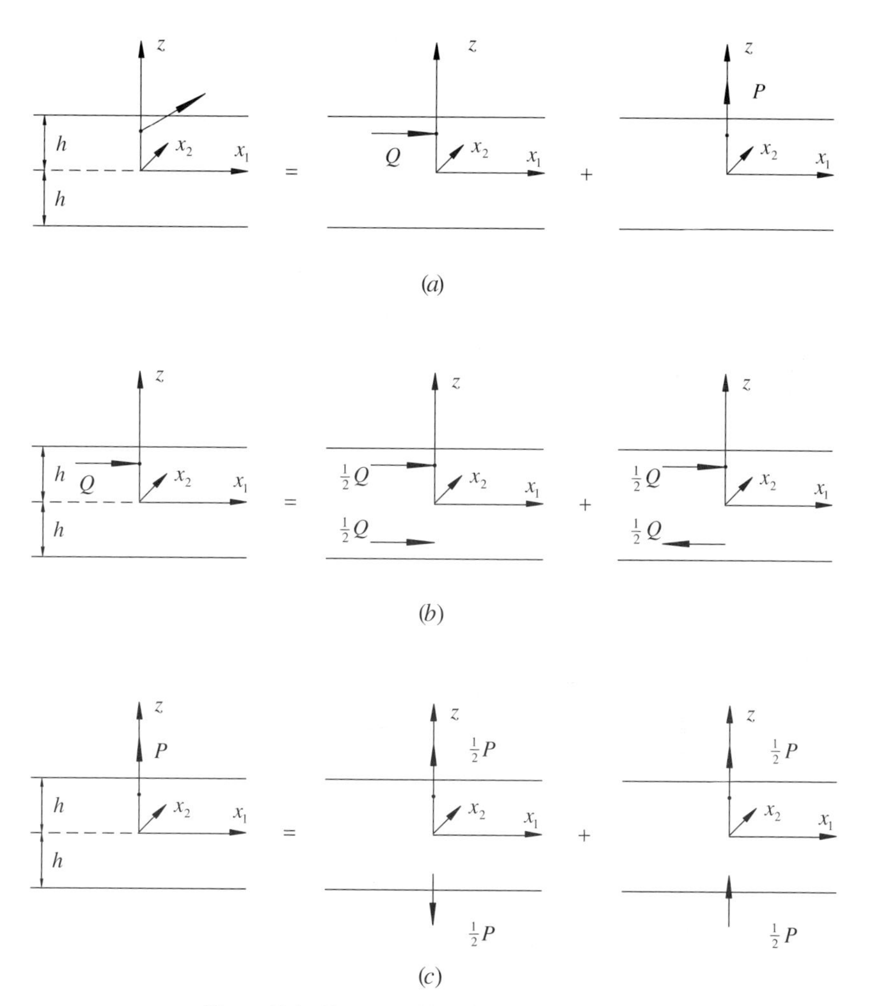

Figure 10.1 Decomposition of point-load problem.

of interest to compare the full three-dimensional results with expressions that can be obtained from simple approximations. This is done in Section 10.6 for comparison with results from plane-stress plate theory and in Section 10.7 for comparison with Kirchhoff plate theory.

10.2 Elastic layer subjected to a time-harmonic anti-plane line load

The elastic layer is referred to a Cartesian coordinate system where the x_1x_2-plane coincides with the mid-plane of the layer, and the layer is defined by

$|z| \leq h,\ -\infty < x_1,\ x_2 < \infty$. We consider an anti-plane line load of the form

$$\boldsymbol{F} = F_0 e^{-i\omega t}\delta(x_1)\delta(z)\boldsymbol{i}_2, \tag{10.2.1}$$

which is applied along the x_2-coordinate axis. The wave motion generated by this line load is clearly two-dimensional, in that the only displacement component, $u_2(x_1, z)\exp(-i\omega t)$, is independent of the x_2-coordinate. It can also be noted that $u_2(x_1, z)$ must be symmetric relative to $z = 0$.

Intuitively it is to be expected that the wave motion generated by $\boldsymbol{F}$ can be efficiently expressed as a summation over the symmetric modes of anti-plane wave motion in the layer. These modes were discussed in Section 9.2. Thus we write for $x_1 \geq 0$:

$$u_2^{FS} = \sum_{m=0,2,4,\ldots}^{\infty} A_m^S \cos(\overline{q}_m z)\ e^{il_m x_1}. \tag{10.2.2}$$

Similarly we write for $x_1 \leq 0$:

$$u_2^{FS} = \sum_{m=0,2,4,\ldots}^{\infty} A_m^S \cos(\overline{q}_m z)\ e^{-il_m x_1}. \tag{10.2.3}$$

Here $\overline{q}_m$ follows from Eq. (7.5.11) and the corresponding l_m can be obtained from (7.5.5). The corresponding stresses follow from Hooke's law. The displacements given by Eqs. (10.2.2) and (10.2.3) represent wave motions propagating away from the point of application of the load, in the positive and negative x_1-direction, respectively.

Now we consider again an application of the reciprocity relation, but over a domain defined by $a \leq x_1 \leq b,\ -h \leq z \leq h$. This domain includes the origin of the coordinate system, where the line load is applied. We then obtain from Eq. (6.2.8)

$$\int_{-h}^{h} \left(\tau_{12}^B u_2^A - \tau_{12}^A u_2^B\right)_{x_1=b} dz - \int_{-h}^{h} \left(\tau_{12}^B u_2^A - \tau_{12}^A u_2^B\right)_{x_1=a} ds = F_0 u_2^B(0,0).$$

For state A we take the displacements defined by Eqs. (10.2.2) and (10.2.3) and their corresponding stresses, while for state B, the virtual wave, we take a single symmetric mode that propagates in the negative x_1-direction:

$$u_2^B = B_n^S \cos(\overline{q}_n z)\ e^{-il_n x_1},$$
$$\tau_{12}^B = -il_n \mu B_n^S \cos(\overline{q}_n z)\ e^{-il_n x_1}.$$

Substitution of the relevant expressions yields after some manipulation

$$-\sum_{m=0,2,4,\ldots}^{\infty} i\mu A_m^S I_{mn}^S (l_m + l_n) e^{i(l_m - l_n)b} - \sum_{m=0,2,4,\ldots}^{\infty} i\mu A_m^S I_{mn}^S (l_m - l_n) e^{-i(l_m + l_n)a}$$
$$= F_0, \tag{10.2.4}$$

where I^S_{mn} is defined by Eq. (9.2.7). The terms in the second summation on the left-hand side vanish both for $n = m$ and for $n \neq m$. The terms in the first summation vanish for $n \neq m$. For $n = m$ we obtain from Eq. (10.2.4)

$$i\mu A^S_m I^S_{mm} 2l_m = -F_0,$$

which is an equation for A^S_m. Using (9.2.9) and (9.2.10) we find

$$m \neq 0: \quad A^S_m = -\frac{F_0}{2il_m \mu h}, \tag{10.2.5}$$

$$m = 0: \quad A^S_0 = -\frac{F_0}{4il_m \mu h}. \tag{10.2.6}$$

For $x_1 \geq 0$ the displacement now follows from (10.2.2) as

$$u_2^{FS} = \sum_{m=0,2,4..}^{\infty} C^S_m \cos\left(\frac{m\pi z}{2h}\right) \frac{1}{il_m} e^{-i\omega t + il_m x_1}, \tag{10.2.7}$$

where

$$C^S_m = il_m A^S_m.$$

In (10.2.2) we have eliminated $\overline{q}$ by the use of Eq. (7.5.11).

Now let us consider the case where the anti-plane line load is applied in a pulse-like fashion. Thus, instead of (10.2.1) we consider

$$\boldsymbol{F} = F_0 f(t)\delta(x_1)\delta(z)\boldsymbol{i}_2.$$

Using Fourier integrals, we write

$$f(t) = \frac{1}{2\pi}\int_{-\infty}^{\infty} F(\omega)\, e^{-i\omega t}\, d\omega,$$

where

$$F(\omega) = \int_{-\infty}^{\infty} f(t) e^{i\omega t} dt.$$

It may be verified that for this case the displacement may be written as

$$u_2^{FS} = \frac{1}{2\pi}\sum_{m=0,2,4,...}^{\infty} C^S_m I_m \cos\left(\frac{m\pi z}{2h}\right),$$

where

$$I_m = \int_{-\infty}^{\infty} F(\omega) \frac{\exp\left\{-i\omega t + i[(\omega/c_T)^2 - (m\pi/2h)^2]^{1/2} x_1\right\}}{i[(\omega/c_T)^2 - (m\pi/2h)^2]^{1/2}}\, d\omega.$$

In this equation I_m has been eliminated by the use of (9.2.3). For the special case

$$f(t) = \delta(t), \qquad \text{we have} \quad F(\omega) = 1,$$

and the integral can be evaluated to yield:

$$I_m = 2\pi c_T J_0 \left[\left(t^2 - \frac{x_1^2}{c_T^2} \right)^{1/2} \frac{m\pi c_T}{2h} \right] H\left(t - \frac{x_1}{c_T} \right),$$

where $J_0(\ \ )$ is the ordinary Bessel function. This result can be found in a table of exponential Fourier transforms. The displacement then follows as

$$u_2^{FS} = c_T \sum_{m=0,2,4,\ldots} C_m^S \cos\left(\frac{m\pi z}{2h} \right) J_0 \left[\left(t^2 - \frac{x_1^2}{c_T^2} \right)^{1/2} \frac{m\pi c_T}{2h} \right] H\left(t - \frac{x_1}{c_T} \right).$$

10.3 Response to a line load normal to the plate faces

A line load in the z-direction, applied along the x_2-axis, may be expressed in the form

$$f = Pe^{-i\omega t}\delta(x_1)\delta(z)\mathbf{i}_z. \qquad (10.3.1)$$

The wave motion generated by this line load is clearly two-dimensional and in a state of plane strain, with displacements $u_1(x_1, z)\exp(-i\omega t)$ and $u_z(x_1, z)\exp(-i\omega t)$ that do not depend on the x_2-coordinate. Because of the direction of the applied load, and its point of application in the plane of symmetry of the layer, the displacements will be antisymmetric relative to the mid-plane of the layer.

It is again to be expected that the wave motion can be expressed as a summation over antisymmetric Lamb-wave modes. At some distance from the point of application of the load, only the modes with real-valued wavenumbers have to be included in the summation. For specified frequencies, these real-valued wavenumbers can be computed from the Rayleigh–Lamb frequency equation for antisymmetric modes, which is given by Eq. (7.6.18). As discussed earlier, the frequency equation also yields imaginary and complex-valued wavenumbers, but the corresponding modes are ignored because they propagate with a decaying amplitude, and it may be assumed that sufficiently far from the load their amplitudes are negligible.

The carrier wave for the mth mode is

$$\varphi(x_1) = e^{ik_m x_1}. \qquad (10.3.2)$$

For $x_1 > 0$ the expansions for the displacements and the stress components may be written as

$$u_z = \sum_{m=0}^{\infty} u_z^m = \sum_{m=0}^{\infty} A_m^A W_A^m(z) e^{ik_m x_1} \tag{10.3.3}$$

$$u_1 = \sum_{m=0}^{\infty} u_1^m = \sum_{m=0}^{\infty} A_m^A V_A^m(z) i e^{ik_m x_1} \tag{10.3.4}$$

$$\tau_{1z} = \sum_{m=0}^{\infty} \tau_{1z}^m = \sum_{m=0}^{\infty} A_m^A T_{1z}^{Am}(z) i e^{ik_m x_1} \tag{10.3.5}$$

$$\tau_{11} = \sum_{m=0}^{\infty} \tau_{11}^m = \sum_{m=0}^{\infty} A_m^A T_{11}^{Am}(z) e^{ik_m x_1}, \tag{10.3.6}$$

where $V_A^m(z)$, $W_A^m(z)$, $T_{1z}^{Am}(z)$ and $T_{11}^{Am}(z)$ are given by Eqs. (7.6.26), (7.6.27), (9.4.15) and (9.4.17), respectively. The constants A_m^A are determined by use of the reciprocity theorem, where Eqs. (10.3.3)–(10.3.6) define state A. Even though the summations are for an infinite number of modes, the actual number is finite since only modes with real-valued wavenumbers are included. For state B, the virtual wave, a single mode is selected, which propagates in the negative x_1-direction:

$$u_z^n = B_n^A W_A^n(z) e^{-ik_n x_1}, \tag{10.3.7}$$

$$u_1^n = -B_n^A V_A^n(z) i e^{-ik_n x_1}, \tag{10.3.8}$$

$$\tau_{1z}^n = -B_n^A T_{1z}^{An}(z) i e^{-ik_n x_1}, \tag{10.3.9}$$

$$\tau_{11}^n = B_n^A T_{11}^{An}(z) e^{-ik_n x_1}. \tag{10.3.10}$$

As the domain for the reciprocity theorem we again take the region defined by $a \leq x_1 \leq b, -h \leq z \leq h$. As discussed several times in earlier applications of the reciprocity theorem, the interaction of state A with the virtual wave only yields a contribution when the two wave systems are counter-propagating. For the present application this occurs at $x_1 = b$. Hence the reciprocity theorem reduces to the following relation between the contribution from the body force at $x_1 = 0, z = 0$ and the integral at $x_1 = b$:

$$P u_z^n(0) = \sum_{m=0}^{\infty} \int_{-h}^{h} \left(\tau_{11}^n u_1^m + \tau_{1z}^n u_z^m - \tau_{11}^m u_1^n - \tau_{1z}^m u_z^n\right)_{x_1=b} dz.$$

Substitution of Eqs. (10.3.3)–(10.3.6) and (10.3.7)–(10.3.10) yields

$$P B_n^A W_A^n(0) = i \sum_{m=0}^{\infty} A_m^A B_n^A \left(I_{mn}^A + I_{nm}^A\right) e^{i(k_m - k_n)b}, \tag{10.3.11}$$

where I^A_{mn} and I^A_{nm} are defined by Eq. (9.4.21) and (9.4.22). Since I^A_{mn} and I^A_{nm} both vanish for $m \neq n$, the only terms that survive in the summation are for $m = n$. It then follows that

$$A^A_n = -\frac{i}{2}\frac{W^n_A(0)}{I^A_{nn}}P. \tag{10.3.12}$$

10.4 Response to a point load parallel to the plate faces

Without loss of generality we can consider a time-harmonic point load applied in the direction of the x_1-axis. In cylindrical coordinates ($x_1 = r\cos\theta, x_2 = r\sin\theta, z$), we write equivalently to Eq. (7.4.10)–(7.4.12):

$$u_r = \frac{1}{k_n}V^n(z)\frac{\partial\varphi}{\partial r}(r,\theta), \tag{10.4.1}$$

$$u_\theta = \frac{1}{k}V^n(z)\frac{1}{r}\frac{\partial\varphi}{\partial\theta}(r,\theta), \tag{10.4.2}$$

$$u_z = W^n(z)\varphi(r,\theta), \tag{10.4.3}$$

where

$$\frac{\partial^2\varphi}{\partial r^2} + \frac{1}{r}\frac{\partial\varphi}{\partial r} + \frac{1}{r^2}\frac{\partial^2\varphi}{\partial\theta^2} + k_n^2\varphi = 0. \tag{10.4.4}$$

The first point of consideration now is to select the right solution of Eq. (10.4.4) for $\varphi(r,\theta)$. For this we take guidance from the displacement solutions for a time-harmonic point force in an unbounded solid applied at the origin in the x_1-direction. These solutions are given by Eqs. (8.3.7)–(8.3.9). In Chapter 8 it was argued that the presence of a surface parallel to the $r\theta$-plane does not change the dependence on θ. The same argument holds when the body has two surfaces parallel to the $r\theta$-plane, i.e., when the body is a layer. Thus $\varphi(r,\theta)$ is given by Eq. (8.3.11). For convenience the $r\theta$-plane is taken as the mid-plane of the layer.

At some distance from the applied load, the wave motion generated by Q can be expressed as a summation over the symmetric and antisymmetric modes. Let us first consider the *symmetric* problem, and write the displacement solutions as sums of symmetric Lamb-wave and horizontally polarized modes:

$$u_r^{QS} = \sum_{m=0}^{\infty} A^S_m V^n_S(z)\Phi'(k_m r)\cos\theta + \sum_{m=0}^{\infty} B^S_m U^m_S(z)\frac{1}{l_m r}\Psi(l_m r)\cos\theta, \tag{10.4.5}$$

$$u_z^{QS} = \sum_{m=0}^{\infty} A_m^S W_S^n(z)\Phi(k_m r)\cos\theta, \tag{10.4.6}$$

$$u_\theta^{QS} = \sum_{m=0}^{\infty} A_m^S V_S^n(z)\left(\frac{-1}{k_m r}\right)\Phi(k_m r)\sin\theta - \sum_{m=0}^{\infty} B_m^S U_S^m(z)\Psi'(l_m r)\sin\theta, \tag{10.4.7}$$

where $V_S^n(z)$ and $W_S^n(z)$ are defined by Eqs. (7.6.22) and (7.6.23), and $U_S^m(z) = \cos[m\pi z/(2h)]$. For a specified value of ω, k_m and l_m are the solutions of Eqs. (7.6.15) and (7.5.18), respectively. It is of interest to note that for the radial displacement the contribution due to the horizontally polarized waves is $O(1/r)$ compared with the Lamb-wave contribution. For the circumferential displacement the Lamb-wave contribution is $O(1/r)$ compared with the horizontally polarized part. Only the Lamb waves contribute to the displacement in the z-direction.

The coefficients A_m^S in Eqs. (10.4.5)–(10.4.7) can be determined by another use of the reciprocal identity. This time we consider the identity for the domain $0 \le r \le b$, $-h \le z \le h$, $0 \le \theta \le 2\pi$. For state A we choose the displacement and stress fields generated by the force Q, for simplicity applied at $z_0 = 0$, $r = 0$, $\theta = 0$. The corresponding displacements are represented by Eqs. (10.4.5)–(10.4.7). For state B we choose a virtual wave in the form of a single symmetric Lamb-wave mode consisting of the sum of an outgoing and a converging wave:

$$u_r^n = \tfrac{1}{2} C_n^S V_S^n(z)\left[\Phi'(k_n r) + \overline{\Phi}'(k_n r)\right]\cos\theta, \tag{10.4.8}$$

$$u_z^n = \tfrac{1}{2} C_n^S W_S^n(z)[\Phi(k_n r) + \overline{\Phi}(k_n r)]\cos\theta, \tag{10.4.9}$$

$$u_\theta^n = \tfrac{1}{2} C_n^S V_S^n(z)\left(\frac{-1}{k_n r}\right)\left[\Phi(k_n r) + \overline{\Phi}(k_n r)\right]\sin\theta, \tag{10.4.10}$$

where $\Phi(k_n r)$ and $\overline{\Phi}(k_n r)$ are defined by Eqs. (9.5.3) and (9.5.4), respectively.

The displacements in Eqs. (10.4.8)–(10.4.10) are bounded at $r = 0$. It can be verified that the left-hand side of the reciprocity relation (8.4.9) becomes

$$Qu_r(0, 0, 0) = \tfrac{1}{2} QC_n^S V_S^n(0). \tag{10.4.11}$$

To evaluate the right-hand side at $r = b$, we recall the observation made after Eq. (10.4.7) that the corresponding field quantities for the Lamb-wave modes and the horizontally polarized modes are of different orders for large r. This implies that for sufficiently large b the product terms of the two kinds of mode can be neglected because they are $O(1/b)$. The consequence is that the virtual wave (10.4.8)–(10.4.10) only provides contributions in conjunction with the Lamb-wave terms in Eqs. (10.4.5)–(10.4.7). In that light, the right-hand side of Eq. (8.4.9) can be evaluated by using the result for $r = b$, Eq. (9.5.23), for the first term in Eqs. (10.4.8)–(10.4.10). The second term yields similar expressions, and the complete result is

$$QV_S^n(0) = \sum_{m=0}^{\infty} A_m^S\big[Q_{mn}^S(b) + \overline{Q}_{mn}^S(b)\big], \tag{10.4.12}$$

where $Q_{mn}^S(b)$ follows from Eq. (9.5.23), and $\overline{Q}_{mn}^S$ from (9.5.27) as

$$\overline{Q}_{mn}^S(b) = \pi b\big[I_{nm}^S \Phi'(k_m b)\overline{\Phi}(k_n b) - I_{mn}^S \overline{\Phi}'(k_n b)\Phi(k_m b)\big]. \tag{10.4.13}$$

Inspection of Eq. (9.4.25) shows that $Q_{mn}^S(b)$ vanishes for both $m \neq n$ and $m = n$. However, $\overline{Q}_{mn}^S(b)$ vanishes only for $m \neq n$. For $m = n$, Eq. (10.4.12) can be simplified further by using the following identity for Hankel functions (McLachlan, 1961, p. 198):

$$\left[\frac{d}{d\xi}H_v^{(1)}(\xi)\right]H_v^{(2)}(\xi) - H_v^{(1)}(\xi)\left[\frac{d}{d\xi}H_v^{(2)}(\xi)\right] = \frac{4i}{\pi\xi}.$$

We obtain

$$T_{nn}^S = \frac{4i}{k_n} I_{nn}^S, \tag{10.4.14}$$

and Eq. (10.4.13) yields

$$A_n^S = \frac{k_n}{4i}\frac{V_S^n(0)}{I_{nn}^S}Q, \tag{10.4.15}$$

where I_{nn}^S follows from Eq. (9.5.24) as

$$I_{nn}^S = \mu\big[c_1^S \cos^2(ph) + c_2^S \cos^2(qh)\big]. \tag{10.4.16}$$

The constants c_1^S and c_2^S are

$$c_1^S = \frac{(k_n^2 - q^2)(k_n^2 + q^2)}{2q^3k_n^3}\left[2qh(k_n^2 - q^2) - (k_n^2 + 7q^2)\sin(2qh)\right] \tag{10.4.17}$$

$$c_2^S = \frac{k_n^2 + q^2}{pk_n^3}\left[4k_n^2 ph + 2(k_n^2 - 2p^2)\sin(2ph)\right]. \tag{10.4.18}$$

Next we go through the same procedure, but now we choose for state B a virtual wave in the form of a single symmetric horizontally polarized mode consisting of the sum of an outgoing and a converging wave:

$$u_r^n = \frac{1}{2} D_n^S \cos\left(\frac{n\pi z}{2h}\right) \frac{1}{l_n r} \left[\Psi(l_n r) + \overline{\Psi}(l_n r)\right] \cos\theta \tag{10.4.19}$$

$$u_\theta^n = -\frac{1}{2} D_n^S \cos\left(\frac{n\pi z}{2h}\right) \left[\Psi'(l_n r) + \overline{\Psi}'(l_n r)\right] \sin\theta, \tag{10.4.20}$$

where $\Psi(l_n r)$ follows from Eq. (9.3.6), and

$$\overline{\Psi}(l_n r) = H_1^{(2)}(l_n r). \tag{10.4.21}$$

Proceeding through the same steps as for the Lamb-wave modes, which means that we only use the second term of Eq. (10.4.7) and Eq. (10.4.20), we find

$$B_n^S = \frac{1}{4i} \frac{Q}{\mu J_{nn}}, \qquad n = 0, 2, 4, \ldots \tag{10.4.22}$$

The analogous result for the antisymmetric modes is

$$B_n^A = \frac{1}{4i} \frac{Q}{\mu J_{nn}}, \qquad n = 1, 3, 5, \ldots. \tag{10.4.23}$$

Finally we consider the case where the horizontal point load is applied at a position $z = z_0$. It is then convenient to split the problem in two problems, a symmetric and an antisymmetric one, as illustrated in Fig. 10.1. For the symmetric problem the left-hand side of Eq. (9.4.21) must then be replaced by two contributions, each involving $Q/2$ and, considering the Lamb-wave modes, either $V_S^n(z_0)$ or $V_S^n(-z_0)$. However, $V_S^n(z_0)$ and $V_S^n(-z_0)$, and thus the two contributions, are equal. It can then be verified that Eq. (10.4.15) becomes

$$A_n^S = \frac{k_n}{4i} \frac{V_S^n(z_0)}{I_{nn}^S} Q. \tag{10.4.24}$$

For the antisymmetric problem the two forces of magnitude $Q/2$ are in opposite directions, but so are the normal displacements, and it then follows that

$$A_n^A = \frac{k_n}{4i} \frac{V_A^n(z_0)}{I_{nn}^A} Q, \tag{10.4.25}$$

$$I_{nn}^A = \mu\left[c_1^A \sin^2(ph) + c_2^A \sin^2(qh)\right], \tag{10.4.26}$$

where

$$c_1^A = \frac{\left(k_n^2 - q^2\right)\left(k_n^2 + q^2\right)}{2q^3 k_n^3} \left[2qh\left(k_n^2 - q^2\right) + \left(k_n^2 + 7q^2\right)\sin(2qh)\right] \tag{10.4.27}$$

$$c_2^A = \frac{k_n^2 + q^2}{pk_n^3} \left[4k_n^2 ph - 2\left(k_n^2 - 2p^2\right)\sin(2ph)\right]. \tag{10.4.28}$$

The constants for the horizontally polarized modes can be modified analogously from Eqs. (10.4.22) and (10.4.23) as

$$B_n^S = \frac{1}{4i} \frac{\cos[n\pi z_0/(2h)]}{\mu J_{nn}} Q \tag{10.4.29}$$

$$B_n^A = \frac{1}{4i} \frac{\sin[n\pi z_0/(2h)]}{\mu J_{nn}} Q. \tag{10.4.30}$$

The response to a normal point load P can also be split up into a symmetric and an antisymmetric response, as shown in Fig. 10.1. The details of the computation of the axially symmetric displacements $u_r^n(r, z)$ and $u_z^n(r, z)$ can be found in the paper by Achenbach and Xu (1999b). A summary of the results is given in the next section.

10.5 Summary of solutions

As illustrated in Fig. 10.1, the displacement due to a point load of arbitrary direction can be expressed as the superposition of displacements due to the vertical component, P, and the horizontal component, Q:

$$\mathbf{u}(r, \theta, z) = \mathbf{u}^P(r, \theta, z) + \mathbf{u}^Q(r, \theta, z). \tag{10.5.1}$$

Each of these solutions can be obtained as the superposition of a symmetric and an antisymmetric solution:

$$\mathbf{u}^P = \mathbf{u}^{PS} + \mathbf{u}^{PA}, \tag{10.5.2}$$

$$\mathbf{u}^Q = \mathbf{u}^{QS} + \mathbf{u}^{QA}. \tag{10.5.3}$$

The displacements due to a horizontal point load Q have been derived in some detail in the preceding sections. The displacements due to the vertical point load can be obtained in a similar manner. Here we summarize all results.

Horizontal point load

$$u_r^{QS} = \sum_{m=0}^{\infty} A_m^S V_S^m(z) \left[H_0^{(1)}(k_m r) - \frac{1}{k_m r} H_1^{(1)}(k_m r) \right] \cos\theta + \sum_{m=0}^{\infty} B_m^S \cos\left(\frac{m\pi z}{2h}\right) \frac{1}{l_m r} H_1^{(1)}(l_m r) \cos\theta, \tag{10.5.4}$$

$$u_z^{QS} = \sum_{m=0}^{\infty} A_m^S W_S^m(z) H_1^{(1)}(k_m r) \cos\theta, \tag{10.5.5}$$

$$u_\theta^{QS} = \sum_{m=0}^{\infty} A_m^S V_S^m(z) \left(\frac{-1}{k_m r} \right) H_1^{(1)}(k_m r) \sin\theta$$
$$- \sum_{m=0}^{\infty} B_m^S \cos\left(\frac{m\pi z}{2h} \right) \left[H_0^{(1)}(l_m r) - \frac{1}{l_m r} H_1^{(1)}(l_m r) \right] \sin\theta, \quad (10.5.6)$$

where A_m^S follows from Eq. (10.4.24) and B_m^S from Eq. (10.4.29). Also,

$$u_r^{QA} = \sum_{m=0}^{\infty} A_m^A V_A^m(z) \left[H_0^{(1)}(k_m r) - \frac{1}{k_m r} H_1^{(1)}(k_m r) \right] \cos\theta$$
$$+ \sum_{m=0}^{\infty} B_m^A \sin\left(\frac{m\pi z}{2h} \right) \frac{1}{l_m r} H_1^{(1)}(l_m r) \cos\theta, \quad (10.5.7)$$

$$u_z^{QA} = \sum_{m=0}^{\infty} A_m^A W_A^m(z) H_1^{(1)}(k_m r) \cos\theta, \quad (10.5.8)$$

$$u_\theta^{QA} = \sum_{m=0}^{\infty} A_m^A V_S^m(z) \left(\frac{-1}{k_m r} \right) H_1^{(1)}(k_m r) \sin\theta$$
$$- \sum_{m=0}^{\infty} B_m^A \sin\left(\frac{m\pi z}{2h} \right) \left[H_0^{(1)}(l_m r) - \frac{1}{l_m r} H_1^{(1)}(l_m r) \right] \sin\theta, \quad (10.5.9)$$

where A_m^A follows from Eq. (10.4.25) and B_m^A from Eq. (10.4.30).

Vertical point load

$$u_r^{PS} = -\sum_{m=0}^{\infty} C_m^S V_S^m(z) H_1^{(1)}(k_m r), \quad (10.5.10)$$

$$u_z^{PS} = \sum_{m=0}^{\infty} C_m^S W_S^m(z) H_0^{(1)}(k_m r), \quad (10.5.11)$$

where

$$C_m^S = \frac{k_m}{4i} \frac{W_S^m(z_0)}{I_{mm}^S} P. \quad (10.5.12)$$

Also,

$$u_r^{PA} = -\sum_{m=0}^{\infty} C_m^A V_A^m(z) H_1^{(1)}(k_m r), \quad (10.5.13)$$

$$u_z^{PA} = \sum_{m=0}^{\infty} C_m^A W_A^m(z) H_0^{(1)}(k_m r), \quad (10.5.14)$$

where

$$C_m^A = \frac{k_m}{4i}\frac{W_A^m(z_0)}{I_{mm}^A}P. \tag{10.5.15}$$

The terms I_{mm}^S and I_{mm}^A are given by Eqs. (10.4.16) and (10.4.26), respectively.

It is of interest to check the reciprocity of the solutions given by Eqs. (10.5.4)–(10.5.15). The simplest check of reciprocity is for the case where P is applied at $r = 0, z = 0$, and Q is applied at $r = r, z = 0$. In this case the displacement u_r at $r = r, z = 0, \theta = 0$ due to P should be just the same as the displacement u_z at $r = r, z = 0, \theta = \pi$. It can be verified that this equality is indeed satisfied.

10.6 Comparison with plane-stress solution

Equations (8.3.7)–(8.3.9) give the displacements in an unbounded solid for a point load applied at the origin of the coordinate system and pointing in the x_1-direction. Similarly to a derivation given by Achenbach, Gautesen and McMaken (1982, p. 28), it can be shown that the corresponding displacements due to a line load can be obtained from these expressions by replacing

$$G(k_\gamma R) \qquad \text{by} \qquad -\frac{\mathrm{i}}{4}H_0^{(1)}(k_\gamma r), \tag{10.6.1}$$

where $G(k_\gamma R)$ is defined by Eq. (8.3.6). This replacement results in displacement expressions for the case of plane strain. The other two-dimensional case, the case of plane stress, only differs in that different elastic constants must be used. For a comparison of the two-dimensional formulations for plane strain and plane stress, we refer to Achenbach (1973, p. 59). From the two sets of equations it can be seen that solutions for plane stress follow from those for plane strain on replacing

$$\lambda + 2\mu \qquad \text{by} \qquad \frac{2\mu\lambda}{\lambda + 2\mu} \tag{10.6.2}$$

or, equivalently,

$$c_L^2 \qquad \text{by} \qquad c_0^2, \qquad \text{where} \quad c_0^2 = 4\frac{\kappa^2 - 1}{\kappa^2}c_T^2. \tag{10.6.3}$$

Here c_0 is the so-called plate velocity, and

$$\kappa = c_L/c_T. \tag{10.6.4}$$

The plane-stress case is an approximation that is valid for a thin plate at low frequencies and thus small wavenumbers. For simplicity we will take r sufficiently large that terms of $O(1/kr)$ can be ignored. For a thin plate with a line load uniform through the thickness of the plate, the displacements in the r- and θ- directions then follow from (8.3.7) and (8.3.8) as

$$u_r \cong -\frac{F}{\mu}\frac{c_T^2}{c_0^2}\frac{i}{4}H_0^{(1)}(k_0 r)\cos\theta, \tag{10.6.5}$$

$$u_\theta \cong \frac{F}{\mu}\frac{i}{4}H_0^{(1)}(k_T r)\sin\theta. \tag{10.6.6}$$

Here the plate velocity c_0 is defined by Eq. (10.6.3), and

$$k_0 = \omega/c_0. \tag{10.6.7}$$

Now let us consider the lowest modes in the expansions (10.4.5) and (10.4.7) in the limit $h \to 0$. For small wavenumbers the frequency and wavenumber of the lowest symmetric mode are related by (see Achenbach, 1973)

$$\omega \cong 2\left(\frac{\kappa^2-1}{\kappa^2}\right)^{1/2} c_T k_0. \tag{10.6.8}$$

After some manipulation we then find for small h

$$I_{00}^S \cong 128\mu\left(\frac{\kappa^2-1}{\kappa^2}\right)^3 k_0 h, \tag{10.6.9}$$

$$V_S^0 \cong 4\frac{\kappa^2-1}{\kappa^2}. \tag{10.6.10}$$

From (10.4.24) the constant A_0^S is then obtained as

$$A_0^S \cong \frac{1}{64i}\frac{1}{\mu}\left(\frac{\kappa^2}{\kappa^2-1}\right)^2\frac{Q}{2h}, \tag{10.6.11}$$

and thus from (10.5.4)

$$u_r \cong -\frac{Q}{2h}\frac{1}{\mu}\frac{\kappa^2}{(\kappa^2-1)}\frac{i}{16}H_0^{(1)}(k_0 r)\cos\theta. \tag{10.6.12}$$

Using $c_0 = 2[(\kappa^2-1)/\kappa^2]^{1/2}c_T$, Eq. (10.6.5) can be rewritten as

$$u_r \cong -\frac{Q}{2h}\frac{1}{\mu}\frac{c_T^2}{c_0^2}\frac{i}{4}H_0^{(1)}(k_0 r)\cos\theta. \tag{10.6.13}$$

Equations (10.6.5) and (10.6.13) show that the two radial displacements agree, since $Q/(2h)$ is equivalent to F.

For the circumferential displacement, Eq. (10.4.7) yields for the lowest mode, as $h \to 0$,

$$u_\theta \cong \frac{Q}{2h}\frac{1}{\mu}\frac{i}{4}H_0^{(1)}(k_T r)\sin\theta. \tag{10.6.14}$$

Again we have agreement, here of Eq. (10.6.14) with Eq. (10.6.6).

10.7 Comparison with plate theory for the normal-point-load problem

In this section we will show that for a normal load applied in the mid-plane of the layer the limiting case, that of the lowest antisymmetric mode, gives the same result as the classical Kirchhoff plate theory. The displacement response to point excitation of an infinite Kirchhoff plate was given by Junger and Feit (1972) as

$$w(r,t) = \frac{iP}{8\omega(2\rho h D)^{1/2}}\left[H_0^{(1)}(kr) + \frac{2i}{\pi}K_0(kr)\right]e^{-i\omega t}, \tag{10.7.1}$$

where $K_0(kr)$ is the modified Bessel function, which decays exponentially with kr, $2h$ is the *total* thickness of the plate and D is its bending stiffness. For a plate of thickness $2h$, as considered here, the bending stiffness is

$$D = \frac{8Eh^3}{12(1-\upsilon^2)}, \tag{10.7.2}$$

where E is Young's modulus and υ is Poisson's ratio.

For large values of the argument kr, Eq. (10.7.1) yields the asymptotic form

$$w(r,t) = \frac{iP}{4\omega(2\rho h D)^{1/2}}\frac{e^{i(kr-\pi/4)}}{(2\pi kr)^{1/2}}e^{-i\omega t}. \tag{10.7.3}$$

Next we eliminate ω by its relation to k for the lowest antisymmetric mode at small values of k (see Achenbach, 1973, p. 228)

$$\omega = k^2 h\left[\frac{E}{3\rho(1-\upsilon^2)}\right]^{1/2}, \tag{10.7.4}$$

and we introduce D as given by Eq. (10.7.2), to obtain the displacement as

$$w = -\frac{iP}{4}\frac{3}{4}\frac{1-\upsilon^2}{Ek^2h^3}\left(\frac{2}{\pi kr}\right)^{1/2}e^{i(kr-\pi/4)}. \tag{10.7.5}$$

According to Eq. (10.5.14) the displacement for the lowest antisymmetric mode is

$$u_z^{PA} = C_0^A[a_3\cos(pz) + a_4\cos(qz)]H_0^{(1)}(kr), \tag{10.7.6}$$

where a_3 and a_4 are defined by Eqs. (7.6.29), and C_0^A follows from Eq. (10.5.15). By using Eqs. (7.6.29) we obtain

$$a_3 + a_4 = 2\frac{p}{k}\sin(qh) + \frac{k^2 - q^2}{qk}\sin(ph). \tag{10.7.7}$$

For small values of kh we have $\sin(qh) \sim qh$ and $\sin(ph) \sim ph$, and Eq. (10.7.7) becomes

$$a_3 + a_4 \approx \frac{h}{k}\frac{p}{q}\frac{\omega^2}{c_T^2}, \tag{10.7.8}$$

where the definitions of p and q have been used. Next we introduce from Eq. (10.7.4)

$$\frac{\omega^2 h^2}{c_T^2} = \gamma(kh)^4, \qquad \text{where} \qquad \gamma = \frac{2}{3}\frac{1}{1-\upsilon}. \tag{10.7.9}$$

To the lowest order in kh we then obtain

$$a_3 + a_4 = \gamma(kh)^3. \tag{10.7.10}$$

Next we determine C_0^A, Eq. (10.5.15), for very small kh. This is done by replacing $\sin(2ph)$ and $\sin(2qh)$ in the constants C_1^A and C_2^A by two-term expansions in kh. These constants are subsequently substituted in the form for I_{00}^A at small values of kh, which follows from Eq. (10.4.26), again by using two-term expansions for $\sin(ph)$ and $\sin(qh)$. Expanding $1/q^2$ and $1/p^2$ in powers of kh, and using Eq. (10.7.9), we obtain upon collecting the lowest-order terms

$$I_{00}^A = 4\mu\gamma^3(kh)^9. \tag{10.7.11}$$

Lower-order terms in kh have canceled out. By the use of (10.7.10) and (10.7.11) it then follows from Eq. (10.5.15) that

$$C_0^A = \frac{k}{4i}\frac{\hat{W}_A(0)P}{I_{00}^A} = \frac{k}{4i}\frac{P}{4\mu}\frac{1}{\gamma^2}\frac{1}{(kh)^6}. \tag{10.7.12}$$

Now we are ready to calculate u_z^{PA} for small kh from Eq. (10.5.14). We also take the asymptotic value of $H_0^{(1)}(kr)$ at large kr. The result is

$$u_z^{PA} = C_0^A(a_3 + a_4)\left(\frac{2}{\pi kr}\right)^{1/2} e^{i(kr-\pi/4)}. \tag{10.7.13}$$

Substitution of (10.7.10) and (10.7.12) yields

$$u_z^{PA} = \frac{k}{4i}\frac{P}{4\mu}\frac{1}{\gamma}\frac{1}{(kh)^3}\left(\frac{2}{\pi kr}\right)^{1/2} e^{i(kr-\pi/4)}. \tag{10.7.14}$$

Elimination of γ by Eq. (10.7.9) then gives

$$u_z^{PA} = \frac{P}{4i}\frac{3}{4}\frac{1-\upsilon^2}{Ek^2h^3}\left(\frac{2}{\pi kr}\right)^{1/2} e^{i(kr-\pi/4)}, \tag{10.7.15}$$

which is just the same as the plate deflection according to the Kirchhoff plate theory, given by Eq. (10.7.5).

11
Integral representations and integral equations

11.1 Introduction

An important application of the reciprocity relation is its use to generate integral representations. With the aid of the basic singular elastodynamic solution for an unbounded solid, an integral representation can be derived that provides the displacement field at a point of observation in terms of the displacements and tractions on the boundary of a body. In the limit as the point of observation approaches the boundary, a boundary integral equation is obtained. This equation can be solved numerically for the unknown displacements or tractions. The calculated boundary values are subsequently substituted in the original integral representation to yield the desired field variables at an arbitrary point of observation.

The boundary element method is often used for the numerical solution of boundary integral equations. The advantage of the boundary element method for solving boundary integral equations is that the dimensionality of the problem is reduced by one. Rather than calculations in a two- or three-dimensional discretized space, we have calculations for discretized curves or surfaces. For detailed discussions of boundary element methods in elastodynamics we refer to the review papers by Beskos (1987) and Kobayashi (1987). These papers contain numerous additional references. We also mention the book edited by Banerjee and Kobayashi (1992), and a recent book by Bonnet (1995) that has several sections on dynamic problems.

We start this chapter with an exposition of the basic ideas for the simpler, two-dimensional, case of anti-plane strain in Section 11.2. For the three-dimensional case, detailed derivations and discussions of integral representations and integral equations are given in Sections 11.3 and 11.4, respectively. A specific case, scattering by a cavity, is also discussed in Section 11.4. Scattering by a crack is considered in Section 11.5. Interesting reciprocity relations for the scattering of plane elastic waves by obstacles of arbitrary shape are presented in Section 11.6. An approximate solution for the scattering by a crack based on the Kirchhoff approximation is discussed in Section 11.7. Finally, Section 11.8 is concerned with the use of reciprocity considerations to obtain an approximate far-field solution for the diffraction of a signal generated by a source near the tip of a crack, for the case of anti-plane strain.

11.2 Radiation of anti-plane shear waves from a cavity

To provide a simple introduction to the material of this chapter we first consider an integral representation for radiation from a cavity for the case of anti-plane strain. The geometry is shown in Fig. 11.1. The figure shows a cylindrical cavity of area V and boundary S, and a point P outside S. The origin of the coordinate system is placed inside the cavity.

For this two-dimensional case, we apply the reciprocity relation to the domain $\overline{V}$ bounded by S and a circle S_r centered at the origin; the circle has a very large radius r and contains both V and the point P. We obtain

$$\int_{\overline{V}} \left(f_z^A w^B - f_z^B w\right) dV = \int_S \left(w^A \tau_{\alpha z}^B - w^B \tau_{\alpha z}^A\right) n_\alpha \, dS + \int_{S_r} \left(w^A \tau_{\alpha z}^B - w^B \tau_{\alpha z}^A\right) n_\alpha \, dS. \quad (11.2.1)$$

Here the normals on S and S_r are taken as being into the cavity and towards the outside from S_r, respectively.

For state A we choose the radiated field, $w^{\text{rad}}(x;\omega)$, generated by the loading of the cavity's surface S. Thus $f_z^A \equiv 0$. State B, which is the virtual wave considered for the present problem, is taken as the field generated in a solid without a cavity by an anti-plane line load at point P. The line load is defined by

$$f_z^P(\mathbf{x}) = F\delta(\mathbf{x} - \mathbf{y}). \quad (11.2.2)$$

Here $\mathbf{x}$ defines a field point, while $\mathbf{y}$ is the point of application of the line load. It follows from Section 6.6 that the line load generates displacements of the form

$$W(\mathbf{x}, \mathbf{y}; \omega) = \frac{i}{4}\frac{F}{\mu} H_0^{(1)}(k_T|\mathbf{x} - \mathbf{y}|), \quad (11.2.3)$$

where

$$|\mathbf{x} - \mathbf{y}| = [(x_1 - y_1)^2 + (x_2 - y_2)^2]^{1/2}. \quad (11.2.4)$$

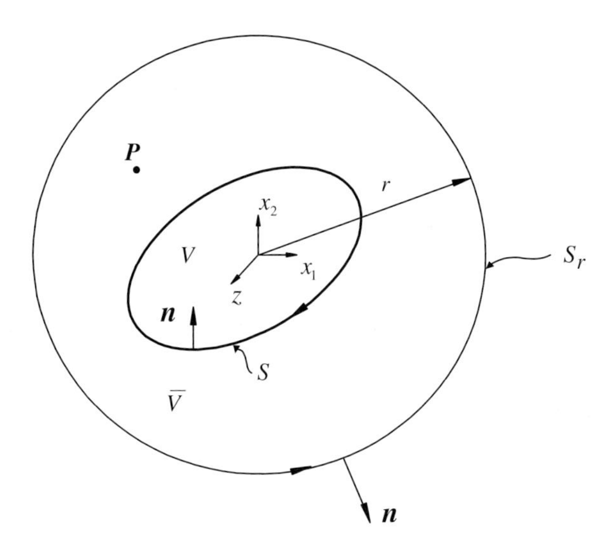

Figure 11.1 A cylindrical cavity in a homogeneous, isotropic, linearly elastic solid with an anti-plane line load applied at point P.

The stresses corresponding to Eq. (11.2.3) are

$$T_{\alpha z}(\boldsymbol{x}, \boldsymbol{y}; \omega) = \frac{i}{4} F \frac{\partial}{\partial x_\alpha} H_0^{(1)}(k_T\, |\boldsymbol{x} - \boldsymbol{y}|), \tag{11.2.5}$$

where $\alpha = 1, 2$. Thus w^B can be defined using Eq. (11.2.3), while for the surface tractions of states B and A we use the notation

$$T = T_{\alpha z} n_\alpha \qquad \text{and} \qquad T^{\mathrm{rad}} = \tau_{\alpha z}^{\mathrm{rad}} n_\alpha, \tag{11.2.6}$$

where $T_{\alpha z}$ is defined by Eq. (11.2.5). The reciprocity relation (11.2.1) then becomes

$$\begin{aligned} w^{\mathrm{rad}}(\boldsymbol{y}; \omega) F = \int_S &[W(\boldsymbol{x}, \boldsymbol{y}; \omega) T^{\mathrm{rad}}(\boldsymbol{x}; \omega) \\ &- w^{\mathrm{rad}}(\boldsymbol{x}; \omega) T(\boldsymbol{x}, \boldsymbol{y}; \omega)]\, dS(\boldsymbol{x}) + I(\boldsymbol{y}), \end{aligned} \tag{11.2.7}$$

where

$$I(\boldsymbol{y}) = \int_{S_r} [W(\boldsymbol{x}, \boldsymbol{y}; \omega) T^{\mathrm{rad}}(\boldsymbol{x}; \omega) - w^{\mathrm{rad}}(\boldsymbol{x}; \omega) T(\boldsymbol{x}, \boldsymbol{y}; \omega)]\, dS(\boldsymbol{x}).$$

The directions of the integrations along S and S_r are indicated in Fig. 11.1.

To determine the limit of $I(\boldsymbol{y})$ as $r \to \infty$, we use the Sommerfeld radiation condition in the form given by Courant and Hilbert (1962, p. 315),

$$\lim_{r\to\infty} \int_{S_r} \left|\tau_{\alpha z}^{\text{rad}} n_\alpha - ik_T \mu w^{\text{rad}}\right|^2 dS(\boldsymbol{x}) = 0. \tag{11.2.8}$$

For large values of $k_T|\boldsymbol{x} - \boldsymbol{y}|$ we obtain from (11.2.3) and (11.2.5)

$$W(\boldsymbol{x}, \boldsymbol{y}; \omega) = \frac{i}{4}\frac{F}{\mu}\left(\frac{2}{\pi k_T |\boldsymbol{x}|}\right)^{1/2} e^{i(k_T|\boldsymbol{x}|-\pi/4)} e^{-ik_T \hat{\boldsymbol{x}}\cdot\boldsymbol{y}},$$

$$T_{\alpha z}(\boldsymbol{x}, \boldsymbol{y}; \omega) = \frac{i}{4}F\left(\frac{2}{\pi k_T |\boldsymbol{x}|}\right)^{1/2} ik_T \hat{x}_\alpha e^{i(k_T|\boldsymbol{x}|-\pi/4)} e^{-ik_T \hat{\boldsymbol{x}}\cdot\boldsymbol{y}}.$$

Substitution of these expressions into Eq. (11.2.8) yields

$$\begin{aligned} I(y) &= \frac{i}{4}\frac{F}{\mu}\left(\frac{2}{\pi k_T}\right)^{1/2} e^{-i\pi/4} \int_{S_r} \frac{e^{ik_T|\boldsymbol{x}|}}{|\boldsymbol{x}|^{1/2}} \left(\tau_{\alpha z}^{\text{rad}} - ik_T \mu w^{\text{rad}} \hat{x}_\alpha\right) \\ &\qquad \times n_\alpha e^{-ik_T \hat{\boldsymbol{x}}\cdot\boldsymbol{y}}\, dS(\boldsymbol{x}) \\ &= \frac{i}{4}\frac{F}{\mu}\left(\frac{2}{\pi k_T}\right)^{1/2} e^{-i\pi/4} \frac{e^{ik_T r}}{r^{1/2}} \int_{S_r} \left(\tau_{\alpha z}^{\text{rad}} n_\alpha - ik_T \mu w^{\text{rad}}\right) \\ &\qquad \times e^{-ik_T \boldsymbol{n}\cdot\boldsymbol{y}}\, dS(\boldsymbol{x}), \end{aligned} \tag{11.2.9}$$

where we have made the substitutions $\hat{\boldsymbol{x}} = \boldsymbol{n}$, $|\boldsymbol{x}| = r$ and $\hat{x}_\alpha n_\alpha = 1$.

For integration of a function $f(\boldsymbol{x})$ over a circle of radius r, an application of the Schwarz inequality yields

$$\left|\int_{S_r} f(\boldsymbol{x})\, dS(\boldsymbol{x})\right| \leq (2\pi r)^{1/2} \left[\int_{S_r} |f(\boldsymbol{x})|^2\, dS(\boldsymbol{x})\right]^{1/2}. \tag{11.2.10}$$

By making the identification

$$f(\boldsymbol{x}) = \left(\tau_{\alpha z}^{\text{rad}} n_\alpha - ik_T \mu w^{\text{rad}}\right) e^{-ik_T \hat{\boldsymbol{x}}\cdot\boldsymbol{y}},$$

it now follows from the radiation condition Eq. (11.2.8) and the Schwarz inequality Eq. (11.2.10) that $I(\boldsymbol{y})$ vanishes in the limit $r \to \infty$. Consequently we have obtained the following integral representation:

$$w^{\text{rad}}(\boldsymbol{y}; \omega) = \int_S \left[W(\boldsymbol{x}, \boldsymbol{y}; \omega) T^{\text{rad}}(\boldsymbol{x}; \omega) - w^{\text{rad}}(\boldsymbol{x}; \omega) T(\boldsymbol{x}, \boldsymbol{y}; \omega)\right] dS(\boldsymbol{x}), \tag{11.2.11}$$

where we have taken F as equal to one force unit, and $\boldsymbol{y}$ is in the region $\overline{V}$ outside the boundary S.

Equation (11.2.11) is a representation integral for the radiated field at an arbitrary point P defined by position vector $\boldsymbol{y}$. The radiated field is generated by surface loading of the cavity. This representation integral requires complete information on the loading of the cavity, more than is generally directly available since both the traction, $T^{\mathrm{rad}}(\boldsymbol{x};\omega)$, and the displacement, $w^{\mathrm{rad}}(\boldsymbol{x};\omega)$, must be known. If, say, only the traction is known in the first instance, the representation integral given by Eq. (11.2.11) can be changed into an integral equation for the unknown displacement in the limit as $\boldsymbol{y}$ approaches the boundary S. In this limiting process, suitable accommodations have to be made for the singularity that appears on the boundary due to the singular behavior of $T(\boldsymbol{x},\boldsymbol{y};\omega)$ as $\boldsymbol{y}$ approaches S. We will consider two ways of dealing with this singularity. The first way is to regularize the integral equation by introducing a difference form with a corresponding static traction term that has the same singularity as the dynamic traction term. The second way of dealing with the singularity on S is by integrating along a small indentation around the singular point. This procedure provides a bounded term and a Cauchy principal value of a singular integral.

To regularize the integral equation we use the static solution for the line-load problem in an unbounded solid. Whereas the dynamic displacements and tractions are indicated by $W(\boldsymbol{x},\boldsymbol{y};\omega)$ and $T(\boldsymbol{x},\boldsymbol{y};\omega)$, the corresponding static solutions are indicated by omitting ω, i.e., by $W(\boldsymbol{x},\boldsymbol{y})$ and $T(\boldsymbol{x},\boldsymbol{y})$. For a line load of one force unit per unit length in the z-direction, static equilibrium of the region V in an unbounded body then implies

$$\gamma + \int_S T(\boldsymbol{x},\boldsymbol{y})\,dS(\boldsymbol{x}) = 0 \qquad \text{where} \qquad \begin{cases} \gamma = 1, & \boldsymbol{x} \in V, \\ \gamma = 0, & \boldsymbol{x} \notin V. \end{cases}$$

On multiplying this result by $w^{\mathrm{rad}}(\boldsymbol{y};\omega)$ and then adding the result for a point outside V to the integral representation for the domain outside S, given by Eq. (11.2.11), we can write

$$\begin{aligned} w^{\mathrm{rad}}(\boldsymbol{y};\omega) = & \int_S W(\boldsymbol{x},\boldsymbol{y};\omega)T^{\mathrm{rad}}(\boldsymbol{x};\omega)\,dS(\boldsymbol{x}) \\ & - \int_S \left[w^{\mathrm{rad}}(\boldsymbol{x};\omega)T(\boldsymbol{x},\boldsymbol{y};\omega) - w^{\mathrm{rad}}(\boldsymbol{y};\omega)T(\boldsymbol{x},\boldsymbol{y})\right] dS(\boldsymbol{x}). \end{aligned} \tag{11.2.12}$$

Equation (11.2.12) can now be changed to an integral equation. For small values of $\rho = |\boldsymbol{x}-\boldsymbol{y}|$, the first integral in Eq. (11.2.12) converges because

$$W(\boldsymbol{x},\boldsymbol{y};\omega) = O(\ln\rho).$$

The terms in the second integral require more careful attention. We have

$$\begin{aligned} T(\boldsymbol{x},\boldsymbol{y}) &= -\frac{1}{2\pi}\frac{1}{\rho}\rho_{,i}\,n_i, \\ T(\boldsymbol{x},\boldsymbol{y};\omega) &= -\frac{ik_T}{4}H_1^{(1)}(k_T\rho)\rho_{,i}\,n_i. \end{aligned}$$

In the limit $\rho \to 0$ we can write

$$H_1^{(1)}(k_T \rho) = \frac{2i}{\pi k_T \rho} + O(\rho \ln \rho).$$

Using the expressions for $T(\boldsymbol{x}, \boldsymbol{y}; \omega)$ and $T(\boldsymbol{x}, \boldsymbol{y})$ it is now immediately seen that

$$T(\boldsymbol{x}, \boldsymbol{y}; \omega) - T(\boldsymbol{x}, \boldsymbol{y}) = O(\rho \ln \rho).$$

In the limit $\rho \to 0$ this expression is bounded. We then conclude that the second integral can be evaluated as $\boldsymbol{y}$ approaches S, and the boundary integral equation (11.2.12) can be solved numerically.

In the second method we let the point $\boldsymbol{y}$ approach S, but we consider a second boundary $\overline{S}$ that surrounds both S and the point $\boldsymbol{y}$. Instead of Eq. (11.2.11) we then have

$$0 = \int_{\overline{S}} \left[W(\boldsymbol{x}, \boldsymbol{y}; \omega) T^{\mathrm{rad}}(\boldsymbol{y}; \omega) - w^{\mathrm{rad}}(\boldsymbol{y}; \omega) T(\boldsymbol{x}, \boldsymbol{y}; \omega) \right] d\overline{S}(\boldsymbol{y}).$$

The left-hand side is zero, because the line load is not outside $\overline{S}$. We now let $\boldsymbol{y}$ approach the boundary, and simultaneously we shrink $\overline{S}$ to S. When $\boldsymbol{y}$ has reached the boundary S, the surface $\overline{S}$ has shrunk to S except for a small semicircle of radius ρ where $\overline{S}$ goes around the point $\boldsymbol{y}$. In the limit $\rho \to 0$ we then obtain the following boundary integral equation:

$$e(\boldsymbol{y}) w^{\mathrm{rad}}(\boldsymbol{y}; \omega) = ⨍_{\overline{S}} \left[w^{\mathrm{rad}}(\boldsymbol{x}; w) T(\boldsymbol{x}; \boldsymbol{y}; \omega) - W(\boldsymbol{x}, \boldsymbol{y}; \omega) T^{\mathrm{rad}}(\boldsymbol{x}; \omega) \right] dS(x), \tag{11.2.13}$$

where the kernel $T(\boldsymbol{x}, \boldsymbol{y}; \omega)$, although singular of order ρ^{-1}, is integrable as a Cauchy principal value, as indicated by the slash through the integration symbol. The free term is

$$e(\boldsymbol{y}) = \lim_{\rho \to 0} \int_{S_\rho} T(\boldsymbol{x}, \boldsymbol{y}; \omega) \, dS(\rho),$$

where S_ρ is a semicircle of radius ρ centered at $\boldsymbol{y}$. For a smooth boundary at $\boldsymbol{y}$ we find $e = 0.5$.

It is well known that, in the frequency domain, boundary integral equations, like Eq. (11.2.13) for a cavity, fail to yield a unique solution at certain wavenumbers that correspond to the eigenfrequencies of the interior region bounded by S; see Schenk (1968). Since similar behavior occurs for three-dimensional elastodynamics, we will postpone a listing of remedies until the next section.

11.3 Integral representation in three dimensions

For the three-dimensional case, a cavity of arbitrary shape in a homogeneous isotropic linearly elastic solid is subjected to time-harmonic surface tractions and/or surface displacements. As for the anti-plane shear case the displacements elsewhere in the solid are expressed in terms of an integral over the surface of the cavity. Let V be the bounded domain of the cavity, and let S be its boundary. Let S_R be a sphere with radius R centered at the origin, and let V_R denote the domain inside S_R. The radius R is chosen sufficiently large that S_R completely surrounds S.

We now consider the reciprocal theorem Eq. (6.2.8) for the domain $V_R - V$ and for the following two states,

$$\left\{u_i^A(\boldsymbol{x},\boldsymbol{y}),\, \tau_{ij}^A(\boldsymbol{x},\boldsymbol{y})\right\} = \{u_i(\boldsymbol{x}),\, \tau_{ij}(\boldsymbol{x})\}$$

and

$$\left\{u_i^B(\boldsymbol{x},\boldsymbol{y}),\, \tau_{ij}^B(\boldsymbol{x},\boldsymbol{y})\right\} = \left\{u_{i;k}^G(\boldsymbol{x}-\boldsymbol{y}),\, \tau_{ij;k}^G(\boldsymbol{x}-\boldsymbol{y})\right\}.$$

Here state A corresponds to the field generated by the tractions and/or displacements on the surface of the cavity, while state B is the basic singular solution for a point force of one force unit, f^B, in the direction i_k, applied at the point of observation, P, whose position is defined by $x = y$:

$$\boldsymbol{f}^B = \boldsymbol{i}_k \delta(\boldsymbol{x}-\boldsymbol{y}).$$

In accordance with the definitions of $\boldsymbol{x}$ and $\boldsymbol{y}$ given in the previous section, we stipulate that the field point is defined by $\boldsymbol{x}$ and the point of application of the load by $\boldsymbol{y}$. These definitions conform with the definitions that are generally used in the literature. Also, even though the problem is dynamic in the frequency domain, we do not yet indicate the dependence on ω, to keep the expressions shorter and simpler. For $\boldsymbol{y} \in V_R - V$, application of (6.2.8) now yields

$$u_k(\boldsymbol{y}) = \int_S \left[u_{i;k}^G(\boldsymbol{x},\boldsymbol{y})\tau_{ij}(\boldsymbol{x}) - u_i(\boldsymbol{x})\tau_{ij;k}^G(\boldsymbol{x},\boldsymbol{y})\right] n_j(\boldsymbol{x})\, dS(\boldsymbol{x}) + I_k(\boldsymbol{y}), \tag{11.3.1}$$

where

$$I_k(\boldsymbol{y}) = \int_{S_R} \left[u_{i;k}^G(\boldsymbol{x},\boldsymbol{y})\tau_{ij}(\boldsymbol{x}) - u_i(\boldsymbol{x})\tau_{ij;k}^G(\boldsymbol{x},\boldsymbol{y})\right] n_j(\boldsymbol{x})\, dS(\boldsymbol{x}). \tag{11.3.2}$$

Here the normals $\boldsymbol{n}$ are as indicated in Fig. 11.1. In particular it should be noted that the normal to the surface S of the cavity points into the cavity. The displacements $u_{i;k}^G$ and stresses $\tau_{ij;k}^G$ are given by Eqs. (3.9.21) and (3.9.23).

On physical grounds it is to be expected that the integral in (11.3.2) over S_R is independent of R, since the particle displacement at $\boldsymbol{y}$ should be uniquely determined by the boundary conditions on the cavity surface S. We will show that I_k vanishes as $R \to \infty$. This can be done on the basis of the elastodynamic equivalent of the Sommerfeld radiation condition. This condition states that at large distances from the sources of radiation the interrelation between the displacements and the stresses is locally the same as for out-going plane waves.

Mathematically the radiation conditions of three-dimensional elastodynamic radiation theory may be stated in the form

$$\lim_{R\to\infty} \int_{S_R} \left|[\tau_{ij} n_i - i k_L(\lambda + 2\mu) u_j] n_j\right|^2 dS(\boldsymbol{x}) = 0, \tag{11.3.3}$$

$$\lim_{R\to\infty} \int_{S_R} |(\tau_{ij} n_i - i k_T \mu u_j)(\delta_{jk} - n_j n_k)|^2 dS(\boldsymbol{x}) = 0. \tag{11.3.4}$$

Equation (11.3.3) implies a plane-wave relation between the normal traction and the normal displacement in the far field, and (11.3.4) does the same for the transverse traction and the transverse displacement. It may be verified that (11.3.3) and (11.3.4) are satisfied if the displacement potentials for the radiated waves satisfy the usual Sommerfeld radiation condition for solutions of the reduced wave equation.

We will now introduce some of the far-field expressions presented in (3.10.1)–(3.10.12). For the present purpose they will be recast in the following forms:

$$u_{i;k} = u_{i;k}^{G;L} + u_{i;k}^{G;T}, \tag{11.3.5}$$

where $|\boldsymbol{x}| \gg |\boldsymbol{y}|$, and

$$u_{i;k}^{G;\alpha}(\boldsymbol{x}, \boldsymbol{y}) \approx A_{i;k}^{G;\alpha} \frac{\exp(i k_\alpha |\boldsymbol{x}|)}{4\pi |\boldsymbol{x}|} \exp(-i k_\alpha \hat{\boldsymbol{x}} \cdot \boldsymbol{y}). \tag{11.3.6}$$

Here $\alpha = L$ or T, and the constants are

$$A_{i;k}^{G;L} = \frac{1}{\lambda + 2\mu} \hat{x}_i \hat{x}_k, \tag{11.3.7}$$

$$A_{i;k}^{G;T} = \frac{1}{\mu}(\delta_{ik} - \hat{x}_i \hat{x}_k). \tag{11.3.8}$$

Similarly,

$$\tau_{ij;k}^{G} = \tau_{ij;k}^{G;L} + \tau_{ij;k}^{G;T}, \tag{11.3.9}$$

where according to (3.10.13)–(3.10.14)

$$\tau_{ij;k}^{G;L}(\boldsymbol{x}, \boldsymbol{y}) \hat{x}_j \approx i k_L (\lambda + 2\mu) u_{i;k}^{G;L}, \tag{11.3.10}$$

$$\tau_{ij;k}^{G;T}(\boldsymbol{x}, \boldsymbol{y}) \hat{x}_j \approx i k_T \mu u_{i;k}^{G;L}. \tag{11.3.11}$$

Substitution of these results into (11.3.2) yields

$$I_k = \frac{1}{4\pi(\lambda+2\mu)} \frac{\exp(ik_L R)}{R}$$
$$\times \int_{S_R} [\tau_{ij} n_i - ik_L(\lambda+2\mu)u_j] n_j n_k \exp(-ik_L \boldsymbol{n}\cdot\boldsymbol{y})\, dS(\boldsymbol{x})$$
$$+ \frac{1}{4\pi\mu} \frac{\exp(ik_T R)}{R}$$
$$\times \int_{S_R} (\tau_{ij} n_i - ik_T \mu u_j)(\delta_{jk} - n_j n_k) \exp(-ik_T \boldsymbol{n}\cdot\boldsymbol{y})\, dS(\boldsymbol{x}),$$

where we have also made the substitutions $\hat{\boldsymbol{x}} = \boldsymbol{n}$ and $R = (x_j x_j)^{1/2}$.

For integration of a function $f(\boldsymbol{x})$ over a sphere of radius R an application of Schwarz' inequality yields

$$\left| \int_{S_R} f(\boldsymbol{x})\, dS(\boldsymbol{x}) \right| \le 2R\sqrt{\pi} \left(\int_{S_R} |f(\boldsymbol{x})|^2\, dS(\boldsymbol{x}) \right)^{1/2}.$$

By combining this result with (11.3.3) and (11.3.4) we find, similarly to the case of anti-plane shear,

$$\lim_{R\to\infty} I_k = 0.$$

It follows from the above result that for a point of observation not inside the cavity or on its boundary the displacement may be expressed simply as the integral on the right-hand side of Eq. (11.3.1).

In the literature the following simplified notation is frequently used in Eq. (11.3.1):

replace	$u^G_{i;k}(\boldsymbol{x}, \boldsymbol{y})$	by	$U_{ik}(\boldsymbol{x}, \boldsymbol{y}; \omega)$,
replace	$\tau^G_{ij;k}(\boldsymbol{x}, \boldsymbol{y}) n_j(\boldsymbol{y})$	by	$T_{ik}(\boldsymbol{x}, \boldsymbol{y}; \omega)$.

Here $U_{ik}(\boldsymbol{x}, \boldsymbol{y}; \omega)$ is given by Eq. (3.9.24) with $F_0 = 1$ (force unit), while $T_{ik}(\boldsymbol{x}, \boldsymbol{y}, \omega)$ can be obtained from (3.9.27) in the form

$$T_{ik}(\boldsymbol{x}, \boldsymbol{y}; \omega) = \left(\frac{\lambda}{\mu} R_{,k} n_i + R_{,j} n_j \delta_{ik} + R_{,i} n_k \right) \frac{dU_1}{dR}$$
$$- \left(\frac{\lambda}{\mu} R_{,k} n_i + 2R_{,i} R_{,k} R_{,j} n_j \right) \frac{dU_2}{dR}$$
$$- \left[2\frac{\lambda}{\mu} R_{,k} n_i + 2(n_i - 2R_{,j} n_j R_{,i}) R_{,k} \right.$$
$$\left. + (R_{,i} n_k + R_{,j} n_j \delta_{ik}) \right] \frac{U_2}{R}, \qquad (11.3.12)$$

where $R = |\boldsymbol{x} - \boldsymbol{y}|$. It should be noted that the following relation holds:

$$R,_j n_j = \frac{\partial R}{\partial n}.$$

The expression for $u_k(\boldsymbol{y};\omega)$ now takes the form

$$u_k(\boldsymbol{y};\omega) = \int_S [U_{ik}(\boldsymbol{x},\boldsymbol{y};\omega)t_i(\boldsymbol{x}) - T_{ik}(\boldsymbol{x},\boldsymbol{y};\omega)u_i(\boldsymbol{x})]\, dS(\boldsymbol{x}), \qquad (11.3.13)$$

where $t_i(\boldsymbol{x})$ are the traction components corresponding to $\tau_{ij}(\boldsymbol{x})$.

It is useful to write the equivalent expression for large distances from the cavity, i.e., when $|\boldsymbol{y}| \gg |\boldsymbol{x}|$. For large values of $|\boldsymbol{y}|$, $G(k_\gamma R)$, which is defined by Eq. (3.9.15), becomes

$$G(k_\alpha R) \approx \frac{\exp(ik_\alpha |\boldsymbol{y}|}{4\pi |\boldsymbol{y}|} \exp(-ik_\alpha \boldsymbol{x}\cdot\hat{\boldsymbol{y}}).$$

Also,

$$R,_k \approx -\hat{y}_k.$$

The expressions for U_{ik} and T_{ik} then reduce to

$$U_{ik} \approx \sum_{\alpha=L,T} A_{ik}^{G;\alpha}(\hat{\boldsymbol{y}}) \frac{\exp(ik_\alpha|\boldsymbol{y}|)}{4\pi|\boldsymbol{y}|} e^{-ik_\alpha \boldsymbol{x}\cdot\hat{\boldsymbol{y}}}$$

and

$$T_{ik} \approx \sum_{\alpha=L,T} ik_\alpha B_{ik}^{G;\alpha}(\hat{\boldsymbol{y}}) \frac{\exp(ik_\alpha|\boldsymbol{y}|)}{4\pi|\boldsymbol{y}|} e^{-ik_\alpha \boldsymbol{x}\cdot\hat{\boldsymbol{y}}},$$

where, for $F_0 = 1$ (force unit), $A_{ik}^{G;\alpha}(\hat{\boldsymbol{y}})$ and $B_{ik}^{G;\alpha}(\hat{\boldsymbol{y}})$ follow from Eqs. (3.10.4), (3.10.5) and (3.10.9), (3.10.10):

$$A_{ik}^{G;L}(\hat{\boldsymbol{y}}) = \frac{1}{\lambda + 2\mu}\hat{y}_i\hat{y}_k,$$

$$A_{ik}^{G;T}(\hat{\boldsymbol{y}}) = \frac{1}{\mu}(\delta_{ki} - \hat{y}_i\hat{y}_k),$$

$$B_{ik}^{G;L}(\hat{\boldsymbol{y}}) = -\left(\frac{2\mu}{\lambda+2\mu}\hat{y}_i\hat{y}_j n_j + \frac{\lambda}{\lambda+2\mu}n_i\right)\hat{y}_k,$$

$$B_{ik}^{G;T}(\hat{\boldsymbol{y}}) = -(\delta_{ik}\hat{y}_j n_j + \hat{y}_i n_k - 2\hat{y}_i\hat{y}_j\hat{y}_k n_j).$$

Substitution in Eq. (11.3.13) yields

$$u_k^\alpha(\boldsymbol{y};\omega) \approx U_k^\alpha(\hat{\boldsymbol{y}})\frac{\exp(ik_\alpha|\boldsymbol{y}|)}{4\pi|\boldsymbol{y}|} \qquad \text{as} \qquad |\boldsymbol{y}| \to \infty, \qquad (11.3.14)$$

where $\alpha = L$ or T. We find

$$U_k^L(\hat{\boldsymbol{y}}) = \frac{1}{\lambda + 2\mu}\hat{y}_k \int_S (\hat{y}_i t_i) e^{-ik_L \boldsymbol{x}\cdot\hat{\boldsymbol{y}}}\, dS(\boldsymbol{x}) + \frac{ik_L}{\lambda + 2\mu}\hat{y}_k \int_S [\lambda u_j n_j + 2\mu(u_i \hat{y}_i)(\hat{y}_j n_j)] e^{-ik_L \boldsymbol{x}\cdot\hat{\boldsymbol{y}}}\, dS(\boldsymbol{x}), \tag{11.3.15}$$

$$U_k^T(\hat{\boldsymbol{y}}) = \frac{1}{\mu}\int_S [t_k - (\hat{y}_i t_i)\hat{y}_k] e^{-ik_T \boldsymbol{x}\cdot\hat{\boldsymbol{y}}}\, dS(\boldsymbol{x}) + ik_T \int_S [(\hat{y}_j n_j) u_k + (u_i \hat{y}_i) n_k - 2(u_i \hat{y}_i)(\hat{y}_j n_j)\hat{y}_k] e^{-ik_T \boldsymbol{x}\cdot\hat{\boldsymbol{y}}} dS(\boldsymbol{x}). \tag{11.3.16}$$

As before we have

$$u_k(\boldsymbol{y};\omega) = u_k^L(\boldsymbol{y};\omega) + u_k^T(\boldsymbol{y};\omega).$$

Returning to Eq. (11.3.13), we write in a more general form

$$e(\boldsymbol{y}) u_k(\boldsymbol{y};\omega) = \int_S U_{ik}(\boldsymbol{x},\boldsymbol{y};\omega) t_i(\boldsymbol{x};\omega)\, dS(\boldsymbol{x}) - \int_S T_{ik}(\boldsymbol{x},\boldsymbol{y};\omega) u_i(\boldsymbol{x};\omega)\, dS(\boldsymbol{x}), \tag{11.3.17}$$

where

$$e(\boldsymbol{y}) = \begin{cases} 1, & \boldsymbol{y} \in V_+, \\ 0, & \boldsymbol{y} \in V. \end{cases}$$

Here V_+ is the domain outside the boundary S. Similarly, for a point $\boldsymbol{y} \in V$ we write

$$e(\boldsymbol{y}) u_k(\boldsymbol{y};\omega) = \int_S U_{ik}(\boldsymbol{x},\boldsymbol{y};\omega) t_i(\boldsymbol{x};\omega)\, dS(\boldsymbol{x}) - \int_S T_{ik}(\boldsymbol{x},\boldsymbol{y};\omega) u_i(\boldsymbol{x};\omega)\, dS(\boldsymbol{x}), \tag{11.3.18}$$

where

$$e(\boldsymbol{y}) = \begin{cases} 1, & \boldsymbol{y} \in V, \\ 0, & \boldsymbol{y} \in V_+. \end{cases}$$

11.4 Integral equation

The integral representations derived in the previous section can be used to derive an integral equation for unknown field variables on the surface of a cavity or another inhomogeneity. Again we consider a cavity occupying a domain V with boundary S. The domain outside V is denoted by V_+. The normal $\boldsymbol{n}$ points into the surrounding solid. We follow the approach of letting the point $\boldsymbol{x}$ tend

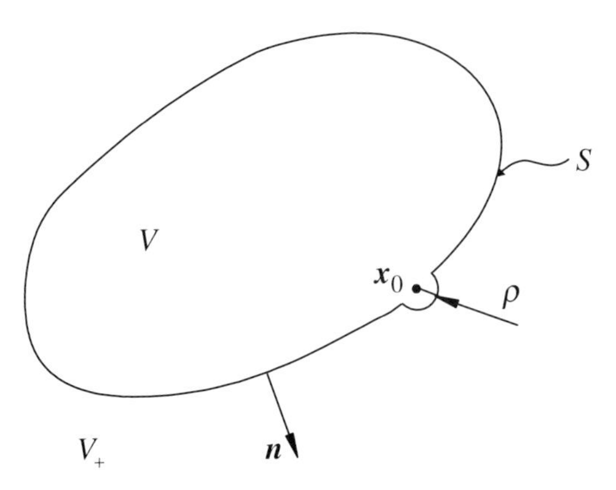

Figure 11.2 Boundary point on boundary S of V.

towards S. As discussed in Section 11.2 this approach gives rise to a singular integral equation which requires the computation of a Cauchy principal value.

First we derive some results that apply when the point of application of a concentrated load tends towards S. The following behavior near the point of application of the load follows directly from the expressions for $U_{ik}(\boldsymbol{x}-\boldsymbol{y})$ and $T_{ik}(\boldsymbol{x}-\boldsymbol{y})$:

$$U_{ik}(\boldsymbol{x},\boldsymbol{y};\omega) = O(1/\rho),$$
$$T_{ik}(\boldsymbol{x},\boldsymbol{y};\omega) = O(1/\rho^2),$$

as $\rho = |\boldsymbol{x}-\boldsymbol{y}| \to 0$. We can then evaluate the limiting value of the surface integral when $\rho \to 0$ on the part of a sphere $s(\boldsymbol{x}_0;\rho)$, with radius ρ and center $\boldsymbol{x}_0$, that is included in V, as shown in Fig. 11.2. The results are

$$\lim_{\rho\to 0}\int_{s(\boldsymbol{x}_0;\rho)\in V} U_{ki}(\boldsymbol{x}_0-\boldsymbol{x})h_i(\boldsymbol{x})\,dS(\boldsymbol{x}) = 0,$$

$$\lim_{\rho\to 0}\int_{s(\boldsymbol{x}_0;\rho)\in V} T_{ki}(\boldsymbol{x}_0-\boldsymbol{x})g_i(\boldsymbol{x})\,dS(\boldsymbol{x}) = C_{ik}(\boldsymbol{x}_0)g_i(\boldsymbol{x}_0),$$

where $h_i(\boldsymbol{x})$ and $g_i(\boldsymbol{x})$ are regular in V and on S. In these expressions $C_{ik}(\boldsymbol{x}_0) = \delta_{ik}$ for $\boldsymbol{x}_0 \in V$ and $\frac{1}{2}\delta_{ik}$ for $\boldsymbol{x}_0 \in S$. By the use of these relations we can write the integral equation for $\boldsymbol{y} \in V$ as $\boldsymbol{y} \to \boldsymbol{x}_0$.

As an example we will derive the boundary integral equation for the problem of scattering of an incident wave by a cavity of volume V and boundary S. The cavity is free of surface tractions. The incident wave is defined by $\boldsymbol{u}^{\text{in}}(\boldsymbol{x})$. In order to render the surface of the cavity free of tractions we define a second field, the scattered field, whose surface tractions on S are equal in magnitude but opposite in sign to those of the incident field. The scattered field is defined by

$$\boldsymbol{u}^{\text{tot}}(\boldsymbol{x};\omega) = \boldsymbol{u}^{\text{sc}}(\boldsymbol{x};\omega) + \boldsymbol{u}^{\text{in}}(\boldsymbol{x};\omega),$$

where $\boldsymbol{u}^{\text{tot}}(\boldsymbol{x};\omega)$ is the total field generated by the scattering of the incident wave. The traction-free boundary conditions on the surface of the scatterer are satisfied by $\boldsymbol{u}^{\text{tot}}(\boldsymbol{x};\omega)$. For the scattered field the results of Section 11.3 are valid for the domain V_+ outside the boundary, S, of the scatterer:

$$\int_S T_{ik}(\boldsymbol{x},\boldsymbol{y};\omega)u_i^{\text{sc}}(\boldsymbol{x};\omega)\,dS(\boldsymbol{x}) - \int_S U_{ik}(\boldsymbol{x},\boldsymbol{y};\omega)t_i^{\text{sc}}(\boldsymbol{x};\omega)\,dS(\boldsymbol{x}) = \begin{cases} u_k^{\text{sc}}(\boldsymbol{x};\omega), & \boldsymbol{x}\in V_+, \\ 0, & \boldsymbol{x}\in V, \end{cases} \tag{11.4.1}$$

where the normal on S is now taken as positive when pointing into the domain exterior to V. For the incident wave we can write for the interior domain

$$\int_S U_{ik}\,(\boldsymbol{x},\boldsymbol{y};\omega)t_i^{\text{in}}(\boldsymbol{x};\omega)\,dS(\boldsymbol{x}) - \int_S T_{ik}(\boldsymbol{x},\boldsymbol{y};\omega)u_i^{\text{in}}(\boldsymbol{x};\omega)\,dS(\boldsymbol{x}) = \begin{cases} u_k^{\text{in}}(\boldsymbol{y};\omega), & \boldsymbol{x}\in V, \\ 0, & \boldsymbol{x}\in V_+. \end{cases} \tag{11.4.2}$$

By subtracting these two equations we obtain the following integral representation:

$$\int_S T_{ik}(\boldsymbol{x},\boldsymbol{y};\omega)u_i^{\text{tot}}(\boldsymbol{x};\omega)\,dS(\boldsymbol{x}) - \int_S U_{ik}(\boldsymbol{x},\boldsymbol{y};\omega)t_i^{\text{tot}}(\boldsymbol{x};\omega)\,dS(\boldsymbol{x}) = \begin{cases} u_k^{\text{tot}}(\boldsymbol{y};\omega) - u_k^{\text{in}}(\boldsymbol{y};\omega), & \boldsymbol{x}\in V_+, \\ -u_k^{\text{in}}(\boldsymbol{y};\omega), & \boldsymbol{x}\in V. \end{cases} \tag{11.4.3}$$

It follows that we have the following boundary integral equation for $u_i^{\text{tot}}(\boldsymbol{x}_0)$:

$$\fint_S T_{ik}(\boldsymbol{x},\boldsymbol{x}_0;\omega)u_i^{\text{tot}}(\boldsymbol{x};\omega)\,dS(\boldsymbol{x}) - \int_S U_{ik}(\boldsymbol{x},\boldsymbol{x}_0;\omega)t_i^{\text{tot}}(\boldsymbol{x};\omega)\,dS(\boldsymbol{x}) + u_k^{\text{in}}(\boldsymbol{x}_0;\omega) = C_{ik}(\boldsymbol{x}_0)u_i^{\text{tot}}(\boldsymbol{x}_0;\omega), \qquad \boldsymbol{x}_0\in S. \tag{11.4.4}$$

Equation (11.4.4) can be solved numerically by well-established ways of using the boundary element method.

As noted in the previous section, at certain wavenumbers that correspond to the eigenfrequencies of the interior region the boundary integral equation for scattering by a cavity does not yield a unique solution. These special wavenumbers are called "eigen-" or "characteristic" wavenumbers and the corresponding eigenfrequencies are termed "fictitious" eigenfrequencies. In his review article,

Kobayashi (1987) lists a number of relevant references. When solving dynamic exterior problems special attention must be devoted to the question of how to deal with the fictitious eigenfrequencies.

Kobayashi (1987) listed four techniques to circumvent the fictitious eigenfrequencies.

(i) Use the interior representation for $\boldsymbol{x} \in V$, that is, with null displacements in the interior, or use it as a supplementary condition for the external boundary integral equation.
(ii) Utilize two integral equations, to solve simultaneously both the first and the second kind of integral equation for the same problem, or consider a combined equation with a suitably selected coupling constant, or use a mixed potential.
(iii) Formulate a boundary integral equation using modified fundamental solutions that satisfy a certain vanishing condition inside the interior domain.
(iv) Apply interpolation techniques by using several values of ω near the eigenfrequencies of the interior domain to obtain numerically approximate results corresponding to the fictitious eigenfrequencies.

References were listed in this article for each of these techniques.

11.5 Scattering by a crack

A crack is a slit in a solid body that cannot transmit tractions. Hence when the body is loaded, and when the region around the crack is placed in a state of stress, the crack becomes a surface of displacement discontinuity. The separation of the crack faces for that case is called the crack-opening displacement (COD). In reality a crack can close as well as open up. For mathematical purposes a crack is represented, however, by an infinitesimally thin slit, which can neither transmit tensile nor compressive tractions. Thus it is assumed that the crack faces do not interact, even under compressive loads. A consequence of this representation is, however, that in the mathematical analysis the crack faces may overlap. Physically this is, of course, not acceptable but generally it is considered to give a good approximation for real cracks, because for real cracks the crack faces are often already slightly separated before loads are applied. Thus the elastodynamic displacements due to incident waves or other applied loads are generally assumed not to close the crack.

Let us consider the scattering of a time-harmonic wave by a crack. Even though this problem can be thought of physically as a special case of scattering by an ellipsoidal cavity with its smaller axis tending to zero, its

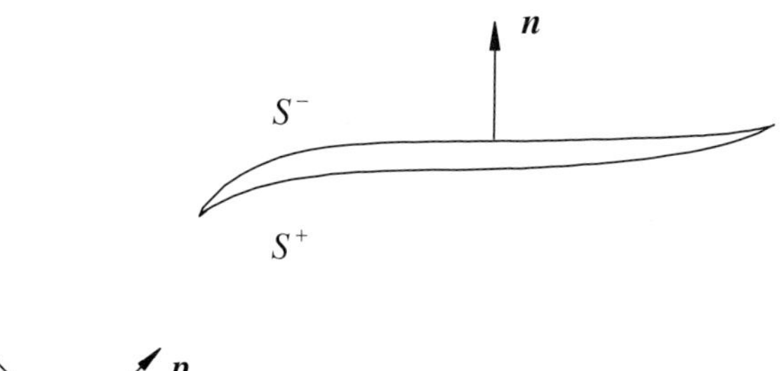

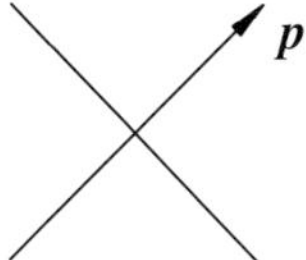

Figure 11.3 Wave incidence on a crack of arbitrary shape.

numerical solution cannot be obtained from the solution corresponding to this flattened cavity.

The geometry is shown in Fig. 11.3. We write $S = S^+ + S^-$, where S^+ is the sonified side of the crack, and S^- is the shadow side. Also, $\boldsymbol{n}$ is the unit normal vector of S, with direction as shown in Fig. 11.3. For the field scattered by the crack the integral representation (11.4.1) can be rewritten as

$$\begin{aligned} u_k^{\mathrm{sc}}(\boldsymbol{y};\omega) = &- \int_{S^+} T_{ik}(\boldsymbol{x},\boldsymbol{y};\omega) u_i^{\mathrm{sc}}(\boldsymbol{x};\omega)\, dS(\boldsymbol{x}) \\ &+ \int_{S^-} T_{ik}(\boldsymbol{x},\boldsymbol{y};\omega) u_i^{\mathrm{sc}}(\boldsymbol{x};\omega)\, dS(\boldsymbol{x}), \end{aligned} \tag{11.5.1}$$

where it has been taken into account that the tractions corresponding to the scattered wave are equal on the two crack faces. They are the negatives of the tractions due to the incident wave. Equation (11.5.1) can be rewritten as

$$u_k^{\mathrm{sc}}(\boldsymbol{y}) = - \int_{S^+} T_{ik}(\boldsymbol{x},\boldsymbol{y};\omega) \Delta u_i^{\mathrm{sc}}(\boldsymbol{x};\omega)\, dS(\boldsymbol{x}), \tag{11.5.2}$$

where

$$\Delta u_i^{\mathrm{sc}}(\boldsymbol{x};\omega) = u_i^{\mathrm{sc}}(\boldsymbol{x}^+;\omega) - u_i^{\mathrm{sc}}(\boldsymbol{x}^-;\omega). \tag{11.5.3}$$

The quantity $\Delta u_i^{\mathrm{sc}}(\boldsymbol{x};\omega)$ is the total crack-opening displacement, since the incident field is continuous across the crack faces.

A system of integral equations cannot be obtained directly from Eq. (11.5.2) because the limiting process $\boldsymbol{y} \to \boldsymbol{x}_0$, where $\boldsymbol{x}_0$ is a point on S^+, will produce a degenerate system of equations. A useful system of equations can be obtained, however, from the surface tractions. Substitution of Eq. (11.5.2) into the general form of Hooke's law, given by Eq. (2.2.10), yields for the traction at position $\boldsymbol{y}$

$$t_p^{\mathrm{sc}}(\boldsymbol{y};\omega) = -C_{pqkl}n_q(\boldsymbol{y})\int_{S^+}\frac{\partial}{\partial x_l}[T_{ik}(\boldsymbol{x},\boldsymbol{y};\omega)]\Delta u_i^{\mathrm{sc}}(\boldsymbol{x};\omega)\,dS^+(\boldsymbol{x}). \quad (11.5.4)$$

This expression is valid when $\boldsymbol{y}$ is not located on S^+. Next we consider the case where $\boldsymbol{y} \to \boldsymbol{x}_0$. This process is done in a few stages, as follows:

$$\begin{aligned}\lim_{y\to x_0} t_p^{\mathrm{sc}}(\boldsymbol{y};\omega) &= t_p^{\mathrm{sc}}(\boldsymbol{x}_0;\omega) = -t_p^{\mathrm{in}}(\boldsymbol{x}_0;\omega)\\ &= -C_{pqkl}\lim_{\boldsymbol{y}\to\boldsymbol{x}_0} n_q(\boldsymbol{y})\int_{S^+}\frac{\partial}{\partial x_l}[T_{ik}(\boldsymbol{x},\boldsymbol{y};\omega)]\Delta u_i^{\mathrm{sc}}(\boldsymbol{x};\omega)\,dS^+(\boldsymbol{x})\\ &= -C_{pqkl}n_q(\boldsymbol{x}_0)\int_{S^+}\frac{\partial}{\partial x_l}[T_{ik}(\boldsymbol{x},\boldsymbol{y};\omega)]\big|_{\boldsymbol{y}=\boldsymbol{x}_0}\Delta u_i^{\mathrm{sc}}(\boldsymbol{x};\omega)\,dS^+(\boldsymbol{x});\end{aligned}$$

finally,

$$-t_p^{\mathrm{in}}(\boldsymbol{x}_0) = -C_{pqkl}n_q(\boldsymbol{x}_0)\int_{S+}\frac{\partial}{\partial x_l}[T_{ki}(\boldsymbol{x},\boldsymbol{y};\omega)]\big|_{\boldsymbol{y}=\boldsymbol{x}_0}\Delta u_i^{\mathrm{sc}}(\boldsymbol{x};\omega)\,dS^+(\boldsymbol{x}). \quad (11.5.5)$$

The system of boundary integral equations given by (11.5.5) is hypersingular when $\boldsymbol{y}$ approaches $\boldsymbol{x}$, since the terms $(\partial/\partial x_l)T_{ik}(\boldsymbol{x}-\boldsymbol{y})$ behave as

$$(\partial/\partial x_l)T_{ik}(\boldsymbol{x}-\boldsymbol{y}) = O(1/\rho^2)$$

for the two-dimensional case, and

$$(\partial/\partial x_l)T_{ik}(\boldsymbol{x}-\boldsymbol{y}) = O(1/\rho^3)$$

for the three-dimensional case, as $\rho \to 0$ where $\rho = |\boldsymbol{x}-\boldsymbol{y}|$.

The presence of a hypersingularity makes it difficult to solve Eq. (11.5.5) numerically, since the integrals arising in this equation do not converge even in the sense of Cauchy principal values. Zhang and Gross (1998) listed three categories of solution procedures. In the first category the hypersingular integrals are regarded as Hadamard finite-part integrals, which are directly integrated analytically or numerically by making certain simplifications on the discretization. In the second category the hypersingular integrals are first reduced to Cauchy principal value integrals by using regularization techniques and then the regularized boundary integral equations are solved numerically. Most regularization techniques use partial integration that shifts the derivative of the stress Green's function to the unknown crack-opening displacement. A third category of solution procedures, discussed in some detail by Zhang and Gross (1998), is based on a two-state conservation integral of elastodynamics. Zhang

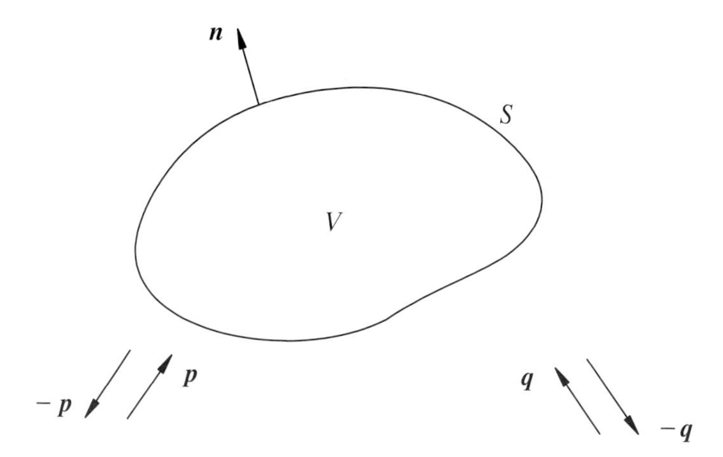

Figure 11.4 Directions of incident and scattered waves for the reciprocity relations.

and Gross (1998) provide an extensive list of references for the three categories of solution procedures.

11.6 Reciprocity relations

Interesting reciprocity relations for the scattering of plane elastic waves by obstacles of arbitrary shape were presented by Tan (1975). These relations are valid in the far field. As shown by Tan, they are most easily derived on the basis of the reciprocity theorem. It will be shown in this section that the far-field scattered wave due to an incident wave propagating in the direction of the unit vector $\boldsymbol{p}$ and observed in the direction $-\boldsymbol{q}$ is related to the far-field scattered wave due to an incident wave propagating in the direction $\boldsymbol{q}$ and observed in the direction $-\boldsymbol{p}$. The geometry is shown in Fig. 11.4. Specific results are stated for the cases of two incident longitudinal waves, two incident transverse waves and an incident longitudinal wave together with an incident transverse wave.

As the starting point of our considerations we write the reciprocal theorem for two scattered fields and for the domain bounded internally by S and externally by S_R. The two scattered fields are defined by

$$\left\{u_i^{\text{sc};A}, \tau_{ij}^{\text{sc};A}\right\} \qquad \text{and} \qquad \left\{u_i^{\text{sc};B}, \tau_{ij}^{\text{sc};B}\right\}.$$

In the limit as $R \to \infty$, the integral along S_R vanishes by virtue of the radiation condition, as discussed in some detail in Section 11.3. Thus, we have

$$\int_S \left(u_i^{\mathrm{sc};A} \tau_{ij}^{\mathrm{sc};B} - u_i^{\mathrm{sc};B} \tau_{ij}^{\mathrm{sc};A}\right) n_j(\boldsymbol{x})\, dS(\boldsymbol{x}) = 0. \tag{11.6.1}$$

In the next step we substitute the relations

$$\tau_{ij}^{\mathrm{sc}} = \tau_{ij} - \tau_{ij}^{\mathrm{in}},$$

and we invoke the boundary conditions of vanishing surface tractions $\tau_{ij} n_j$ on S. Equation (11.6.1) then becomes

$$\int_S \left(u_i^{\mathrm{sc};A} \tau_{ij}^{\mathrm{in};B} - u_i^{\mathrm{sc};B} \tau_{ij}^{\mathrm{in};A}\right) n_j(\boldsymbol{x})\, dS(\boldsymbol{x}) = 0. \tag{11.6.2}$$

Next we consider the two sets of incident waves $\{u_i^{\mathrm{in};A}, u_i^{\mathrm{in};B}\}$ in the solid without a void. For the domain V the reciprocity theorem yields

$$\int_S \left(u_i^{\mathrm{in};A} \tau_{ij}^{\mathrm{in};B} - u_i^{\mathrm{in};B} \tau_{ij}^{\mathrm{in};A}\right) n_j(\boldsymbol{x})\, dS(\boldsymbol{x}) = 0. \tag{11.6.3}$$

Then, by substituting $\tau_{ij}^{\mathrm{in}} n_j = -\tau_{ij}^{\mathrm{sc}} n_j$ into (11.6.3) and subtracting the result from (11.6.2) we obtain

$$\int_S \left(\tau_{ij}^{\mathrm{in};A} u_i^{\mathrm{sc};B} - \tau_{ij}^{\mathrm{sc};B} u_i^{\mathrm{in};A}\right) n_j(\boldsymbol{x})\, dS(\boldsymbol{x})$$
$$= \int_S \left(\tau_{ij}^{\mathrm{in};B} u_i^{\mathrm{sc};A} - \tau_{ij}^{\mathrm{sc};A} u_i^{\mathrm{in};B}\right) n_j(\boldsymbol{x})\, dS(\boldsymbol{x}). \tag{11.6.4}$$

We now consider three cases of reciprocity, all three for two specified cases of wave motion incident on the scatterer. The first of these is concerned with *two incident plane time-harmonic longitudinal waves*. These waves are defined by

$$u_i^{\mathrm{in};L,A} = A^L p_i e^{ik_L \boldsymbol{p}\cdot\boldsymbol{x}},$$
$$u_i^{\mathrm{in};L,B} = B^L q_i e^{ik_L \boldsymbol{q}\cdot\boldsymbol{x}}.$$

By using Hooke's law for plane waves as given by Eq. (3.2.11), i.e.,

$$\tau_{lm} = [\lambda \delta_{lm}(d_j p_j) + \mu(d_l p_m + d_m p_l)] ik A e^{i\eta},$$

we write

$$\tau_{ij}^{\mathrm{in};L,A} = ik_L A^L (\lambda \delta_{ij} + 2\mu p_i p_j) e^{ik_L \boldsymbol{p}\cdot\boldsymbol{x}},$$
$$\tau_{ij}^{\mathrm{in};L,B} = ik_L B^L (\lambda \delta_{ij} + 2\mu q_i q_j) e^{ik_L \boldsymbol{q}\cdot\boldsymbol{x}}.$$

Substitution of these expressions into (11.6.4) yields

$$ik_L A^L \int_S (\lambda\delta_{ij} + 2\mu p_i p_j) n_j(\mathbf{x}) u_i^{\text{sc};B} e^{ik_L \mathbf{p}\cdot\mathbf{x}} \, dS(\mathbf{x})$$

$$- A^L \int_S \tau_{ij}^{\text{sc};B} p_i n_j(\mathbf{x}) e^{ik_L \mathbf{p}\cdot\mathbf{x}} \, dS(\mathbf{x})$$

$$= ik_L B^L \int_S (\lambda\delta_{ij} + 2\mu q_i q_j) n_j(\mathbf{x}) u^{\text{sc};A} e^{ik_L \mathbf{q}\cdot\mathbf{x}} \, dS(\mathbf{x})$$

$$- B^L \int_S \tau_{ij}^{\text{sc};A} q_i n_j(\mathbf{x}) e^{ik_L \mathbf{q}\cdot\mathbf{x}} \, dS(\mathbf{x}). \tag{11.6.5}$$

We wish to prove the relation

$$A^L p_k U_k^{\text{sc};L,B}(-\mathbf{p}) = B^L q_k U_k^{\text{sc};L,A}(-\mathbf{q}). \tag{11.6.6}$$

From Eq. (11.3.15) we obtain

$$A^L p_k U_k^{\text{sc};L,B}(-\mathbf{p}) = \frac{A^L}{\lambda + 2\mu} \int_S \left(p_i t_i^{\text{sc};B}\right) e^{ik_L \mathbf{p}\cdot\mathbf{x}} \, dS(\mathbf{x})$$
$$- \frac{ik_L}{\lambda + 2\mu} A^L \int_S [\lambda u_j n_j + 2\mu (u_i p_i)(p_j n_j)] e^{ik_L \mathbf{p}\cdot\mathbf{x}} \, dS(\mathbf{x}). \tag{11.6.7}$$

Similarly,

$$B^L q_k U_k^{\text{sc};L,A}(-\mathbf{q}) = \frac{B^L}{\lambda + 2\mu} \int_S \left(q_i t_i^{\text{sc};A}\right) e^{ik_L \mathbf{q}\cdot\mathbf{x}} \, dS(\mathbf{x})$$
$$- \frac{ik_L}{\lambda + 2\mu} B^L \int_S [\lambda u_j n_j + 2\mu (u_i q_i)(q_i n_j)] e^{ik_L \mathbf{q}\cdot\mathbf{x}} \, dS(\mathbf{x}). \tag{11.6.8}$$

The equality (11.6.5), when combined with Eqs. (11.6.7) and (11.6.8), then proves the reciprocity relation given by Eq. (11.6.6).

A result analogous to Eq. (11.6.6) can be obtained for the case of two *incident plane time-harmonic transverse waves*. The incident waves are given by

$$u_i^{\text{in};T,A} = A^T d_i e^{ik_T \mathbf{p}\cdot\mathbf{x}},$$
$$u_i^{\text{in};T,B} = B^T d_i e^{ik_T \mathbf{q}\cdot\mathbf{x}},$$

where $d_j p_j = 0$ and $d_j q_j = 0$. By pursuing the approach represented by Eqs. (11.6.5)–(11.6.8) we obtain

$$A^T d_k U_k^{\text{sc};T,B}(-\mathbf{p}) = B^T d_k U_k^{\text{sc};T,A}(-\mathbf{q}). \tag{11.6.9}$$

Finally, for the case of an *incident longitudinal wave and an incident transverse wave*, we can write

$$(\lambda + 2\mu)A^L p_k U_k^{\mathrm{sc};L,B}(-\boldsymbol{p}) = \mu B^T d_k U_k^{\mathrm{sc};T,A}(-\boldsymbol{q}). \qquad (11.6.10)$$

Equations (11.6.5), (11.6.9) and (11.6.10) are reciprocity relations for the amplitudes of the incident and scattered waves along two specific rays.

The exposition of this section has been presented within the context of scattering by a cavity. With modifications it can be shown that the results are also valid for scattering by an inclusion.

11.7 Kirchhoff approximation for scattering by a crack

Solving the boundary integral equation for scattering by a crack requires a considerable effort. It is, therefore, sometimes of interest to obtain a simple approximate solution.

In this section we discuss an approach based on approximating the crack-opening displacement (COD), and the subsequent use of this approximation in the representation integral for the scattered field. For the present purposes the crack is assumed to be flat and located in the plane $x_3 = 0$. The representation integral is given by Eq. (11.5.2) as

$$u_k^{\mathrm{sc}}(\boldsymbol{y}) = -\int_{S^+} T_{ik}(\boldsymbol{x}, \boldsymbol{y}; \omega)\Delta u_i^{\mathrm{sc}}(\boldsymbol{x}; \omega)\, dS(\boldsymbol{x}), \qquad (11.7.1)$$

where $\Delta u_i^{\mathrm{sc}}(\boldsymbol{x}; \omega$ is the crack-opening displacement.

In scattering problems $\Delta \boldsymbol{u}^{\mathrm{sc}}(\boldsymbol{x}; \omega)$ is not known a priori. In two well-known approximations, which are valid at low and high frequencies, respectively, a form of $\Delta \boldsymbol{u}^{\mathrm{sc}}$ is postulated, and $\boldsymbol{u}^{\mathrm{sc}}(\boldsymbol{y})$ is subsequently computed from Eq. (11.7.1). At low frequencies the static COD can be substituted in the integral to give the scattered field according to the so-called quasistatic scattering theory. At very high frequencies the displacements on the sonified crack face, which are taken to be the same as if that face were of infinite extent, can be used as the crack-opening displacement. Subsequent evaluation of (11.7.1) produces the Kirchhoff approximation. An example of the Kirchhoff approximation is discussed in this section.

The representation integral given by Eq. (11.7.1) simplifies considerably for the case where the incident wave is a plane wave propagating in the x_3-direction, the point of observation is taken far from the crack and the origin of the coordinate system is taken in the plane of the crack.

The incident plane wave is represented by

$$u_3^{\mathrm{in};L} = A^L e^{ik_L x_3}.$$

This wave generates only normal displacements on the plane $x_3 = 0$. For a plane free of surface tractions the Kirchhoff approximation implies that

$$\Delta u_3^{\mathrm{sc}} = 2A^L.$$

The only component of T_{ik} appearing in Eq. (11.7.1) then is T_{3k}. For the far field, expressions for T_{3k} were derived in Section 11.3. We have

$$T_{3k} \approx \sum_{\alpha=L,T} ik_\alpha C_{3k}^{G;\alpha} \frac{\exp(ik_\alpha|\mathbf{y}|)}{4\pi|\mathbf{y}|} e^{-ik_\alpha \mathbf{x}\cdot\hat{\mathbf{y}}}, \tag{11.7.2}$$

where, taking into account that $i = 3$ and $j = 3$,

$$C_{3k}^{G;L} = -\left[\frac{2\mu}{\lambda + 2\mu}(\hat{y}_3)^2 + \frac{\lambda}{\lambda + 2\mu}\right]\hat{y}_k,$$

$$C_{3k}^{G;T} = -[\delta_{3k}\hat{y}_3 + \hat{y}_3 - 2(\hat{y}_3)^2\hat{y}_k].$$

Substitution of these results into Eq. (11.7.2) and subsequent substitution in (11.7.1) yields

$$u_k^{\mathrm{sc}}(\mathbf{y};\omega) = u_k^{\mathrm{sc};L}(\mathbf{y};\omega) + u_k^{\mathrm{sc};T}(\mathbf{y};\omega), \tag{11.7.3}$$

where

$$u_k^{\mathrm{sc};\alpha}(\mathbf{y},\omega) = -2A^L C_{3k}^{G;\alpha} \frac{e^{ik_\alpha|\mathbf{y}|}}{4\pi|\mathbf{y}|} I(k_\alpha). \tag{11.7.4}$$

The integral $I(k_\alpha)$ is

$$I(k_\alpha) = ik_\alpha \int_{S^+} e^{-ik_\alpha \mathbf{x}\cdot\hat{\mathbf{y}}}\, dS(\mathbf{x}). \tag{11.7.5}$$

The area integral (11.7.5) can be reduced to a line integral over the crack edge E by noting that $I(k_\alpha)$ can also be expressed in the form

$$I(k_\alpha) = -\frac{1}{\hat{y}_1^2 + \hat{y}_2^2} \int_{S^+} \frac{\partial}{\partial x_i}(\hat{y}_i e^{-ik_\alpha \mathbf{x}\cdot\hat{\mathbf{y}}}) dx_1 dx_2.$$

An application of the divergence theorem subsequently yields

$$I(k_\alpha) = -\frac{1}{\hat{y}_1^2 + \hat{y}_2^2} \int_E \nu_i \hat{y}_i e^{-ik_\alpha \mathbf{x}\cdot\hat{\mathbf{y}}}\, de = -\int_E \nu_i \hat{y}_i e^{-ik_\alpha \mathbf{x}\cdot\hat{\mathbf{y}}}\, de, \tag{11.7.6}$$

where ν_i $(i = 1, 2)$ defines the outward unit normal to the crack edge E, and e is arc length measured along the crack edge. By means of Eqs. (11.7.6) and

(11.7.4) the scattered fields are expressed by a superposition of sources over the edge of the crack.

For a crack of elliptical shape, of major axis b and minor axis a, the integral over the edge E given by Eq. (11.7.6) can be evaluated explicitly. We introduce new variables ϕ and ψ by the relations

$$(x_1, x_2) = [a \cos(\phi + \psi), b \sin(\phi + \psi)] \qquad 0 \leq \phi \leq 2\pi,$$
$$(\cos\psi, \sin\psi) = (a\hat{y}_1, b\hat{y}_2)/\rho,$$

where

$$\rho = \left(a^2\hat{y}_1^2 + b^2\hat{y}_2^2\right)^{1/2}. \tag{11.7.7}$$

It then follows that

$$(\nu_1, \nu_2)\, de = [-b\cos(\phi + \psi), -a\sin(\phi + \psi)]\, d\phi,$$

and the edge integral, Eq. (11.7.6), becomes

$$I(k_\alpha) = \rho \int_0^{2\pi} \left[\frac{b}{a}\cos(\phi + \psi)\cos\psi + \frac{a}{b}\sin(\phi + \psi)\sin\psi\right] e^{-ik_\alpha \rho\cos\phi}\, d\phi.$$

This integral can be evaluated to yield

$$I(k_\alpha) = -\frac{2\pi i a b}{\rho} J_1(k_\alpha \rho), \tag{11.7.8}$$

where $J_1(\ \ )$ is the ordinary Bessel function of order one, and ρ is defined by Eq. (11.7.7). Upon substitution of (11.7.8) into (11.7.4) we obtain the scattered field according to the Kirchhoff approximation for an elliptical crack sonified by a plane longitudinal wave.

A slightly more general analysis for incidence under an arbitrary angle was given by Achenbach, Gautesen and McMaken (1982). A comparison of the Kirchhoff approximation with exact results has shown that at high frequencies the approximation is reasonably good in the direction of specular reflection.

11.8 Reciprocity considerations for a diffraction problem

In the previous section the wave motion induced by a crack of finite dimensions due to the incidence of a plane wave generated far from the crack was investigated in an approximate manner. The problem was termed a scattering problem because the crack as a whole was considered to respond to the incident wave. In this section we investigate the case where in the region of interest the dominant

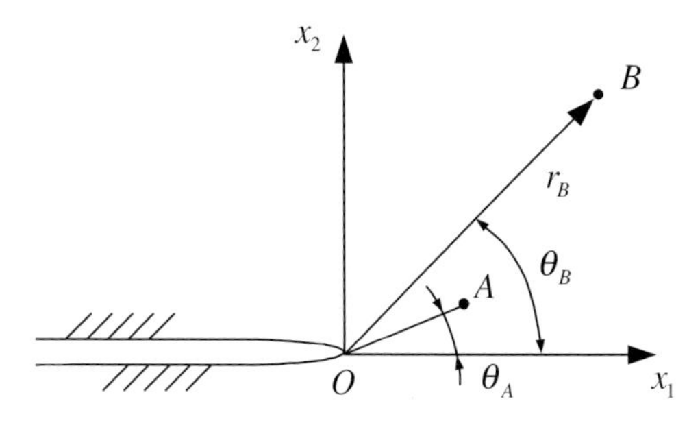

Figure 11.5 Crack tip with source at A and receiver at B.

response from the crack to an incident wave is supposed to come from the crack tip. Such a problem is called a diffraction problem.

The configuration is shown in Fig. 11.5. The deformation is two dimensional in anti-plane strain. A time-harmonic anti-plane line load is applied at point A, in close proximity to the crack tip O. The objective is to determine the diffracted field, w_{BA}^{diff}, at point B, which is far from the crack tip. By a direct approach it is, however, easier to solve the conjugate problem and determine the field at point A, w_{AB}^{diff}, due to a line load at point B. We can subsequently use the reciprocity relation to determine w_{BA}^{diff} from w_{AB}^{diff}.

According to Eqs. (6.6.10), (6.6.18) and (6.6.7) the far-field displacement generated by a line load applied at point B and of amplitude F_B is

$$w_B(r) = \frac{i}{4}\frac{F_B}{\mu}\left(\frac{2}{\pi}\right)^{1/2}\frac{1}{[k_T(r_B - r)]^{1/2}}e^{i[k_T(r_B - r) - \pi/4]}, \tag{11.8.1}$$

where $k_T = \omega/c_T$ and r is measured from the crack tip O. An approximation to the diffracted field at point A may then be written as

$$w_{AB}^{\text{diff}}(r_A, r_B) = w_B(0)(k_T r_A)^{1/2} D_{AB}(\theta_A, \theta_B)e^{ik_T r_A}, \tag{11.8.2}$$

where

$$w_B(0) = \frac{i}{4}\frac{F_B}{\mu}\left(\frac{2}{\pi}\right)^{1/2}\frac{1}{(k_T r_B)^{1/2}}e^{i(k_T r_B - \pi/4)}. \tag{11.8.3}$$

Here $w_B(0)$ is the incident field at the crack tip and $D_{AB}(\theta_A, \theta_B)$ is a diffraction coefficient for the angle of incidence θ_B and the angle of diffraction θ_A. The derivative $(\partial/\partial\theta)D_{AB}(\theta, \theta_B)$ vanishes on the crack faces, to render the crack faces free of surface tractions. Note the appearance of the term $(k_T r_A)^{1/2}$ in Eq. (11.8.2), since we are assuming that point A is very close to the crack tip and hence $k_T r_A \ll 1$. If we considered a point A for $k_T r_A \gg 1$, we would instead have $(k_T r_A)^{-1/2}$ for the far-field solution, in accordance with geometrical diffraction theory.

The total field at point A is

$$w_{AB}^{\text{tot}} = w_{AB}^{\text{in}} + w_{AB}^{\text{diff}}. \tag{11.8.4}$$

Here w_{AB}^{in} is the field at point A if there is no crack. Next we consider two states: state B is the virtual wave due to a line load of amplitude F_B at point B and state A is the wave field due to a line load of amplitude F_A at point A. Since for the total fields of both states A and B the tractions vanish on the crack faces, the reciprocity relation includes only the contributions from body forces. Consequently we obtain

$$F_A \left(w_{AB}^{\text{in}} + w_{AB}^{\text{diff}} \right) = F_B \left(w_{BA}^{\text{in}} + w_{BA}^{\text{diff}} \right). \tag{11.8.5}$$

However, the incident waves satisfy the relations

$$F_A w_{AB}^{\text{in}} = F_B w_{BA}^{\text{in}}. \tag{11.8.6}$$

Hence we have

$$w_{BA}^{\text{diff}}(r) = \frac{F_A}{F_B} w_B(0)(k_T r_A)^{1/2} D_{AB}(\theta_A, \theta_B) e^{i k_T r_A}, \tag{11.8.7}$$

where $w_B(0)$ is defined by Eq. (11.8.3).

12

Scattering in waveguides and bounded bodies

12.1 Introduction

The scattering of elastic waves by defects, such as cracks, voids and inclusions, located in bodies with boundaries is a challenging topic for analytical and numerical studies in elastodynamics. It is, however, also a topic of great practical interest in the field of quantitative non-destructive evaluation (QNDE), because scattering results can be used to detect and size defects. In the context of QNDE, elastodynamics is referred to as ultrasonics, since it is generally necessary to work with wave signals whose principal frequency components are well above the frequency range audible to the human ear.

For realistic defects it is not possible to obtain solutions of scattering problems by rigorous analytical methods. The best numerical technique is generally the one that employs a Green's function to derive a boundary integral equation, as discussed in Chapter 11, which can then be solved by the boundary element method. This process yields the field variables on the surface of the scattering obstacle (the defect). An integral representation can subsequently be used to calculate the scattered field elsewhere. Of course, as an alternative, the fields on the defect can be approximated. Various approximations are available. We mention the quasistatic approximation for the displacement on the surface of a cavity, the Kirchhoff approximation for a crack and the Born approximation for scattering by an inclusion.

In Section 12.2 the interaction of an incident wave motion with a defect in a waveguide is considered. The incident wave is represented by a summation of modes. Since the field scattered by a defect in a waveguide is compact and restricted to the waveguide, we use the terminology "reflection and transmission" rather than scattering. Reflection and transmission coefficients for individual modes of the incident wave motion are calculated in terms of integrals over the surface of the defect, by an elegant use of the reciprocity relation for wave modes in a waveguide. By using an approximation for the field on the defect, approximations for the reflection and transmission coefficients are obtained for a surface-breaking crack and an inclusion. This approach does not require the use of Green functions and integral equations.

In Section 12.3 we consider surface waves in anti-plane strain for a configuration consisting of a thin film bonded to a substrate. The substrate is an inhomogeneous half-space with a shear modulus that varies with depth. The dependence on depth of the surface wave is determined by an approximate solution of the displacement equation of motion for anti-plane strain. Next, the reflection and transmission of surface waves by a disbond at the interface between the film and the substrate are examined, in Section 12.4. By the use of reciprocity considerations, an expression for the reflection coefficient is derived. Section 12.5 is concerned with the reflection of Rayleigh surface waves by a surface-breaking crack for the two-dimensional case of plane strain. Finally, Sections 12.6 and 12.7 discuss scattering matrices and scattering by a crack in a bounded elastic body.

12.2 Interaction of an incident wave with a defect in a layer

In Section 10.2 it was shown that time-harmonic wave motions propagate in a layer as superpositions of wave modes. For an isotropic layer the wave modes can be separated into two classes: modes that are symmetric or antisymmetric relative to the mid-plane of the layer. A wave motion that is symmetric relative to the mid-plane can be fully represented by a superposition of symmetric modes. When the wave motion is incident on a defect in the layer (say a cavity or an inclusion), every single mode will interact with the defect, and this interaction will generate systems of reflected and transmitted wave modes. Clearly the incidence of symmetric wave modes on a symmetric defect, which is also symmetrically located relative to the mid-plane of the layer, gives rise to systems of reflected and transmitted wave modes that are also symmetric.

Elastodynamic reciprocity can quite conveniently be used to investigate the systems of reflected and transmitted wave modes. As an illustration we show that this can be done in a simple and elegant manner for the special case of two-dimensional anti-plane wave motion. The scattering of Lamb waves can be considered in the same way, albeit with some algebraic complications.

A full representation of symmetric wave motion propagating in the positive x_1-direction and hence approaching the defect from $x_1 < 0$ is as

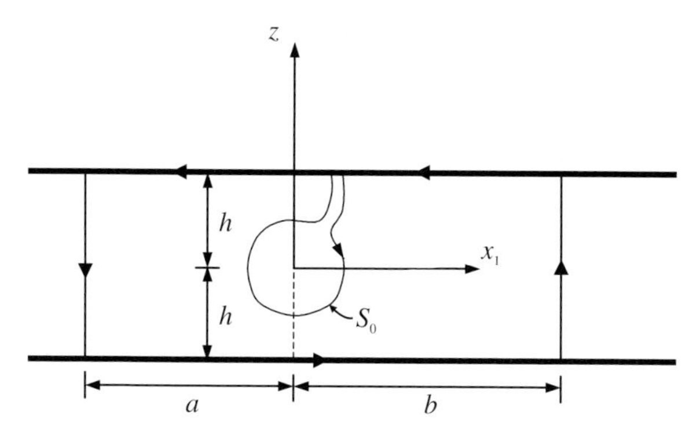

Figure 12.1 Reflection and transmission by a cavity whose boundary is S_0.

follows:

$$u_2 = \sum_{m=0,2,4}^{\infty} u_{2m}(x_1, z) = \sum_{m=0,2,4,\ldots}^{\infty} U_m \cos\left(\overline{q}_m z\right) e^{il_m x_1}, \qquad (12.2.1)$$

where $\overline{q}_m$ and l_m are defined by Eqs. (7.5.5) and (7.5.3), respectively. Here it should be recalled that the time factor is $\exp(-i\omega t)$. It is noted that for the incident wave the coefficients U_m are known. They depend on the shape of the incident signal across the thickness of the layer. The geometry is shown in Fig. 12.1.

The interaction of the incident wave motion with the defect gives rise to reflected and transmitted wave motions, which can be represented by

$$u_2 = \sum_{m=0,2,4,\ldots}^{\infty} R_m U_m \cos(\overline{q}_m z)\, e^{-il_m x_1} \qquad (12.2.2)$$

for $x_1 < 0$, and

$$u_2 = \sum_{m=0,2,4,\ldots}^{\infty} T_m U_m \cos(\overline{q}_m z)\, e^{il_m x_1} \qquad (12.2.3)$$

for $x_1 > 0$. For a defect with specified conditions on its boundary the problem consists in determining the reflection and transmission coefficients R_m and T_m. It is difficult to solve this problem completely, but it is not difficult to derive useful general expressions for R_m and T_m.

The approach follows the technique outlined in previous chapters, namely the use of the reciprocity relation together with a virtual wave. For the two-dimensional geometry being considered here, application of the reciprocity relation is very simple. Referring to Fig. 12.1 we may write in the absence of body forces

$$-\int_{-h}^{h}\left(\tau_{12}^{B}u_{2}^{A}-\tau_{12}^{A}u_{2}^{B}\right)_{x_1=-a}ds+\int_{-h}^{h}\left(\tau_{12}^{B}u_{2}^{A}-\tau_{12}^{A}u_{2}^{B}\right)_{x_1=b}dz$$
$$=-\int_{S_0}\left(t_{2}^{B}u_{2}^{A}-t_{2}^{A}u_{2}^{B}\right)dS, \qquad (12.2.4)$$

where S_0 is the boundary of the defect. In Eq. (12.2.4) we have used surface tractions and displacements over the surface of the defect. It has also been taken into account that the faces of the layer ($z=\pm h$) are free of surface tractions. For the integration over S_0, the normal has been taken as pointing into the defect.

State A consists of the incident wave and the reflected and transmitted waves. Hence we select the sum of the series given by Eqs. (12.2.1) and (12.2.2) for $x_1<0$, and the series given by Eq. (12.2.3) for $x_1>0$. For the defect we consider a cavity that is free of surface tractions, i.e., $t_2^A=0$, and the unknown displacement generated by the incident wave motion on the surface of the cavity is indicated by u_2^A. For state B, the virtual wave, we take a single symmetric mode (mode number n) that propagates in the positive x_1-direction:

$$u_2^B=U_n^B\cos(\overline{q}_n z)\,e^{il_n x_1}, \qquad (12.2.5)$$
$$\tau_{12}^B=il_n\mu U_n^B\cos(\overline{q}_n z)\,e^{il_n x_1}. \qquad (12.2.6)$$

Equations (12.2.5) and (12.2.6) are used to compute expressions for t_2^B and u_2^B on the surface of the cavity. Even though for state B we actually take a virtual wave for another body, namely, a layer without a cavity, the displacements and tractions on the cavity surface are calculated from (12.2.5) and (12.2.6), which ensures that state B is acceptable for a layer with a cavity, since its surface has tractions and displacements corresponding to the virtual-wave mode.

For the incident wave we now consider one wave mode of the summation (12.2.1) at a time, say $U_m^A\cos(\overline{q}_m z)\exp(il_m x_1)$. Substitution of the relevant expressions into Eq. (12.2.4) then yields

$$i\mu U_m^A U_n^B\delta_{mn}^S(l_m-l_n)\,e^{-i(l_m+l_n)a}-i\mu R_m U_m^A U_n^B\delta_{mn}^S(l_m+l_n)\,e^{i(l_m-l_n)a}$$
$$+\,i\mu U_m^A U_n^B\delta_{mn}^S(l_n-l_m)\,e^{i(l_n+l_m)b}-i\mu T_m U_m^A U_n^B\delta_{mn}^S(l_m-l_n)\,e^{i(l_m+l_n)b}$$
$$=-\int_{S_0}t_{2n}^B u_{2m}^A\,dS, \qquad (12.2.7)$$

where $\delta_{mn}^S=0$ for $m\neq n$, $\delta_{mn}^S=h$ for $m=n\neq 0$ and $\delta_{mn}^S=2h$ for $m=n=0$. Also t_{2n}^B is the traction on S_0 due to the virtual mode, while u_{2m}^A is the displacement on S_0 due to the actual incident mode. The first, the third and the fourth term on the left-hand side vanish for $n=m$ and for $n\neq m$. The second term vanishes for $n \neq m$. For $n = m$ we then obtain from Eq. (12.2.7)

$$i\mu R_m U_m^A U_m^B \delta_{mm}^S 2l_m = \int_{S_0} t_{2m}^B u_{2m}^A \, dS. \tag{12.2.8}$$

Using (12.2.7) and (12.2.8) we find for the reflection coefficient of the mth mode of Eq. (12.2.1),

$$m \neq 0: \quad R_m = \frac{1}{2il_m \mu h U_m^A U_m^B} \int_{S_0} t_{2m}^B u_{2m}^A \, dS, \tag{12.2.9}$$

$$m = 0: \quad R_0 = \frac{1}{4il_0 \mu h U_0^A U_0^B} \int_{S_0} t_{2m}^B u_{2m}^A \, dS. \tag{12.2.10}$$

The results given by Eqs. (12.2.9) and (12.2.10) are conceptually interesting but they are not immediately applicable, because u_{2m}^A is as yet unknown. This quantity must be obtained by a separate calculation.

By taking as a virtual wave a single mode that propagates in the negative x_1-direction we can obtain similarly

$$T_m = 1 - R_m. \tag{12.2.11}$$

Reflection and transmission by a surface-breaking crack

As a special case of a cavity we next consider reflection and transmission by a surface-breaking crack. The geometry is shown in Fig. 12.2

As the incident wave we again consider a symmetric wave motion propagating in the x_1-direction, as represented by Eq. (12.2.1). However, since the defect is not symmetric relative to the mid-plane of the layer, it should now be expected that both the reflected and transmitted wave motions contain symmetric as well as antisymmetric modes. For the reflected wave we write

$$\begin{aligned} x_1 < 0: \quad u_2^A = &\sum_{m=0,2,4,\ldots}^{\infty} R_m^S U_m^S \cos(\overline{q}_m z)\, e^{-il_m x_1} \\ &+ \sum_{m=1,3,5,\ldots}^{\infty} R_m^{AS} U_m^{AS} \sin(\overline{q}_m z)\, e^{-il_m x_1}; \end{aligned} \tag{12.2.12}$$

note that on the right-hand side the superscripts S, AS refer to the symmetric and antisymmetric modes, respectively. The transmitted wave is represented by

$$\begin{aligned} x_1 > 0: \quad u_2^A = &\sum_{m=0,2,4,\ldots}^{\infty} T_m^S U_m^S \cos(\overline{q}_m z)\, e^{il_m x_1} \\ &+ \sum_{m=1,3,5,\ldots}^{\infty} T_m^{AS} U_m^{AS} \sin(\overline{q}_m z)\, e^{il_m x_1}. \end{aligned} \tag{12.2.13}$$

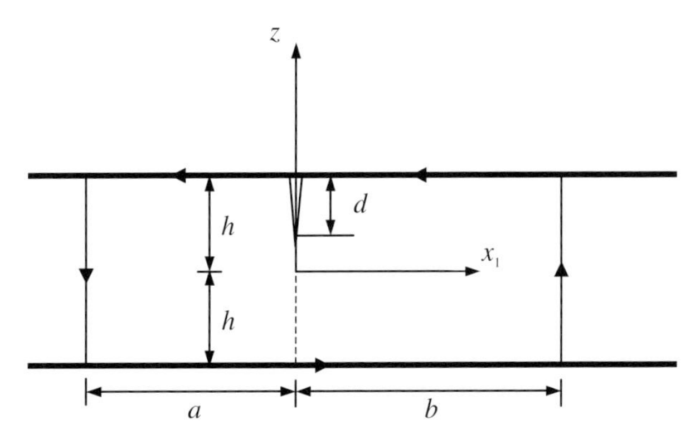

Figure 12.2 Reflection and transmission by a surface-breaking crack.

The calculations are based on a modification of Eq. (12.2.4). With reference to Fig. 12.2, it can be checked that the path integration over S_0 becomes

$$\int_{h}^{h-d} t_2^B u_2^A \, dz + \int_{h-d}^{h} t_2^B u_2^A \, dz = -\int_{h-d}^{h} t_2^B \Delta u_2^A \, dz. \qquad (12.2.14)$$

In Eq. (12.2.14) the normal has been taken as being into the crack, and the quantity Δu_2^A, which is unknown, is usually referred to as the crack-opening displacement: $\Delta u_2^A = u_2^A(x_1^+) - u_2^A(x_1^-)$.

The details of the analysis based on the reciprocity relation now proceed as before. For state B we take two virtual wave modes, a symmetric and an antisymmetric mode, one at a time:

$$\text{symmetric:} \qquad u_2^B = U_n^B \cos(\overline{q}_n z)\, e^{il_n x_1}, \qquad (12.2.15)$$

$$\text{antisymmetric:} \qquad u_2^B = U_n^B \sin(\overline{q}_n z)\, e^{il_n x_1}. \qquad (12.2.16)$$

Equations (12.2.15) and (12.2.16) are used to compute the tractions on the crack faces. These are denoted by

$$t_{2n}^S \quad \text{and} \quad t_{2n}^{AS}.$$

For the case where the virtual wave is a symmetric (antisymmetric) mode, there is no interference with the antisymmetric (symmetric) modes. The interference of modes of the same kind with each other is just as discussed in the previous example. The results can be obtained as

$$m = 2, 4, 6, \ldots, \qquad R_m^S = \frac{1}{2il_m \mu h U_m^S U_m^B} \int_{h-d}^{h} t_{2m}^S \Delta u_{2m}^A \, dz \qquad (12.2.17)$$

$$m = 1, 3, 5, \ldots, \qquad R_m^{AS} = \frac{1}{2il_m \mu h U_m^{AS} U_m^B} \int_{h-d}^{h} t_{2m}^{AS} \Delta u_{2m}^A dz. \quad (12.2.18)$$

As in the example of the cavity, the crack-opening displacement $\Delta u_{2m}^A(z)$ generated by incidence of the mth mode must be calculated separately.

Numerical methods to calculate $\Delta u_{2m}^A(z)$ are available; see Zhang and Gross (1998). Equations (12.2.17) and (12.2.18) are also useful, however, for application, with approximations, to the crack-opening displacement. At low frequencies a quasistatic approximation can be used whereby the incident wave is considered as a static load, and well-known results from static crack analysis can be used for $\Delta u_{2m}^A(z)$. At high frequencies the Kirchoff approximation generally gives acceptable results. In this approximation, the sonified side of the crack is considered to be a specular reflector, while the shadow side is totally unperturbed. For this case the crack-opening displacement is taken as twice the amplitude of the incident wave. Substitution in Eqs. (12.2.17) and (12.2.18) of this approximate crack-opening displacement gives a high-frequency approximation for the reflection coefficients.

Reflection and transmission by an inclusion

As a third example we consider reflection and transmission by an inclusion of a different material. For the details of the analysis we refer to the corresponding cavity case discussed earlier in this section. Analogously to the cavity case we consider a symmetric inclusion that is also symmetrically located relative to the mid-plane of the layer. The reciprocity relation is again given by Eq. (12.2.4). However, for the inclusion the term $t_2^A u_2^B$ does not disappear from the integral over S_0 on the right-hand side of Eq. (12.2.4). In other respects the analysis for the inclusion carries through in the same manner as for the cavity. Instead of Eq. (12.2.9) we now obtain

$$R_m = \frac{1}{2il_m \mu h U_m^A U_m^B} \int_{S_0} \left(t_{2m}^B u_{2m}^A - t_{2m}^A u_{2m}^B \right) dS. \qquad (12.2.19)$$

To determine u_{2m}^A and t_{2m}^A a connection must be established with the different material of the inclusion. For the problem at hand only the difference between the shear moduli of the inclusion and the layer is taken into account. Referring to Fig. 12.1, we write for the inclusion

$$\mu(x_1, z) = \mu + \overline{\mu}(x_1, z), \qquad (12.2.20)$$

where μ is the constant shear modulus of the layer. The equations of motion inside the inclusion then become

$$\frac{\partial}{\partial x_1}\left[\mu(x_1, z)\frac{\partial u_2}{\partial x_1}\right] + \frac{\partial}{\partial z}\left[\mu(x_1, z)\frac{\partial u_2}{\partial z}\right] = -\rho\omega^2 u_2$$

or

$$\mu(x_1, z)\left[\frac{\partial^2 u_2}{\partial x_1^2} + \frac{\partial^2 u_2}{\partial z^2}\right] + \frac{\partial \overline{\mu}}{\partial x_1}\frac{\partial u_2}{\partial x_1} + \frac{\partial \overline{\mu}}{\partial z}\frac{\partial u_2}{\partial z} = -\rho\omega^2 u_2. \quad (12.2.21)$$

The latter equation can be rewritten as

$$\mu\nabla^2 u_2 + f_2 = -\rho\omega^2 u_2, \quad (12.2.22)$$

where

$$f_2 = \frac{\partial \overline{\mu}}{\partial x_1}\frac{\partial u_2}{\partial x_1} + \frac{\partial \overline{\mu}}{\partial z}\frac{\partial u_2}{\partial z} + \overline{\mu}(x_1, z)\nabla^2 u_2. \quad (12.2.23)$$

Thus, the second term in Eq. (12.2.22) can be considered as a body-force term.

For the domain inside the inclusion we next combine in the reciprocity relation the field u_2 from Eq. (12.2.22) as state A with the virtual wave given by Eq. (12.2.5) as state B. The result is

$$\int_{S_0} \left(t_2^A u_2^B - t_2^B u_2^A\right) dS = \int_V f_2^A u_2^B \, dV. \quad (12.2.24)$$

Here the integration over S_0 is counter-clockwise and the normal points into the layer.

A comparison of Eqs. (12.2.4) and (12.2.24) shows that the integration over S_0 can be replaced by the right-hand side of Eq. (12.2.24), to yield

$$R_m = -\frac{1}{2il_m \mu h U_m^A U_m^B}\int_V f_{2m}^A u_{2m}^B \, dV \quad (12.2.25)$$

where f_{2m}^A is given by Eq. (12.2.23). Considerable computation will obviously be required to determine the actual field inside the inclusion from Eq. (12.2.25). This equation does, however, lend itself to a simple approximation known as the Born approximation.

Let us first simplify f_{2m}^A by considering the case where μ is uniform inside the inclusion. Then the derivatives in f_{2m}^A vanish and Eq. (12.2.25) becomes

$$R_m = -\frac{\overline{\mu}}{\mu}\frac{1}{2il_m h U_m^A U_m^B}\int_V \left(\nabla^2 u_{2m}^A\right) u_{2m}^B \, dV. \quad (12.2.26)$$

In the Born approximation the field u_{2m}^A is replaced by an incident wave mode and the integral is evaluated. This approximation is generally thought to be acceptable for $\overline{\mu}/\mu \ll 1$.

Reflection and transmission by an infinite planar array

The results already derived in this section can easily be extended to the reflection and transmission of an incident wave by an infinite planar array of identical inhomogeneities in an unbounded elastic solid. To that end we place an infinite number of layers, each with the same inhomogeneity, next to each other, with conditions of continuity of relevant stresses and displacements at the common boundaries. The centers of the inhomogeneities are in a single plane.

It is not difficult to show that a periodic array of identical compact inhomogeneities, whose geometrical centers are located in a plane interior to an elastic solid, acts as a homogeneous plane for the reflection and transmission of incident waves. A typical example is provided by the incidence of a plane elastic wave on an array of periodically spaced spherical cavities. Reflection and transmission coefficients for that configuration were obtained by Achenbach and Kitahara (1986). Their results show that for an arbitrary angle of incidence, an incident plane wave gives rise to an infinite number of reflected and transmitted longitudinal and transverse wave modes. The higher-order modes have cut-off frequencies, below which these modes are evanescent. Below the first cut-off frequency only the zeroth-order modes are propagating modes, and these modes correspond to reflected and transmitted homogeneous plane waves.

The analysis of Achenbach and Kitahara (1986) exploits the periodicity of an array of spherical cavities. Achenbach *et al.* (1988) generalized that formulation to a periodic array of identical inhomogeneities that are of arbitrary shape. The reflection and transmission problem was formulated rigorously. An infinite number of reflected and transmitted wavemodes was again identified, each mode with its own cut-off frequency. Reflection and transmission coefficients were defined in terms of the coefficients of the zeroth-order wavemodes. These coefficients were computed by application of the reciprocity relation. They were expressed as integrals over the surface of a single inhomogeneity, in terms of the field variables on that surface. The fields on the surface of the inhomogeneity are governed by a system of boundary integral equations, which were derived in some detail. This system was solved numerically by the use of the boundary element method. Numerical results were presented for spherical cavities and spherical inclusions.

The interaction of an incident wave with a single layer of cracks was considered by Angel and Achenbach (1985) for a two-dimensional configuration of equally spaced collinear cracks. A more general formulation for an array of inclined cracks was provided by Mikata and Achenbach (1988). A brief summary and some additional results are included in a paper by Achenbach *et al.* (1988).

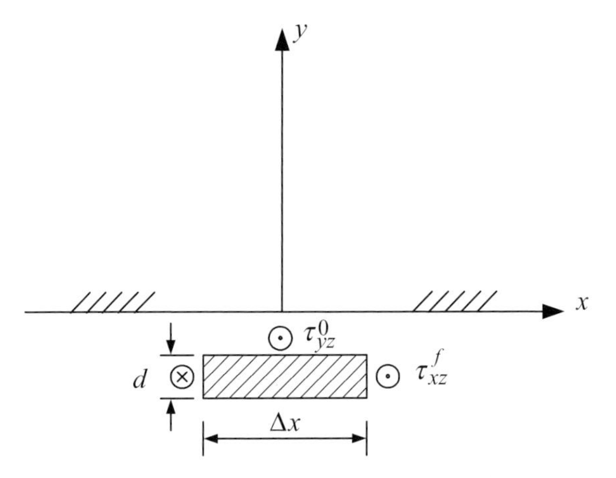

Figure 12.3 Inhomogeneous half-space with thin-film element.

12.3 Thin film on an inhomogeneous substrate

We consider the configuration of a thin film perfectly adhering to an inhomogeneous substrate. Figure 12.3 shows the free-body diagram of an element of the film. The case of anti-plane strain, which only involves out-of-plane displacements $w(x, y, t)$, is considered in this section. The displacement and the stresses of the thin film are identified by a superscript f.

The first order of business is the determination of appropriate solutions for the displacement and the stresses of the substrate. The relevant equations are

$$\frac{\partial \tau_{xz}}{\partial x} + \frac{\partial \tau_{yz}}{\partial y} = \rho \frac{\partial^2 w}{\partial t^2}, \tag{12.3.1}$$

$$\tau_{xz} = \mu \frac{\partial w}{\partial x}, \tag{12.3.2}$$

$$\tau_{yz} = \overline{\mu}(y) \frac{\partial w}{\partial y}. \tag{12.3.3}$$

It is noted that the substrate is homogeneous in the x-direction, with shear modulus μ, but inhomogeneous in the y-direction with shear modulus $\overline{\mu}(y)$. Substitution of (12.3.2) and (12.3.3) into (12.3.1) yields

$$\mu \frac{\partial^2 w}{\partial x^2} + \frac{\partial}{\partial y} \left[\overline{\mu}(y) \frac{\partial w}{\partial y} \right] = \rho \frac{\partial^2 w}{\partial t^2}. \tag{12.3.4}$$

Now we consider a time-harmonic displacement wave propagating in the x-direction, with an as yet unknown variation in the y-direction:

$$w = Wf(y)\, e^{i(kx-\omega t)}, \qquad \text{where} \quad f(0) = 1. \tag{12.3.5}$$

Here k is a wavenumber that is to be determined as part of the analysis. Substitution of this expression for the displacement into the equation of motion, Eq. (12.3.4), yields

$$\frac{d}{dy}\left[\overline{\mu}(y)\frac{df}{dy}\right] - (k^2\mu - \rho\omega^2) f(y) = 0 \tag{12.3.6}$$

or

$$\overline{\mu}(y)\frac{d^2 f}{dy^2} + \frac{d\overline{\mu}}{dy}\frac{df}{dy} - (k^2\mu - \rho\omega^2) f = 0. \tag{12.3.7}$$

This equation does not seem to have an analytical solution $f(y)$ for any reasonable choice of $\overline{\mu}(y)$. An approximate solution can be obtained by replacing $\overline{\mu}(y)$ and $d\overline{\mu}(y)/dy$ by their values at $y = 0$, μ_0 and $-\mu_1$, respectively. The relevant solution of Eq. (12.3.7) can then be written as

$$f(y) = e^{-\alpha y}, \tag{12.3.8}$$

where

$$\alpha = -\frac{1}{2}\frac{\mu_1}{\mu_0} + \frac{1}{2\mu_0}\left[\mu_1^2 + 4\mu_0(k^2\mu - \rho\omega^2)\right]^{1/2}. \tag{12.3.9}$$

The approximation used here might apply near $y = 0$ for the case

$$\overline{\mu}(y) = \mu_0 e^{-(\mu_1/\mu_0)y}. \tag{12.3.10}$$

For a surface $y = 0$ free of surface tractions the boundary condition $\tau_{yz}(0) = 0$ becomes

$$\alpha = 0. \tag{12.3.11}$$

This equation implies that for a half-space free of surface tractions, an anti-plane surface wave of the form (12.3.5) is not possible.

It remains to obtain an equation for the wavenumber k when the half-space is covered by a thin film. This can be done by considering appropriate conditions at the interface of the film and the substrate. Since the film is very thin, an element of the film is taken to consist of its complete thickness, d, in the y-direction, and a length Δx in the x-direction. On the basis of Eqs. (12.3.5) and (12.3.8), continuity of displacement then requires

$$w^f = w(x, 0, t) = We^{i(kx-\omega t)}. \tag{12.3.12}$$

The free-body diagram of the film element, shown in Fig. 12.3, then requires the following equation of motion for the film element:

$$\frac{\partial \tau_{xz}^f}{\partial x} d + \tau_{yz}^0 = (\rho^f d) \frac{\partial^2 w^f}{\partial t^2}, \tag{12.3.13}$$

where ρ^f and d are the film's mass density and thickness, respectively, and τ_{yz}^0 follows from (12.3.3) as

$$\tau_{yz}^0 = -\alpha \mu_0 W e^{i(kx-ct)}. \tag{12.3.14}$$

Equation (12.3.13) then yields

$$(\rho^f \omega^2 - k^2 \mu^f) d + \tfrac{1}{2}\mu_1 - \tfrac{1}{2}\left[\mu_1^2 + 4\mu_0(k^2\mu - \rho\omega^2)\right]^{1/2} = 0, \tag{12.3.15}$$

where u^f is the film's shear modulus. Equation (12.3.15) is a dispersion equation that will yield k as a function of ω.

12.4 Scattering by a disbond

A surface wave incident on a disbond of length $2l$ between a thin film and a substrate generates a reflected and a transmitted surface wave. Leaving out the term $\exp(-i\omega t)$ these three waves may be represented by

$$x < -l, \text{ incident:} \qquad w^{\text{in}} = Ae^{-\alpha y + ikx}, \tag{12.4.1}$$

$$x < -l, \text{ reflected:} \qquad w^{\text{re}} = RAe^{-\alpha y - ikx}, \tag{12.4.2}$$

$$x > l, \text{ transmitted:} \qquad w^{\text{tr}} = TAe^{-\alpha y + ikx}. \tag{12.4.3}$$

State A is defined by

$$w^A = w^{\text{in}} + w^{\text{re}} + w^{\text{tr}},$$

where the ranges of validity of the three terms are defined by Eqs. (12.4.1)–(12.4.3). State B, the virtual wave, is taken as if there is no disbond:

$$w^B = e^{-\alpha y + ikx}, \tag{12.4.4}$$

$$-\infty < x < \infty: \qquad \tau_{xz}^B = ik\mu e^{-\alpha y + ikx}, \tag{12.4.5}$$

$$\tau_{yz}^B = -\alpha \overline{\mu}(y) e^{-\alpha y + ikx}. \tag{12.4.6}$$

The reciprocity relation is of the form

$$-\int_0^\infty F_{AB}\big|_{x=-a} dy - \int_{-l}^{l} G_{AB}\big|_{y=0} dx + \int_0^\infty F_{AB}\big|_{x=b} dy = 0, \tag{12.4.7}$$

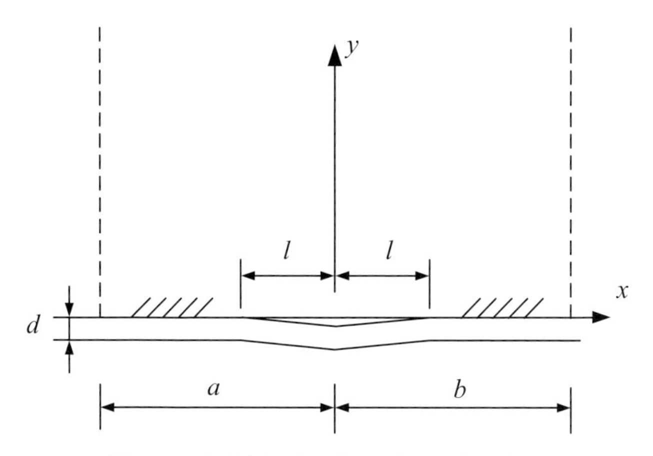

Figure 12.4 Disbond at film–substrate interface.

where

$$F_{AB} = w^A \tau_{xz}^B - w^B \tau_{xz}^A, \tag{12.4.8}$$

$$G_{AB} = w^A \tau_{yz}^B. \tag{12.4.9}$$

Figure 12.4 shows the contour with integrations along $x = -a$, $\infty > y \geq 0$ and $x = b$, $0 \leq y < \infty$. The part of the contour at $y \to \infty$, $-a \leq x \leq b$, is left out since the surface waves have exponentially decayed. Except along the disbond part, $-l \leq x \leq l$, the contribution along $y = 0$ vanishes because $\partial w^A/\partial y$ has the same relation to w^A as $\partial w^B/\partial y$ to w^B. Along $-l \leq x \leq l$ the shear traction τ_{yz}^A vanishes, but w^A is unknown and must be calculated separately. Both τ_{yz}^B and w^B are known along the disbond. The disbond also generates body waves of anti-plane shear, but they decay and are not included in the present considerations.

Just as in other cases that have been considered, the integrations over states A and B vanish when the waves propagate in the same direction. Thus only the counter-propagating waves, i.e., the virtual wave combined with the reflected wave, provide a contribution to the first integral in Eq. (12.4.7). The third integral does not yield a contribution to Eq. (12.4.7). After some manipulation we obtain

$$R = \frac{\alpha \mu_0}{2ik\mu A} \frac{\int_{-l}^{l} w^A(x)\, e^{ikx} dx}{\int_0^\infty e^{-2\alpha y} dy}, \tag{12.4.10}$$

where α is defined by Eq. (12.3.9). This expression for the reflection coefficient is valid for a specified frequency. The value of k corresponding to that frequency must be determined from the dispersion equation given by Eq. (12.3.15).

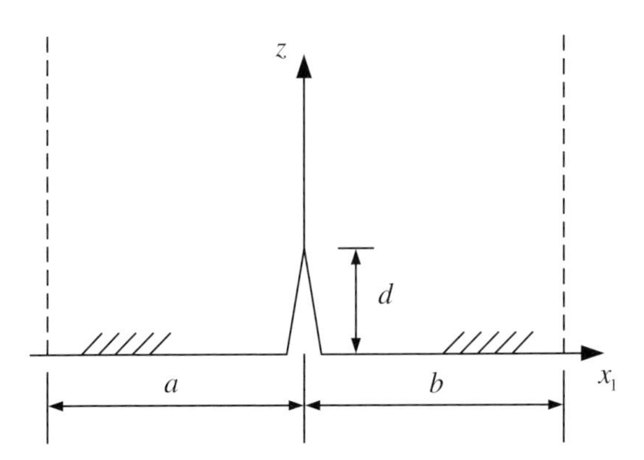

Figure 12.5 Surface-breaking crack in a homogeneous half-space.

12.5 Interaction of a Rayleigh surface wave with a crack

Surface waves are often used to detect surface-breaking cracks and near-surface defects. The principle is simple: a transducer produces a surface wave, which interacts with the defect and is reflected as a surface-wave signal to the transducer or transmitted across the defect to a second transducer. The reflected and/or transmitted signal indicates that there is a defect, and it often also provides information on the type and the size of the defect.

In this section we consider a two-dimensional (plane-strain) situation. The configuration of a surface-breaking crack of depth d is indicated in Fig. 12.5. In conjunction with the time factor $\exp(-i\omega t)$, the relevant solutions of Eq. (7.2.6) for $\varphi(x_1, x_2)$ are

$$\varphi(x_1) = e^{ik_R x_1} \tag{12.5.1}$$

for surface waves propagating in the positive x_1-direction, such as the incident and the transmitted waves, and

$$\varphi(x_1) = e^{-ik_R x_1} \tag{12.5.2}$$

for propagation in the negative x_1-direction, such as for the reflected wave. Here $k_R = \omega/c_R$, where c_R is the solution of Eq. (7.3.8), i.e., the phase velocity of Rayleigh surface waves.

As shown in Section 7.3, Eqs. (7.3.9) and (7.3.10), the displacements may be written as

$$u_1(x_1, z) = \frac{A}{k_R} V^R(z) \frac{d\varphi}{dx_1}, \tag{12.5.3}$$

$$u_z(x_1, z) = A W^R(z) \varphi(x_1), \tag{12.5.4}$$

while

$$\tau_{11}(x_1, z) = AT_{11}^R(z)\varphi(x_1), \tag{12.5.5}$$

$$\tau_{1z}(x_1, z) = AT_{1z}^R(z)\frac{1}{k_R}\frac{d\varphi}{dx_1}. \tag{12.5.6}$$

In these expressions $\exp(-i\omega t)$ has been omitted, and $V^R(z)$, $W^R(z)$, $T_{11}^R(z)$, $T_{1z}^R(z)$ are given by Eqs. (7.3.12), (7.3.13) and (7.3.19), (7.3.20).

State A is defined by the sum of the incident, reflected and transmitted waves. For state B, the virtual wave, we take a wave propagating in the positive x_1-direction of the form given by Eqs. (12.5.3) and (12.5.4), with $\varphi(x_1)$ defined by Eq. (12.5.1).

Since the surface $z = 0$ is free of surface tractions and since the surface waves vanish as $z \to \infty$, the reciprocity relation becomes

$$-\int_0^\infty F_{AB}\Big|_{x=-a} dz - \int_0^d G_{AB}\Big|_{x=0} dz + \int_0^\infty F_{AB}\Big|_{x=b} dz = 0, \tag{12.5.7}$$

where we refer to Fig. 12.5 for the configuration, and

$$F_{AB}(z) = \tau_{11}^B u_1^A + \tau_{1z}^B u_z^A - \tau_{11}^A u_1^B - \tau_{1z}^A u_z^B, \tag{12.5.8}$$

$$G_{AB}(z) = \tau_{11}^B \Delta u_1^A + \tau_{1z}^B \Delta u_z^A. \tag{12.5.9}$$

In Eq. (12.5.9),

$$\Delta u_1^A = u_1^A\Big|_{x_1=0^+} - u_1^A\Big|_{x_1=0^-},$$

with an analogous expression for the crack-opening displacement $\Delta u_z(z)$.

Just as in previous cases, only the counter-propagating waves provide a contribution to Eq. (12.5.7). This means that the combination of the virtual wave and the reflected wave provides the only contribution to the integration at $x_1 = -a$, while there is no contribution from the integration at $x_1 = b$. For the reflected wave we have

$$u_1^A(x_1, z) = -iRAV^R(z)\,e^{-ik_Rx_1},$$

$$u_z^A(x_1, z) = RAW^R(z)\,e^{-ik_Rx_1},$$

$$\tau_{11}^A(x_1, z) = RAT_{11}^R(z)\,e^{-ik_Rx_1},$$

$$\tau_{1z}^A(x_1, z) = -iRAT_{1z}^R(z)\,e^{-ik_Rx_1}.$$

The corresponding expressions for the virtual wave are

$$u_1^B(x_1, z) = iBV^R(z)\,e^{ik_Rx_1},$$

$$u_z^B(x_1, z) = BW^R(z)\,e^{ik_Rx_1},$$

$$\tau_{11}^B(x_1, z) = BT_{11}^R(z)\,e^{ik_Rx_1},$$

$$\tau_{1z}^B(x_1, z) = iBT_{1z}^R(z)\,e^{ik_Rx_1}.$$

Substitution of these results into the first integral of Eq. (12.5.7) yields

$$\int_0^\infty F_{AB}\Big|_{x=-a} dz = -2iRA\int_0^\infty \left[T_{11}^R(z)V^R(z) - T_{1z}^R(z)W^R(z)\right]dz. \tag{12.5.10}$$

After a similar evaluation of the second term in Eq. (12.5.7) it then follows from Eq. (12.5.7) that

$$R = -\frac{i}{2A}\frac{\int_0^d \left[T_{11}^R(z)\Delta u_1^A(z) + iT_{1z}^R(z)\Delta u_z^A(z)\right]dz}{\int_0^\infty \left[T_{11}^R(z)V^R(z) - T_{1z}^R(z)W^R(z)\right]dz}. \tag{12.5.11}$$

12.6 Scattering matrix

We again consider a solid body of volume V bounded by a surface S. All field quantities in V exhibit a time dependence $\exp(-i\omega t)$, which is omitted. The Betti–Rayleigh reciprocity relation connects the stress components τ_{ij} and the displacements u_i of the two states A and B.

Now let us consider the case where two ultrasonic contact transducers are placed on S, as shown in Fig. 12.6. These transducers can act both as transmitters and receivers of ultrasound. The contact surfaces between the transducers and the body are S_1 and S_2, respectively. The remaining part of S is denoted by S_e. Thus

$$S = S_1 + S_2 + S_e. \tag{12.6.1}$$

First, the transducer used for the transmission of energy into the solid is considered. The displacement at the transducer–specimen interface may be taken as

$$u_i^{(T)} = A_T u_i^{(0)}, \tag{12.6.2}$$

where A_T is an amplitude constant for transmission and $u_i^{(0)}$ is a reference displacement. If the transmission into the specimen is locally assumed to be plane-wave, it must be of the form $\exp(-ikn)$ since the normal is directed outward and the time dependence is taken as $\exp(-i\omega t)$. The corresponding traction on the specimen surface is then

$$\tau_{ij}^T n_j = -A_T \tau_{ij}^{(0)} n_j, \tag{12.6.3}$$

where, as before, n_j is the outward normal to S and $\tau_{ij}^{(0)}$ is the reference stress corresponding to $u_i^{(0)}$.

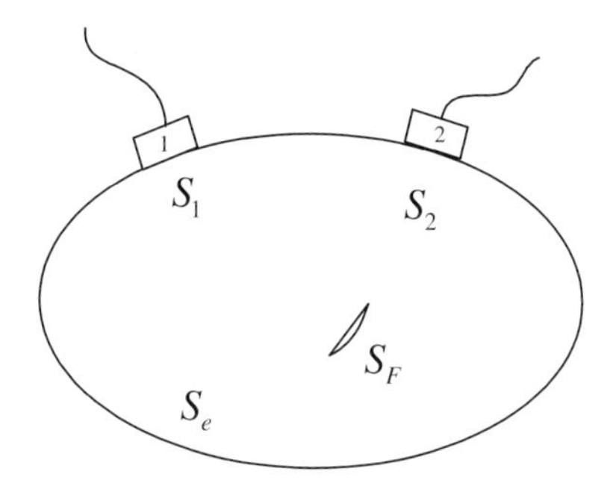

Figure 12.6 Bounded body with two transducers and a flaw.

Next let us consider the case where the transducer is used as a receiver. At the transducer–specimen interface we have

$$u_i^R = A_R u_i^{(0)}. \tag{12.6.4}$$

For this case the energy is propagating out of the body, and hence locally the displacement is of the form $\exp(ikn)$. It follows that

$$\tau_{ij}^R n_j = A_R \tau_{ij}^{(0)} n_j. \tag{12.6.5}$$

Next we introduce the concept of the scattering matrix, similarly to Auld (1979) and Kino (1978). The scattering matrix relates the outputs, A_{1R} and A_{2R} at transducers 1 and 2, to the inputs, A_{1T} and A_{2T}, at the two transducers:

$$\begin{bmatrix} A_{1R} \\ A_{2R} \end{bmatrix} = \begin{bmatrix} M_{11} & M_{12} \\ M_{21} & M_{22} \end{bmatrix} \begin{bmatrix} A_{1T} \\ A_{2T} \end{bmatrix}. \tag{12.6.6}$$

In the coefficients $M_{\alpha\beta}$ of the scattering matrix $[M]$ the subscript α defines the receiver and β the transmitter. A simple relationship exists between M_{12} and M_{21}. To derive this relationship we consider two states A and B:

state A: transducer 1 transmits, transducer 2 receives;

state B: transducer 2 transmits, transducer 1 receives.

For both cases the surface S is free of tractions, except on the areas of contact with the transducers.

For state A, Eq. (12.6.6) reduces to

$$\begin{bmatrix} A_{1R}^A \\ A_{2R}^A \end{bmatrix} = \begin{bmatrix} M_{11} & M_{12} \\ M_{21} & M_{22} \end{bmatrix} \begin{bmatrix} A_{1T}^A \\ 0 \end{bmatrix} \tag{12.6.7}$$

or

$$A_{2R}^A = M_{21} A_{1T}^A. \tag{12.6.8}$$

Analogously, we obtain for state B

$$A_{1R}^{B} = M_{12} A_{2T}^{B}. \tag{12.6.9}$$

The two elastodynamic states are now entered in the reciprocal identity, with the surface S subdivided according to Eq. (12.6.1). Since S_e is free of surface tractions, the integral vanishes over that area and the reciprocal identity yields

$$\int_{S_1} \left(\tau_{ij}^{A} u_i^{B} - \tau_{ij}^{B} u_i^{A}\right) n_j \, dS + \int_{S_2} \left(\tau_{ij}^{A} u_i^{B} - \tau_{ij}^{B} u_i^{A}\right) n_j \, dS = 0. \tag{12.6.10}$$

By the use of Eqs. (12.6.2)–(12.6.9) we obtain the following relations:

$$\text{on } S_1: \quad u_i^{A} = A_{1T}^{A} u_i^{(0)}, \quad \tau_{ij}^{A} n_j = -A_{1T}^{A} \tau_{ij}^{(0)} n_j,$$
$$u_i^{B} = A_{1R}^{B} u_i^{(0)}, \quad \tau_{ij}^{B} n_j = A_{1R}^{B} \tau_{ij}^{(0)} n_j;$$

$$\text{on } S_2: \quad u_i^{A} = A_{2R}^{A} u_i^{(0)}, \quad \tau_{ij}^{A} n_j = A_{2R}^{A} \tau_{ij}^{(0)} n_j,$$
$$u_i^{B} = A_{2T}^{B} u_i^{(0)}, \quad \tau_{ij}^{B} n_j = -A_{2T}^{B} \tau_{ij}^{(0)} n_j.$$

Substitution of these relations in Eq. (12.6.10) yields

$$A_{1T}^{A} A_{1R}^{B} I_1 - A_{2R}^{A} A_{2T}^{B} I_2 = 0, \tag{12.6.11}$$

where

$$I_\alpha = \int_{S_\alpha} \tau_{ij}^{(0)} u_i^{(0)} n_j \, dS, \qquad \alpha = 1, 2. \tag{12.6.12}$$

Subsequent use of Eqs. (12.6.8) and (12.6.9) yields

$$M_{12} A_{1T}^{A} A_{2T}^{B} I_1 - M_{21} A_{1T}^{A} A_{2T}^{B} I_2 = 0,$$

or

$$M_{12} I_1 = M_{21} I_2. \tag{12.6.13}$$

For the case where the two transducers are exactly the same, i.e., $I_1 = I_2$, Eq. (12.6.13) further reduces to

$$M_{12} = M_{21}. \tag{12.6.14}$$

12.7 Transducer response to scattering by a flaw

If the body of volume V has both an external surface, S, and an internal surface, S_F, as shown in Fig. 12.6, the reciprocity relation for the volume V bounded by both S and S_F becomes

$$\int_S \left(\tau_{ij}^A u_i^B - \tau_{ij}^B u_i^A\right) n_j \, dS - \int_{S_F} \left(\tau_{ij}^A u_i^B - \tau_{ij}^B u_i^A\right) n_j \, dS = 0, \tag{12.7.1}$$

where the normal $\boldsymbol{n}$ on S_F is taken as into the solid body. Again two elastodynamic states are considered:

state A: transducer 1 transmits, transducer 2 receives and the flaw is present, i.e., $\tau_{ij}^{1A} n_j = 0$ on S_F;

state B: transducer 2 transmits, transducer 1 receives and no flaw is present, i.e., $\tau_{ij}^{2B} n_j \neq 0$ on S_F.

It should be noted that a superscript 1 or 2 has been added to identify the transmitting transducer. Equations (12.6.7), (12.6.8) again apply, and we may write

$$A_{1T}^A A_{1R}^B I_1 - A_{2R}^A A_{2T}^B I_2 = -\int_{S_F} \tau_{ij}^{2B} u_i^{1A} n_j \, dS, \tag{12.7.2}$$

where it has been taken into account that $\tau_{ij}^{1A} n_j = 0$ on S_F. Since there is a further distinction between states A and B with regard to the conditions on S_F, we rewrite (12.6.8) and (12.6.9) as

$$A_{2R}^B = M_{21}^A A_{1T}^A \qquad \text{and} \qquad A_{1R}^B = M_{12}^B A_{2T}^B. \tag{12.7.3}$$

Substitution of these results into (12.7.2) yields

$$M_{12}^B I_1 - M_{21}^A I_2 = \frac{-1}{A_{1T}^A A_{2T}^B} \int_{S_F} \tau_{ij}^{2B} u_i^{1A} n_j \, dS. \tag{12.7.4}$$

To discuss Eq. (12.7.4) further, we recall that state A is for a body with a flaw and state B for a body without a flaw. By virtue of linear superposition we may write

$$M_{21}^A = M_{21}^{NF} + M_{21}^F, \tag{12.7.5}$$

where the superscripts F and NF indicate that the body contains or does not contain a flaw. By Eq. (12.6.13) we then have

$$M_{21}^{NF} I_2 = M_{12}^{NF} I_1 = M_{12}^B I_1. \tag{12.7.6}$$

It follows that

$$M_{12}^B I_1 - M_{21}^A I_2 = M_{12}^B I_1 - M_{21}^{NF} I_2 - M_{21}^F I_2 = -M_{21}^F I_2.$$

Substitution of this result into Eq. (12.7.4) gives

$$M_{21}^F = \frac{1}{A_{1T}^A A_{2T}^B I_2} \int_{S_F} \tau_{ij}^{2B} u_i^{1A} n_j \, dS. \tag{12.7.7}$$

By interchanging transducers 1 and 2 in states A and B we find analogously

$$M_{12}^F = \frac{1}{A_{2T}^A A_{1T}^B I_1} \int_{S_F} \tau_{ij}^{1B} u_i^{2A} n_j \, dS. \tag{12.7.8}$$

For pulse-echo situations, where a single transducer both transmits and receives, we find analogously

$$M_{11}^F = \frac{1}{A_{1T}^A A_{1T}^B I_1} \int_{S_F} \tau_{ij}^{1A} u_i^{1A} n_j \, dS, \tag{12.7.9}$$

$$M_{22}^F = \frac{1}{A_{2T}^B A_{2T}^B I_2} \int_{S_F} \tau_{ij}^{2B} u_i^{2B} n_j \, dS. \tag{12.7.10}$$

The expressions for the scattering coefficients given by Eqs. (12.7.7)–(12.7.10) have the disadvantage that they are in terms of properties of the transducers which are not a priori available and which are not necessarily easily and accurately obtainable by calibration procedures. A significant simplification can be obtained by considering the following ratio:

$$C = \frac{M_{11}^F \, M_{22}^F}{M_{12}^F \, M_{21}^F}. \tag{12.7.11}$$

It is easily verified that all terms appearing in the denominators of Eqs. (12.7.7)–(12.7.10) cancel out in this ratio, and only the integrals containing τ_{ij}^{1A}, u_i^{1A} etc. are maintained. The ratio of these integrals can be simplified further by the observation that for small flaws the ultrasonic wave incident on the flaw can be assumed to be locally very well approximated by a plane wave. Suppose that the amplitudes of these plane waves at the crack locations are denoted by P^{1A}, P^{1B}, P^{2A} and P^{2B}. It then follows that the stresses and displacements may be written as

$$\tau_{ij}^{1A} = P^{1A} \overline{\tau}_{ij}^{1A}, \qquad u_i^{1A} = P^{1A} \overline{u}_j^{A}, \tag{12.7.12}$$

where $\overline{\tau}_{ij}^{(1)}$ and $\overline{u}_i^{(1)}$ are the stress and displacement components due to a plane incident wave of unit amplitude from the direction of transducer 1. Expressions analogous to Eq. (12.7.12) can be written for the other stress and displacement components in Eqs. (12.7.7)–(12.7.10). It is then easily verified that the constant amplitude factors P^{1A} etc. will also cancel out in Eq. (12.7.11), to yield the following result:

$$C = \frac{\int_{S_F} \overline{\tau}^1_{ij}\overline{u}^1_i n_j \, dS \times \int_{S_F} \overline{\tau}^2_{ij}\overline{u}^2_i n_j \, dS}{\int_{S_F} \overline{\tau}^1_{ij}\overline{u}^2_i n_j \, dS \times \int_{S_F} \overline{\tau}^2_{ij}\overline{u}^1_i n_j \, dS}. \tag{12.7.13}$$

Here $\overline{\tau}^1_{ij}$ are the stress components at the location of the flaw in the absence of the flaw and $\overline{u}^1_i$ is the displacement on the flaw's surface, both due to an incident plane wave coming from the direction of transducer 1. Similar interpretations hold for the other stress and displacement components in Eq. (12.7.13). The expression for the self-compensated scattering factor C requires the results of two pulse-echo and two pitch–catch measurements, but it has the significant advantage that C is independent of the transducer properties and the coupling of the transducers to the sample. Clearly C is just as representative of the geometrical aspects of the flaw as a single scattering coefficient. The theoretical scattering coefficients $M_{\alpha\beta}$ and the ratio C still have to be connected to the measured voltages. This connection is provided by relations of the type

$$V_{21} = S_2 M^F_{21} A_{1T}, \tag{12.7.14}$$

where V_{21} is the voltage measured at transducer 2 due to excitation at transducer 1, and $S_2(\omega)$ is a response function for transducer 2 that relates voltage to displacement. Unfortunately, $S_2(\omega)$ depends significantly on the coupling between the transducer and the body. This coupling is generally achieved with a thin layer of oil, whose coupling properties are difficult to determine quantitatively since transducers are often moved about in order to obtain the best measurement. To avoid these difficulties with coupling, a self-compensating technique has been introduced which involves the combination of four measurements, similar to that in Eq. (12.7.11). Analogously to Eq. (12.7.14) we can write

$$\begin{aligned} V_{12} &= S_1 M^F_{12} A_{2T}, \\ V_{11} &= S_1 M^F_{11} A_{1T}, \\ V_{22} &= S_2 M^F_{22} A_{2T}. \end{aligned}$$

It follows that

$$\frac{V_{11}\,V_{22}}{V_{12}\,V_{21}} = \frac{M^F_{11}\,M^F_{22}}{M^F_{12}\,M^F_{21}} = C. \tag{12.7.15}$$

We note that the response functions have cancelled out and that Eq. (12.7.15) provides direct information on the flaw.

13
Reciprocity for coupled acousto-elastic systems

13.1 Introduction

In many applications, waves in an acoustic medium such as water are coupled to wave motion in submerged elastic bodies. There are examples in the area of structural acoustics, which is concerned with the generation of sound in a surrounding acoustic medium by time-variable forces in submerged structures as well as with the detection of submerged bodies by the scattering of incident sound waves. Structural acoustics has been discussed in considerable detail in books by Junger and Feit (1972), Fahy (1985), and Cremer, Heckl and Ungar (1973). Another class of coupled acousto-elastic systems is defined by seismic problems in an oceanic environment, where acoustic waves are used to probe the geological strata under the ocean floor.

In this chapter and the next we distinguish the coupling of wave phenomena as either configurational or physical. For the configurational case, coupling comes about because bodies of different constitutive behaviors are in contact, as is the case for acousto-elastic systems, as described above. However, for physical coupling the wave interaction takes place in the same body when different physical phenomena are coupled by the constitutive equations. That is the case for electromagneto-elastic coupling, specifically piezoelectricity, which will be discussed in the next chapter.

Reciprocity for acousto-elastic systems was already anticipated by Rayleigh (1873), when in his statement of the reciprocity theorem for acoustics, see Chapter 4, he allowed for a space filled with air that is partly bounded by finitely extended fixed bodies.

After a general statement of reciprocity for configurationally coupled systems in the next section, reciprocity between fields generated by sources in the acoustic and elastic regions is discussed in Section 13.3. Of particular interest is

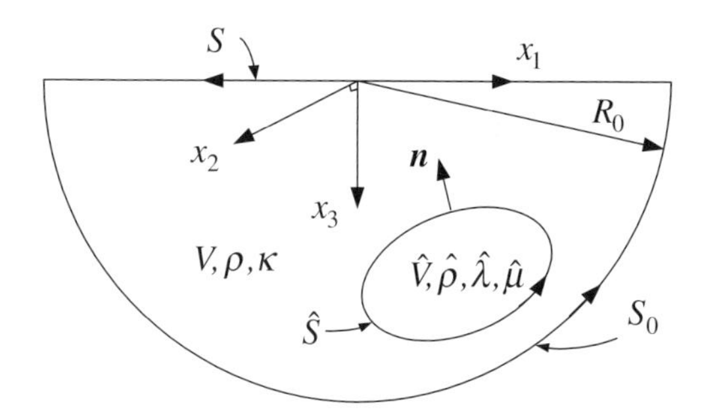

Figure 13.1 Compact inhomogeneity in an acoustic half-space.

reciprocity for free-field sources and sources located on rigid and elastic baffles, which is discussed in Sections 13.5 and 13.7.

13.2 Reciprocity for configurationally coupled systems

The configuration that will be considered consists of an acoustic medium in a hemispherical domain D of radius R_0, located in the half-space $x_3 \geq 0$. The domain contains a compact elastic inhomogeneity defined by $\hat{V}$, with boundary $\hat{S}$. The geometry is shown in Fig. 13.1. In the domain $V = D - \hat{V}$ the acoustic medium is homogeneous with mass density ρ and compressibility κ. The compact inhomogeneity is itself also homogeneous with mass density $\hat{\rho}$ and elastic constants $\hat{\lambda}$ and $\hat{\mu}$. There may be source distributions f_i and q in V and body forces $\hat{f}_i$ in $\hat{V}$.

For steady-state time-harmonic solutions we now write the acoustic reciprocity theorem for the domain $V = D - \hat{V}$. For two time-harmonic states of the inhomogeneous configuration, we have, analogously to Eq. (4.4.1),

$$\int_{S+S_0} (p^A v_i^B - p^B v_i^A) n_i \, dS - \int_{\hat{S}} (p^A v_i^B - p^B v_i^A) n_i \, dS$$
$$= \int_V \left(f_i^A v_i^B + q^B p^A - f_i^B v_i^A - q^A p^B \right) dV. \qquad (13.2.1)$$

The minus sign in front of the integral over $\hat{S}$ occurs because of the choice of clockwise integration with the normal pointing into V, as shown in Fig. 13.1. For the domain $\hat{V}$ the usual elastodynamic reciprocity theorem yields, see

Eq. (6.2.8),

$$\int_{\hat{S}} \left(\hat{\tau}_{ij}^B \hat{v}_j^A - \hat{\tau}_{ij}^A \hat{v}_j^B\right) n_i d\hat{S} = \int_{\hat{V}} \left(\hat{f}_j^A \hat{v}_j^B - \hat{f}_j^B \hat{v}_j^A\right) d\hat{V}, \qquad (13.2.2)$$

where the carets define field quantities in $\hat{V}$ and on $\hat{S}$. Note that in the reciprocity theorem (6.2.8) we have replaced the displacements by the particle velocities. The theorem is for the time-harmonic case, and thus the terms in Eqs. (13.2.1) and (13.2.2) depend only on position $\boldsymbol{x}$.

Next we consider the interface conditions on $\hat{S}$. They are that the normal traction is continuous, but the shear tractions vanish since the acoustic medium is inviscid. In addition the normal displacement is continuous. Thus, for both states A and B we have on $\hat{S}$

$$-p = \hat{t}_n, \qquad (13.2.3)$$

$$t_{ns} = t_{nt} = 0, \qquad (13.2.4)$$

$$\frac{1}{i\omega\rho_0}\frac{\partial p}{\partial n} = \hat{v}_n. \qquad (13.2.5)$$

Here $\hat{t}_n$, $\hat{t}_{ns}$ and $\hat{t}_{nt}$ are tractions, and $\hat{v}_n$ is the normal velocity in the elastic solid relative to an orthogonal coordinate system whose *st*-plane is locally tangential to the inhomogeneity; $\boldsymbol{n}$ is the outward normal. The expression for the particle velocity in terms of the pressure follows by combining Eqs. (2.6.25) and (2.6.15).

By virtue of Eq. (13.2.4), the reciprocity relation for the domain $\hat{V}$, Eq. (13.2.2), may be rewritten as

$$\int_{\hat{S}} \left(\hat{t}_n^A \hat{v}_n^B - \hat{t}_n^B \hat{v}_n^A\right) d\hat{S} = \int_{\hat{V}} \left(\hat{f}_j^A \hat{v}_j^B - \hat{f}_j^B \hat{v}_j^A\right) d\hat{V}. \qquad (13.2.6)$$

Next we subtract this relation from the reciprocity relation for the acoustic medium given by Eq. (13.2.1). Using the interface conditions that $-p = t_n$, Eq. (13.2.3), and that the normal particle velocity is continuous, we find that the integrals over $\hat{S}$ cancel each other to yield the result

$$\int_{S+S_0} \left(p^A v_i^B - p^B v_i^A\right) n_i \, dS = \int_V \left(f_i^A v_i^B + q^B p^A - f_i^B v_i^A - q^A p^B\right) dV$$
$$+ \int_{\hat{V}} \left(\hat{f}_j^A \hat{v}_j^B - \hat{f}_j^B \hat{v}_j^A\right) d\hat{V}. \qquad (13.2.7)$$

This relation is included in a more complicated reciprocity relation given by de Hoop (1995).

13.3 Reciprocity between source solutions

For an unbounded medium, Eq. (13.2.7) can be used to derive reciprocity relations between point-source solutions. If the sources are only in the acoustic medium, Eq. (13.2.7) reduces to Eq. (4.7.3), and the relations will be just the same as the ones presented in Section 4.7. Similarly, if the sources are only in the region of the solid material, the results become the same as those given in Section 6.10 for fields generated by point forces in a solid body.

A case of interest occurs when two source functions, one in the acoustic medium and one in the solid inhomogeneity, are considered. For example, suppose we have

$$q^A(\boldsymbol{x}) = Q^A \delta(\boldsymbol{x} - \boldsymbol{x}^A), \tag{13.3.1}$$

$$\hat{f}_k^B(\boldsymbol{x}) = \hat{F}^B \delta(\boldsymbol{x} - \hat{\boldsymbol{x}}^B) e_k, \tag{13.3.2}$$

while all the other source functions in Eq. (13.2.7) are identically zero. We then have

$$Q^A p^B(\boldsymbol{x}^A - \hat{\boldsymbol{x}}^B) = F^B \hat{v}_k^A(\hat{\boldsymbol{x}}^B - \boldsymbol{x}^A). \tag{13.3.3}$$

Clearly Eq. (13.2.7) can produce a few other relations of the type (13.3.3).

13.4 Scattering by a flaw in a submerged solid body

We return to the acousto-elastic reciprocity relation, but we now consider the case where the solid body contains a crack-like flaw. This case is typical of problems of non-destructive testing by the ultrasonic technique, where the objective is to detect the flaw and if possible to determine its location, size, shape and orientation.

Figure 13.2 shows the geometrical configuration for using the reciprocity relation to calculate the scattering coefficient for a crack. The solid body containing the flaw is placed in a water tank. Two identical transducers are also placed in the tank and used in a pitch–catch (pulse-echo is another case) measurement of the signals scattered by the crack.

In the absence of interior sources in the acoustic medium and the solid, the integration over the external surface of the solid body, $\hat{S}$, will now be balanced by an integral over the flaw surface S_F. The integral over $\hat{S}$, in turn, is balanced by an integral over the surface in the acoustic medium defined by $S_w + S_1 + S_2$.

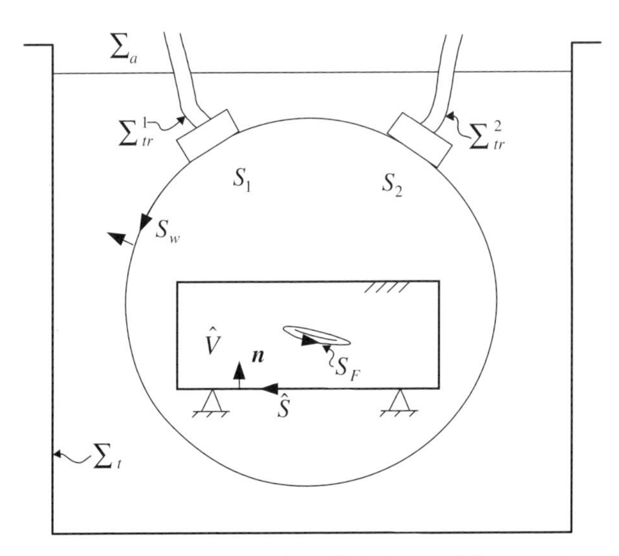

Figure 13.2 Water tank with solid body containing a crack.

Equation (13.2.7) then becomes

$$\int_{S_w+S_1+S_2} \left(p^A u_i^B - p^B u_i^A\right) n_i \, dS = \int_{S_F} \left(\tau_{ij}^A u_i^B - \tau_{ij}^B u_i^A\right) n_j \, dS. \qquad (13.4.1)$$

For an easier connection with the results of Chapter 12 we have replaced the particle velocity v_i by the displacement u_i. This can be done without any problem because the two are related by $v_i = -i\omega u_i$. The surface with the water S_w is indicated in Fig. 13.2; S_1 and S_2 are the faces of the transducers. Other surfaces of interest are Σ_a, the surface between the air and the water, where the pressure may be assumed to vanish; Σ_t, the walls and bottom of the tank where the particle velocity vanishes; and Σ_s^1 and Σ_s^2, the side surfaces of the transducers where the particle velocity also vanishes. For the domain bounded by these surfaces and S_w we have

$$\int_{S_w+\Sigma_s^1+\Sigma_s^2+\Sigma_a+\Sigma_t} (p^A u_i^B - p^B u_i^A) n_i \, dS = 0,$$

since there are no sources in the domain. Because of the conditions on Σ_s^1, Σ_s^2, Σ_a and Σ_t, it follows that the integration over S_w vanishes. Hence the part of Eq. (13.4.1) corresponding to S_w vanishes, and the integral just reduces to integrations over the faces of the two transducers.

Following the development in Section 12.7 we now consider two elastodynamic states:

state A: transducer 1 transmits, transducer 2 receives and the crack is present, i.e., $\tau_{ij}^{1A} n_j = 0$ on S_F,

state B: transducer 2 transmits, transducer 1 receives and no crack is present, i.e., $\tau_{ij}^{2B} n_j \neq 0$ on S_F.

The relevant relations are the same as those presented under Eq. (12.6.10).

For a crack, the scattering coefficient follows from Eq. (12.7.7) as

$$\mathrm{M}_{21}^F = \frac{1}{A_{1T}^A A_{2T}^B I_1} \int_{S_F} \tau_{ij}^{1B} u_i^{\mathrm{sc}} n_j \, dS, \tag{13.4.2}$$

where I is defined by Eq. (12.6.12) but with $\tau_{ij}^{(0)} u_i^{(0)} n_j = -p^0 u_n^0$. Further details for this case when the crack is sonified by a focused transducer can by found in Zhang and Achenbach (1999).

13.5 Reciprocity for free-field sources and sources located on rigid baffles

For the present discussion a baffle is any obstruction that interferes with the unimpeded propagation of sound. Thus a baffle can be a planar element, but equally well a sphere, a cylinder or an obstacle of arbitrary shape. In this section we will be concerned with the presence of sources on rigid baffles and their reciprocity with sources located in the surrounding free space.

To establish the connection with results available in the literature we first slightly reformulate the solution of Eq. (2.6.20). This equation for the pressure perturbation was stated as

$$\frac{1}{c^2} \frac{\partial^2 p}{\partial t^2} - \nabla^2 p = \rho \dot{q}, \tag{13.5.1}$$

where $q(x, t)$ defines the volume source of injection rate, and the prime and subscript on ρ have been omitted for simplicity of notation. We now consider the time-harmonic case and define

$$\rho \dot{q}(\boldsymbol{x}, t) = \rho \ddot{Q} \delta(\boldsymbol{x} - \boldsymbol{x}^A) e^{-i\omega t}. \tag{13.5.2}$$

Here the double dot over Q, and later in this section over W, does not define a conventional second-order time derivative. Rather $\ddot{Q}$ and $\ddot{W}$ are quantities with dimensions length^3/time^2 and length / time^2, respectively. For a point source of magnitude $\rho \ddot{Q}$ at $\boldsymbol{x}^A$, Eq. (13.5.1) now yields

$$p(\boldsymbol{x}) = \frac{\rho \ddot{Q}}{4\pi R} e^{ikR}, \tag{13.5.3}$$

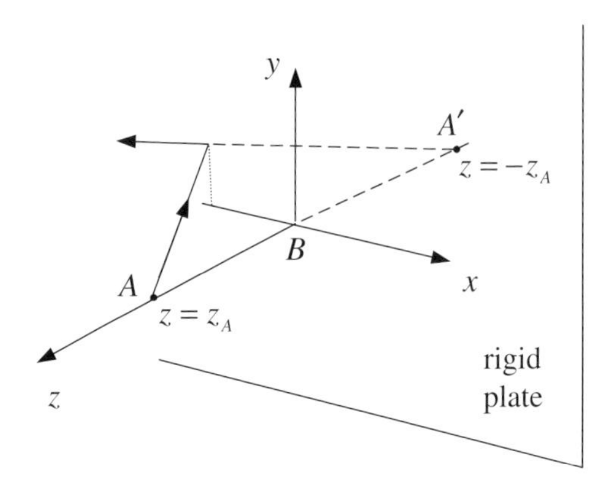

Figure 13.3 Rigid plate with a point source, A, and its image, A'.

where $R = |\boldsymbol{x} - \boldsymbol{x}^A|$. The source function can be thought of as being produced by the surface acceleration of a spherical cavity with a radius small compared with the wavelength. For a given volume acceleration $\ddot{W}$, this approximation is also valid for more general source geometries and for more general acceleration distributions over the source. For a pulsating sphere we obtain the following relation between the constants $\ddot{Q}$ and $\ddot{W}$:

$$\ddot{Q} = 4\pi a^2 \ddot{W}, \tag{13.5.4}$$

where a is the radius of the spherical cavity.

In free space the reciprocity relation for two sources defined by

$$\ddot{Q}^A \delta(\boldsymbol{x} - \boldsymbol{x}^A) \qquad \text{and} \qquad \ddot{Q}^B \delta(\boldsymbol{x} - \boldsymbol{x}^B)$$

follows immediately from the reciprocity relation Eq. (4.7.3) as

$$\ddot{Q}^A p^B(\boldsymbol{x}^A - \boldsymbol{x}^B) = \ddot{Q}^B p^A(\boldsymbol{x}^B - \boldsymbol{x}^A). \tag{13.5.5}$$

Thus, for equal sources the locations of a source and a field point in free space can be interchanged without altering the pressure.

Now let us consider the presence of a rigid baffle, first an infinite rigid plate. The geometry is shown in Figure 13.3. Consider a point source of magnitude $\rho\ddot{Q}$ at point A located at $(x^A = 0, y^A = 0, z^A)$, a distance z^A from the infinite rigid baffle, which is located in the plane $z = 0$. The source generates an incident wave. The presence of the rigid boundary can be simulated by placing a source of equal magnitude, $\rho\ddot{Q}$, at point A' located at $(x^A = 0, y^A = 0, -z^A)$. The combination of the source and its image provides a reflected wave together with conditions of zero particle velocity and a doubling of the pressure on a rigid baffle.

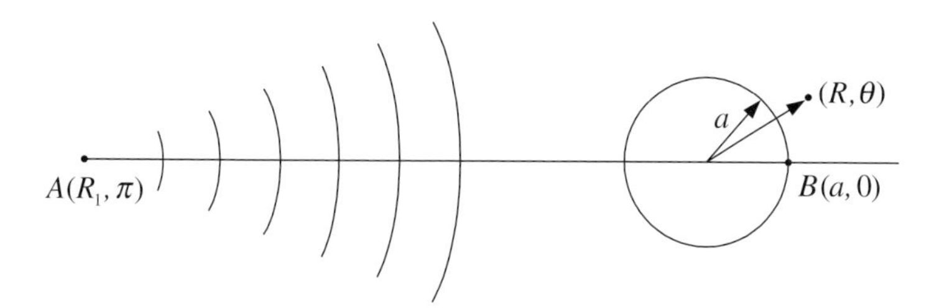

Figure 13.4 Rigid spherical scatterer with sources at A and B.

Now consider a source $\rho\ddot{Q}$ located on the baffle at point $B\,(x=0,\ y=0,\ z=0)$. Since the free-space source at A produces a double pressure at B on the baffle, then, according to reciprocity, a source on the baffle at B should radiate a pressure to A that is twice as large in magnitude as for the source in free space. Thus a rigid baffle doubles the incident pressure signal sensed at a point on the baffle, and it also doubles the pressure signal radiated by a source on the baffle. This result can be explained without the benefit of reciprocity considerations by letting a source and its image approach the rigid baffle along a line normal to the baffle. As the two sources come closer to the baffle, their superposition maintains a zero particle velocity on the rigid baffle as well as a pressure higher than for a single source. In the limit as the two sources reach the rigid baffle from opposite sides, the pressure doubles and a signal of double the magnitude of the single-source signal is radiated by the source on the front side of the baffle.

A more interesting example, that of a spherical baffle, was discussed by Junger and Feit (1972, p. 340). The geometry is shown in Fig. 13.4. A point B is located on spherical baffle of radius a, at a position defined by $\theta = 0$. The far field radiated by this source is given by Junger and Feit (1972, p. 340) as

$$p^B(R_1, \theta = \pi) = \frac{ie^{ikR_1}\rho\ddot{Q}}{4\pi a^2 k^2 R_1}\sum_{n=0}^{\infty}\frac{(2n+1)i^n}{h_n'(ka)}, \qquad kR_1 \gg n^2+1 \quad (13.5.6)$$

where a has been neglected compared with R_1. Here h_n' is the spherical Hankel function of the first kind, also referred to as the spherical Bessel function of the third kind; see Junger and Feit p. 155. The resultant surface pressure produced at the point B at $\theta = 0$ by a plane wave incident from $\theta = \pi$ is also given by Junger and Feit, p. 340, as

$$p(R = a, \theta = 0) = \frac{iP^{\text{in}}}{(ka)^2}\sum_{n=0}^{\infty}\frac{(2n+1)i^n}{h_n'(ka)}. \qquad (13.5.7)$$

It follows from Eq. (13.5.3) that the incident pressure may be written as

$$p^{\text{in}}(a, 0) = \frac{\rho \ddot{Q} e^{ikR_1}}{4\pi R_1}. \tag{13.5.8}$$

Hence in Eq. (13.5.7) we have

$$P^{\text{in}} = \frac{\rho \ddot{Q}}{4\pi R_1} e^{ikR_1}. \tag{13.5.9}$$

Substituting P^{in} into Eq. (13.5.7) we obtain the same expression as Eq. (13.5.6). Hence a source on a rigid spherical baffle at $(a, 0)$ produces a pressure at a field point, located in free space at (R_1, π), that equals, in both amplitude and phase, the pressure generated on the spherical baffle at the field point $(a, 0)$ by a source located in free space at (R_1, π). Junger and Feit (1972) concluded with the following general statement: *Whether scatterers are present or not, the locations of the source and field point can be interchanged without altering the pressure at the field point.*

In general it is easier to determine the pressure field on a scatterer due to a source in free space. The example in this section has illustrated the use of reciprocity to determine the free-field pressure due to a source on a rigid scatterer.

Next we will consider the extension of reciprocity considerations to reflection by elastically supported structures and radiation from vibrating elastic structures.

13.6 Interaction of sound and structural vibrations

In the early years of the development of acoustics, the field was almost exclusively concerned with airborne sound. The effect of the vibrations of structures produced by sound sources in adjoining acoustic media, or of sound in adjoining acoustic media produced by structural vibrations, did not become of interest until early in the twentieth century, during and immediately after World War I. Structural acoustics, which deals with sound–structure interaction problems received, however, ever increasing attention with the advent of submarines, aircraft and the sound-producing machinery of modern times.

In this section and the next we consider problems of sound–structure interaction, with particular interest in reciprocity relations. We first consider the one-dimensional case of the normal incidence of time-harmonic acoustic waves on an elastically supported rigid panel that separates two acoustic media, for example water and air, in a waveguide. The mass density of the acoustic panel is

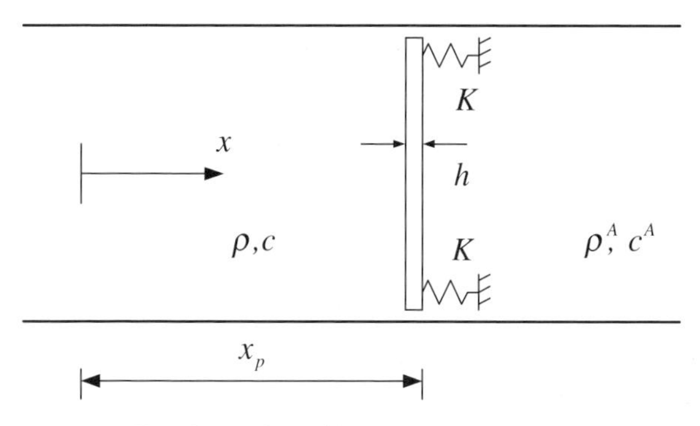

Figure 13.5 Wave interaction with an elastically supported rigid panel.

ρ_p, and the panel is supported by elastic springs of stiffness K per unit area. The geometry is shown in Fig. 13.5. Only the lowest mode, which is like a plane wave in the waveguide, is considered. The panel is located at $x = x_p$. The relevant acoustic properties of the acoustic media are ρ, c for $x < x_p$ and ρ^A, c^A for $x > x_p$, as shown in Fig. 13.5.

We consider an incident wave whose phase is zero at position $x = x_1$, where $x_1 < x_p$. The incident pressure wave may then be expressed as

$$p^{\text{in}} = p_0 e^{ik(x-x_1)}. \tag{13.6.1}$$

Following the usual approach we choose a reflected wave such that the pressure doubles at $x = x_p$:

$$p^{\text{re}} = p_0 e^{ik(x_p-x_1)} e^{-ik(x-x_p)}. \tag{13.6.2}$$

By using either

$$\frac{\partial p}{\partial x} = -\rho \ddot{u} \qquad \text{or} \qquad \frac{\partial p}{\partial x} = \rho\omega^2 u, \tag{13.6.3}$$

we then find that $p^{\text{in}} + p^{\text{re}}$ is commensurate with a vanishing displacement on $x = x_p$, as it is for reflection from a fixed rigid boundary. Naturally the displacement at $x = x_p$ does not actually vanish. The actual displacement is accounted for by the presence of additional pressures on both sides of the plate. These pressures are called radiation pressures, and they are defined by

$$x < x_p: \qquad p^{\text{rad}-} = p^- e^{-ik(x-x_p)}, \tag{13.6.4}$$

$$x > x_p: \qquad p^{\text{rad}+} = p^+ e^{+ik^A(x-x_p)}. \tag{13.6.5}$$

The unknown radiation pressures are obtained from the conditions that the displacement is continuous across $x = x_p$ and that the pressures and elastic forces acting on the plate must satisfy an equation of motion, as shown in what follows.

By virtue of Eq. (13.6.3), displacement continuity yields

$$-\frac{k}{\rho}p^- = \frac{k^A}{\rho^A}p^+. \tag{13.6.6}$$

The equation of motion per unit area becomes

$$2p_0e^{ik(x_p-x_1)} + p^- - p^+ - Ku = h\rho_p(-\omega^2)u, \tag{13.6.7}$$

where h is the thickness of the rigid plate. By using Eq. (13.6.3), the displacement u can be eliminated in favor of p^+. By using Eq. (13.6.6) to eliminate p^- we obtain after some manipulation

$$p^+ = \frac{2k\rho^A p_0 e^{ik(x_p-x_1)}}{k^A\rho + k\rho^A + ikk^A(K/\omega^2 - \rho_p h)}. \tag{13.6.8}$$

From Eqs. (13.6.6) and (13.6.8) we have

$$p^- = -\frac{2k^A\rho p_0 e^{ik(x_p-x_1)}}{k^A\rho + k\rho^A + ikk^A(K/\omega^2 - \rho_p h)}. \tag{13.6.9}$$

The particle velocity at $x = x_2 > x_p$ follows from the relation

$$\frac{\partial p}{\partial x} = i\omega\rho v \tag{13.6.10}$$

and Eq. (13.6.8) as

$$v = -\frac{i}{\omega}\frac{2kk^A p_0 e^{ik(x_p-x_1)+ik^A(x_2-x_p)}}{k^A\rho + k\rho^A + ikk^A(K/\omega^2 - \rho_p h)}. \tag{13.6.11}$$

Equations (13.6.8)–(13.6.11) reveal a resonance effect at the natural frequency of the plate–spring system, which remains, however, bounded owing to the presence of the acoustic media on both sides of the plate. The expressions simplify appropriately when the acoustic media on the two sides are the same, when the rigid plate is not elastically supported ($K = 0$) or when the rigid plate is massless ($\rho_p h = 0$). When all these simplifications are included we find

$$p^- = -e^{ik(x_p-x_1)} \quad \text{and} \quad p^+ = e^{ik(x_p-x_1)}.$$

Now the total pressure at $x = x_p - \varepsilon$ and $x = x_p + \varepsilon$ are the same, namely $p = \exp[ik(x_p - x_1)]$, as should be the case when there is no obstacle at $x = x_p$.

Reciprocity can immediately be verified from the form of the particle velocity given by Eq. (13.6.11). The symmetric dependences on k and k^A and on ρ and ρ^A show that a pressure wave whose phase is zero at $x = x_1$ produces a particle velocity at $x = x_2$, at the other side of the baffle, that is the same as the particle

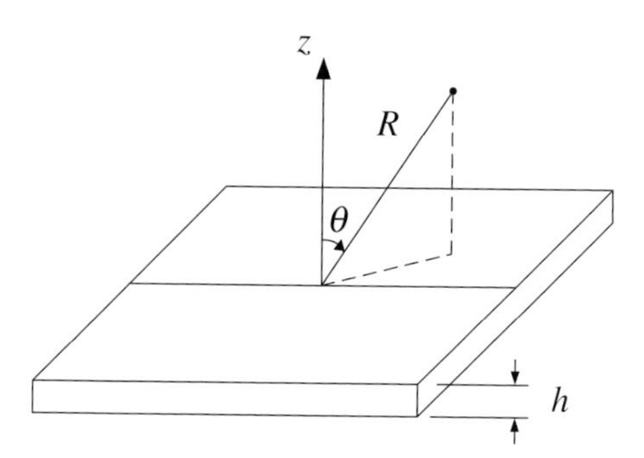

Figure 13.6 Geometry of the plate and coordinates defining a field point.

velocity produced at $x = x_1$ by a pressure wave traveling in the negative x-direction whose phase is zero at $x = x_2$.

13.7 Reciprocity for free-field sources and sources located on elastic baffles

Now let us consider the sound, represented by $p_s(R)$, radiated by a point source located on an elastic plate immersed in an acoustic medium. The far-field pressure is given by Junger and Feit (1972, p. 376) as

$$p(R,\theta) = 2p_s(R)\left\{1 + \frac{1}{ikh(\rho_s/\rho)[1-(\omega/\omega_c)^2\sin^4\theta]\cos\theta - 1}\right\}. \tag{13.7.1}$$

The factor multiplying $p_s(R)$ represents the effect of the elastic plate on the far field. The coordinates R and θ are defined in Fig. 13.6. The mass density of the plate is ρ_s and h is its thickness. The frequency ω_c is the coincidence frequency.

The pressure on an elastic plate insonified by a distant sound source is also given by Junger and Feit (1972, p. 376):

$$p(x,0) = 2P_i\exp(ik\sin\theta)\frac{ikh(\rho_s/\rho)[1-(\omega/\omega_c)^2\sin^4\theta]\cos\theta}{ikh(\rho_s/\rho)[1-(\omega/\omega_c)^2\sin^4\theta]\cos\theta - 1}. \tag{13.7.2}$$

Setting $P_i\exp(ik\sin\theta)$ equal to the pressure $p_s(R)$ generated in the plane of the baffle by a point source located at (R,θ), the two equations (13.7.1) and (13.7.2) become the same.

The reciprocity theorem for arbitrary baffle geometries, attributed by Junger and Feit (1972) to Lyamshev (1959), states: *The pressure generated in an acoustic medium at a field point* $\boldsymbol{x}$ *by a source located at* $\boldsymbol{x}_0$ *on an elastic structure* [here Eq. (13.7.1)] *equals the pressure generated on that structure, at point* x_0, *by a source of the same strength located in the acoustic medium at field point* $\boldsymbol{x}$ [here Eq. (13.7.2)].

Further discussion and examples of reciprocity in structural acoustics can be found in the books by Fahy (1985) and Junger and Feit (1972), and in the papers by Lyamshev (1959) and Belousov and Rimskii-Korsakov (1975).

14
Reciprocity for piezoelectric systems

14.1 Introduction

Fields of classical physics such as electromagnetic wave theory, acoustics and elastodynamics all have their own reciprocity theorems, which for acoustics and elastodynamics have been discussed in the preceding chapters. A comparable discussion of reciprocity in electromagnetic wave theory is outside the scope of this book. Moreover, there are already several books that have dealt in considerable detail with electromagnetic reciprocity; see e.g., Collin (1960), Auld (1973) and de Hoop (1995).

Interesting applications of reciprocity relations for the interactions of electromagnetic and elastodynamic fields to non-destructive evaluation, particularly as it relates to piezoelectricity, have not received the attention that they deserve, with the exception of the work by Auld (1979). In the present chapter we therefore attempt to correct for this lack of exposure by a discussion of reciprocity for piezoelectric systems.

General reciprocity relations involving coupled electromagnetic and elastic waves were first presented by Foldy and Primakoff (1945) and Primakoff and Foldy (1947), who used these relations to demonstrate the interchangeability of source and receiver in electro-acoustic transmission measurements. In the important paper by Auld (1979) these relations were used to analyze elastic wave scattering coefficients from observations *at the electrical terminals* of the electromechanical transducers employed in performing a non-destructive testing experiment. In Auld's paper, an expression was derived that directly relates the electrical signal received by an ultrasonic transducer to the radiation patterns of the transmitting and receiving transducers and to the modified patterns resulting from scattering from a flaw. The reciprocity relation thus provides a formalism to develop a system description of an ultrasonic measurement that includes the influence of all aspects of the measurement system on the observed voltage. The formalism involves implicitly such problems as the efficiency of the transducers, their radiation patterns, the influence on these patterns of the propagation of the beam through interfaces, and the scattering from the flaw. Of course, the reciprocity relation does not provide solutions to each of these problems. Instead, it provides a self-consistent way to interpret the flaw signal if these solutions are known. Obtaining all these solutions can be a complicated problem in elastodynamics.

An interpretation of Auld's electromechanical reciprocity relation via a one-dimensional example was given by Thompson (1994), whose paper also includes a list of references dealing with applications. Examples were also presented by Liang *et al.* (1985) and Block *et al.* (2000).

The chapter starts with a brief discussion of piezoelectricity, which is followed by a statement of the governing equations and a derivation of the reciprocity theorem for a piezoelectric solid. Next we discuss in some detail Auld's adaptation of the reciprocity theorem to calculate scattering coefficients at the electrical terminals of transducers used to generate and detect ultrasonic signals in an elastic body containing a flaw. Two examples are discussed in subsequent sections. The first of these is concerned with measurement of the electrical signal due to the reflection of an ultrasonic signal generated by a transducer mounted on one face of an elastic layer when the signal has been reflected from the opposite face. The second example concerns the backscattering of ultrasound from a fatigue crack.

14.2 Piezoelectricity

For some materials Hooke's law does not fully describe the relation between stress and strain. Certain materials become electrically polarized when they are strained. This effect, called the *direct piezoelectric effect*, manifests itself experimentally by the appearance of bound electrical charges at the surface of a strained medium. It is a linear phenomenon, and the polarization changes sign when the sign of the strain is reversed. Piezoelectricity is related to the microscopic structure of solids and can be explained qualitatively in terms of a rather simple atomic model; see Auld (1973). Briefly, the atoms of a solid (and also the electrons within the atoms themselves) are displaced when the material is deformed. This displacement produces microscopic electrical dipoles within the medium, and in certain crystal structures these dipole moments combine to give an average macroscopic moment (or *electrical polarization*).

The direct piezoelectric effect is always accompanied by the *converse piezoelectric effect*, whereby a solid becomes strained when placed in an electric field.

Like the direct effect, this is also a linear effect and the piezoelectric strain reverses sign with reversal of the applied electric field. Since the piezoelectric strain produced by an electric field will always generate internal stresses, the converse piezoelectric effect must be included in Hooke's law by adding a stress term that is *linearly proportional to the electric field.* This electrically induced stress will be present only in materials with microscopic structures appropriate to the existence of piezoelectricity.

Piezoelectricity provides an effective means for electrically generating and detecting mechanical vibrations by electro-acoustic converters commonly known as transducers.

14.3 Governing equations

The governing equations of a piezoelectric solid include the mechanical equations of motion as well as Maxwell's equations for electromagnetic theory. In indicial notation they may be written as, see Auld (1973),

$$\frac{\partial \tau_{ij}}{\partial x_j} + f_i = \rho \frac{\partial v_i}{\partial t}, \tag{14.3.1}$$

$$e_{ijk} \frac{\partial H_k}{\partial x_j} = \frac{\partial D_i}{\partial t} + J_i, \tag{14.3.2}$$

$$e_{ijk} \frac{\partial E_k}{\partial x_j} = -\frac{\partial B_i}{\partial t}, \tag{14.3.3}$$

$$\frac{\partial D_i}{\partial x_i} = \rho_e, \tag{14.3.4}$$

$$\frac{\partial B_i}{\partial x_i} = 0. \tag{14.3.5}$$

Here v is the particle velocity and e_{ijk} is the alternating tensor:

$$e_{ijk} = \begin{cases} +1 & \text{if } ijk \text{ represents an even permutation of 123,} \\ 0 & \text{if any two of the } ijk \text{ indices are equal,} \\ -1 & \text{if } ijk \text{ represents an odd permutation of 123.} \end{cases}$$

Also, ρ is the mass density and ρ_e (coulombs/m^3) is the volume density of electric charge. Furthermore H (amperes/m) and E (volts/m) represent the magnetic and the electric field strength, respectively. Finally D (coulombs/m^2) and B (tesla) are the electrical and the magnetic flux density, respectively, and J(amperes/m^2) is the volume density of electric current.

The constitutive equations for piezoelectricity are

$$\tau_{ij} = C_{ijkl}\frac{\partial u_k}{\partial x_l} - d_{kij}E_k, \tag{14.3.6}$$

$$D_i = d_{ikl}\frac{\partial u_k}{\partial x_l} + c_{ik}E_k, \tag{14.3.7}$$

$$B_i = \mu H_i, \tag{14.3.8}$$

where the constant tensor components have the following general symmetries:

$$C_{ijkl} = C_{jikl} = C_{ijlk} = C_{klij},$$

$$d_{ijk} = d_{ikj},$$

$$c_{ij} = c_{ji}.$$

14.4 Reciprocity theorem for a piezoelectric solid

Since the reciprocity theorem will be derived in the frequency domain, with the time-dependent term $\exp(-i\omega t)$, Eqs. (14.3.1)–(14.3.3) must be replaced by

$$\frac{\partial \tau_{ij}}{\partial x_j} + f_i = -i\omega\rho v_i, \tag{14.4.1}$$

$$e_{ijk}\frac{\partial H_k}{\partial x_j} = -i\omega D_i + J_i, \tag{14.4.2}$$

$$e_{ijk}\frac{\partial E_k}{\partial x_j} = i\omega B_i. \tag{14.4.3}$$

The total system of equations now consists of Eqs. (14.4.1)–(14.4.3) and (14.3.4)–(14.3.8). The variable field quantities in these equations depend on $\boldsymbol{x}$ only.

In the usual manner we now consider a region of volume V and surface S, and we define states A and B as

$$\text{state } A: \quad \boldsymbol{\tau}_{ij}^A,\ \boldsymbol{v}^A,\ \boldsymbol{f}^A,\ \boldsymbol{B}^A, \boldsymbol{H}^A,\ \boldsymbol{E}^A,\ \boldsymbol{D}^A, \boldsymbol{J}^A,$$

$$\text{state } B: \quad \boldsymbol{\tau}_{ij}^B,\ \boldsymbol{v}^B,\ \boldsymbol{f}^B,\ \boldsymbol{B}^B, \boldsymbol{H}^B,\ \boldsymbol{E}^B,\ \boldsymbol{D}^B, \boldsymbol{J}^B$$

As our point of departure we take the following local-interaction expression:

$$\tau_{kl}^A n_k v_l^B - \tau_{kl}^B n_k v_l^A + e_{ijk}E_j^B H_k^A n_i - e_{ijk}E_j^A H_k^B n_i. \tag{14.4.4}$$

This quantity is subsequently integrated over the surface S to yield a surface integral denoted by I. An application of Gauss' theorem to convert the surface integral into a volume integral yields

$$I = \int_V \frac{\partial}{\partial x_k} \left(\tau_{kl}^A v_l^B - \tau_{kl}^B v_l^A \right) dV + \int_V e_{ijk} \frac{\partial}{\partial x_i} \left(E_j^B H_k^A - E_j^A H_k^B \right) dV. \tag{14.4.5}$$

To simplify this integral further we note that

$$\begin{aligned} \frac{\partial \tau_{kl}^A}{\partial x_k} v_l^B - \frac{\partial \tau_{kl}^B}{\partial x_k} v_l^A &= \left(-f_l^A - i\omega\rho v_l^A \right) v_l^B - \left(-f_l^B - i\omega\rho v_l^B \right) v_l^A \\ &= -f_l^A v_l^B + f_l^B v_l^A, \end{aligned}$$

where the equation of motion Eq. (14.4.1), has been used. Also,

$$\tau_{kl}^A \frac{\partial v_l^B}{\partial x_k} - \tau_{kl}^B \frac{\partial v_l^A}{\partial x_k} = i\omega \left(d_{ikl} E_l^A \frac{\partial u_l^B}{\partial x_k} - d_{ikl} E_l^B \frac{\partial u_l^A}{\partial x_k} \right),$$

where the constitutive equation (14.3.6) has been used, as well as the symmetry properties of C_{ijkl}. Furthermore,

$$e_{ijk} \left(\frac{\partial E_j^B}{\partial x_i} H_k^A - \frac{\partial E_j^A}{\partial x_i} H_k^B \right) = i\omega \left(B_k^B H_k^A - B_k^A H_k^B \right) = 0,$$

since $B_k^A = \mu H_k^A$ and $B_k^B = \mu H_k^B$. Next,

$$e_{ijk} \left(E_j^B \frac{\partial H_k^A}{\partial x_i} - E_j^A \frac{\partial H_k^B}{\partial x_i} \right) = -E_j^B \left(-i\omega D_j^A + J_j^A \right) + E_j^A \left(-i\omega D_j^B + J_j^B \right).$$

But

$$\begin{aligned} E_i^B D_i^A - E_i^A D_i^B &= E_i^B \left(d_{ikl} \frac{\partial u_k^A}{\partial x_l} + c_{ik} E_k^A \right) - E_i^A \left(d_{ikl} \frac{\partial u_k^B}{\partial x_l} + c_{ik} E_k^B \right) \\ &= d_{ikl} E_i^B \frac{\partial u_k^A}{\partial x_l} - d_{ikl} E_i^A \frac{\partial u_k^B}{\partial x_l}, \end{aligned}$$

where $c_{ik} = c_{ki}$ has been used. By using in Eq. (14.4.5) the results derived above, and $d_{ilk} = d_{ikl}$, we finally obtain

$$I = \int_V \left(f_i^B v_i^A - f_i^A v_i^B + J_i^B E_i^A - J_i^A E_i^B \right) dV. \tag{14.4.6}$$

From the definition of I as the integration of (14.4.4) over S and from the result (4.4.6) the reciprocity theorem for a piezoelectric material then follows as

$$\begin{aligned} &\int_S \left(\tau_{kl}^A n_k v_l^B - \tau_{kl}^B n_k v_l^A \right) dS + \int_S \left(e_{ijk} E_j^B H_k^A n_i - e_{ijk} E_j^A H_k^B n_i \right) dS \\ &\quad = \int_V \left(f_l^B v_l^A - f_l^A v_l^B + J_l^B E_l^A - J_l^A E_l^B \right) dV. \end{aligned} \tag{14.4.7}$$

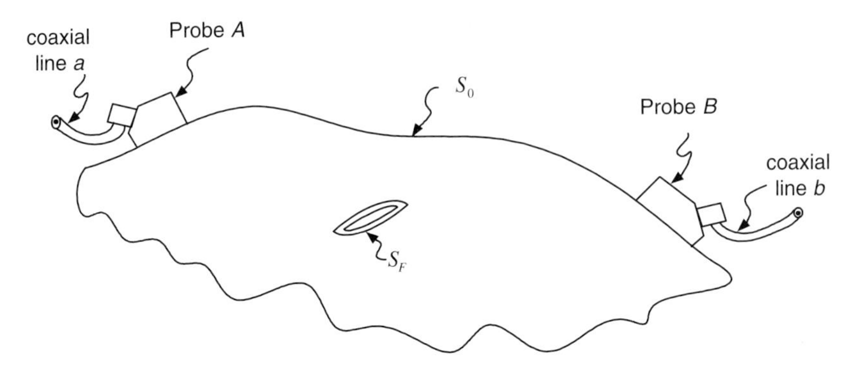

Figure 14.1 Configuration for application of the electromechanical reciprocity relation, showing part of the enclosing surface S.

In vector notation the theorem can be written as

$$\int_S (\boldsymbol{t}^A \cdot \boldsymbol{v}^B - \boldsymbol{t}^B \cdot \boldsymbol{v}^A)\, dS + \int_S (\boldsymbol{E}^B \wedge \boldsymbol{H}^A - \boldsymbol{E}^A \wedge \boldsymbol{H}^B) \cdot \boldsymbol{n}\, dS$$
$$= \int_V (\boldsymbol{f}^B \cdot \boldsymbol{v}^A - \boldsymbol{f}^A \cdot \boldsymbol{v}^B + \boldsymbol{J}^B \cdot \boldsymbol{E}^A - \boldsymbol{J}^A \cdot \boldsymbol{E}^B)\, dV, \qquad (14.4.8)$$

where $\boldsymbol{t}$ defines the surface traction vector on V and $t_i = \tau_{ik} n_k$ has been used. The reciprocity theorem given by Eq. (14.4.8) is specifically formulated for the case of coupling between electromagnetic and elastic fields.

14.5 Application to scattering by a flaw

It was shown by Auld (1979) that the reciprocity theorem for a piezoelectric solid can be used to calculate scattering coefficients at the electrical terminals of transducers used to generate and detect ultrasonic signals in an elastic body containing a flaw. In this section we follow, by and large, the development presented by Auld (1979).

The general configuration is shown in Fig. 14.1. The objective is to determine the transmission from transducer A to transducer B in the presence of a scattering object, the flaw. It is assumed that the only sources present are electrical sources external to the enclosing surface and that the transducers are electromagnetically shielded; that is, the electromagnetic fields are confined within the transducers and the two identical coaxial feeds. In applying the reciprocity theorem given by Eq. (14.4.8), the integration over the bounding surface S comprises an integration over the surface S_0, consisting of the enclosing surface minus the areas covered by the transducers, plus an integral over the surface S_F which surrounds the flaw. The contributions over the transducers vanish, except for those of the cross sections of the coaxial line feeds.

Since the elastic body is nonpiezoelectric and free of either surface tractions or surface displacements, and in the absence of sources, Eq. (14.4.8) becomes

$$\int_{\text{coax } a} (\boldsymbol{E}^B \wedge \boldsymbol{H}^A - \boldsymbol{E}^A \wedge \boldsymbol{H}^B) \cdot \boldsymbol{n}\, dS + \int_{\text{coax } b} (\boldsymbol{E}^B \wedge \boldsymbol{H}^A - \boldsymbol{E}^A \wedge \boldsymbol{H}^B) \cdot \boldsymbol{n}\, dS$$
$$= -\int_{S_F} (\boldsymbol{t}^A \cdot \boldsymbol{v}^B - \boldsymbol{t}^B \cdot \boldsymbol{v}^A)\, dS. \tag{14.5.1}$$

State A is now taken to be the excitation of the system by an incident electromagnetic wave carrying time-averaged power P in coaxial line a for transmission to coaxial line b, *in the absence of the flaw*, and state B corresponds to excitation with the same incident power in coaxial line b for transmission to coaxial line a, *in the presence of the flaw*. The terminal planes in the coaxial lines are assumed to be sufficiently removed from the transducers that any higher-order cut-off waveguide modes of the coaxial line that are excited at the transducer terminals have died away to negligible amplitude. The integrals over the coaxial line cross sections in Eq. (14.5.1) therefore involve only the fundamental mode of propagation of the line. At these points in the line, the electromagnetic fields consist of superposed incident and reflected fundamental modes, and the field quantities appearing in Eq. (14.5.1) may be written as follows. For State A,

$$\text{coaxial line } a: \quad \boldsymbol{E}^A = \left(1 + \Gamma_{aa}^A\right) \boldsymbol{E}^+, \quad \boldsymbol{H}^A = \left(1 - \Gamma_{aa}^A\right) \boldsymbol{H}^+, \tag{14.5.2}$$

$$\text{coaxial line } b: \quad \boldsymbol{E}^A = \Gamma_{ba}^A \boldsymbol{E}^+, \quad \boldsymbol{H}^A = -\Gamma_{ba}^A \boldsymbol{H}^+, \tag{14.5.3}$$

where Γ_{aa}^A is the reflection coefficient in line a in the absence of the scatterer; Γ_{ba}^A is the electrical transmission coefficient, defined as the ratio of the amplitude of the electromagnetic wave transmitted into coaxial line b to the amplitude of the electromagnetic wave incident at coaxial line a in the absence of the scatterer. Also, $\boldsymbol{E}^+$, $\boldsymbol{H}^+$ are the electric and magnetic fields of the fundamental coaxial mode carrying power P toward the transducer. From the power condition, it follows that, Collin (1960),

$$\int_{\text{coaxline}} \boldsymbol{E}^+ \wedge \boldsymbol{H}^+ \cdot \boldsymbol{n}\, dS = -2P, \tag{14.5.4}$$

since $\boldsymbol{n}$ is the *outward* normal to the surface. Similarly, for state B,

$$\text{coaxial line } a: \quad \begin{aligned} \boldsymbol{E}^B &= \Gamma^B_{ab}\boldsymbol{E}^+, \\ \boldsymbol{H}^B &= -\Gamma^B_{ab}\boldsymbol{H}^+, \end{aligned} \tag{14.5.5}$$

$$\text{coaxial line } b: \quad \begin{aligned} \boldsymbol{E}^B &= \left(1+\Gamma^B_{bb}\right)\boldsymbol{E}^+, \\ \boldsymbol{H}^B &= \left(1-\Gamma^B_{bb}\right)\boldsymbol{H}^+, \end{aligned} \tag{14.5.6}$$

where Γ^B_{bb} and Γ^B_{ab} are the reflection and transmission coefficients as perturbed by the presence of the scatterer.

Substitution of Eqs. (14.5.2) and (14.5.5) into Eq. (14.5.1) and use of Eq. (14.5.4) gives

$$4P\left(\Gamma^B_{ba}-\Gamma^A_{ab}\right) = \int_{S_F}(\boldsymbol{t}^B\cdot\boldsymbol{v}^A - \boldsymbol{t}^A\cdot\boldsymbol{v}^B)\,dS. \tag{14.5.7}$$

Because an electromechanical system with only one type of coupling is always reciprocal, we have

$$\Gamma_{ab} = \Gamma_{ba}, \tag{14.5.8}$$

and Eq. (14.5.7) may be written as

$$\begin{aligned} \Delta\Gamma_{ba} &= \Gamma^B_{ba} - \Gamma^A_{ba} \\ &= \frac{1}{4P}\int_{S_F}(\boldsymbol{t}^B\cdot\boldsymbol{v}^A - \boldsymbol{t}^A\cdot\boldsymbol{v}^B)\,dS. \end{aligned} \tag{14.5.9}$$

In Eq. (14.5.9) the field quantities under the integral with superscript A are those excited by incident power P at terminal *a in the absence of the scatterer* and those with superscript B are those excited by incident power P at terminal *b in the presence of the scatterer.* The same formula applies to a one-terminal system if b is replaced everywhere by a. For the latter case, Eq. (14.5.9) gives the change in the input reflection coefficient, observed at a, due to the presence of the scatterer. In a pulse echo experiment $\Delta\Gamma_{aa}$ is the echo return from the scatterer, as observed at the transducer input:

$$\Delta\Gamma_{aa} = \Gamma^B_{aa} - \Gamma^A_{aa} = \frac{1}{4P}\int_{S_F}(\boldsymbol{t}^B\cdot\boldsymbol{v}^A - \boldsymbol{t}^A\cdot\boldsymbol{v}^B)\,dS. \tag{14.5.10}$$

It should be noted that, although the illustration in Fig. 14.1 depicts bulk-wave transducers, the derivation does not depend upon the details of the transducer geometry and is also directly applicable to a transducer of any surface or plate wave geometry (interdigital, wedge etc.) and having any type of coupling between transducer and specimen.

The scattering formula (14.5.9) contains the perturbed field, $\boldsymbol{t}^B$, $\boldsymbol{v}^B$, at the surface of the scatterer. These quantities can be known exactly only from a full solution of the scattering problem and, if this solution were available, it would then be unnecessary to use Eq. (14.5.9) to calculate the scattering. To obtain an approximate solution to the scattering problem one must seek approximate forms for the perturbed field $\boldsymbol{t}^B$, $\boldsymbol{v}^B$. Two commonly used methods are the *Born* and the *quasistatic* approximation. In the first method, valid for scatterers whose properties differ only slightly from those of the surrounding medium, the perturbed field is approximated by the unperturbed field. As discussed in Section 12.2, to apply the Born approximation it is necessary to convert the surface integral over S_F to an integral over the enclosed volume V_F containing the change in material density $\Delta\rho$ and stiffness ΔC within the scatterer. The quasistatic approximation uses the fact that the field around any scatterer that is small compared with a wavelength has approximately the form of an elastostatic solution. For simple shapes many of these solutions are known explicitly.

14.6 Reflection by the bottom surface of an elastic layer

The geometry is shown in Fig. 14.2. A transducer of radius a is placed on the top surface of a homogeneous, isotropic, linearly elastic layer of thickness d. The transducer generates a longitudinal wave in the layer, which is reflected by the bottom surface of the layer. It is assumed that diffraction spreading is negligible. In terms of the wavenumber, $k_L = \omega/c_L$, this approximation is reasonable provided that d does not exceed the Fresnel length $k_L a^2/(2\pi)$ of the transducer.

For this simple example, the reciprocity relation (14.5.10) can be used to write a relation between the measured electrical signal and the mechanical signal incident on the transducer due to reflection from the bottom surface of the elastic layer. A similar illustrative example, for the reflection from a perfectly rigid surface of a normally incident sound beam radiated by a circular transducer in an acoustic medium, was considered by Block *et al.* (2000).

The particle velocity of the incident beam is approximated by

$$v_3^i = Ae^{ik_L(x_3+d)}\chi(\xi), \qquad v_1^i = v_2^i \equiv 0, \tag{14.6.1}$$

where $\xi = (r-a)/a$. The function $\chi(\xi)$ removes the discontinuity at $r = a$. It is defined as

$$\chi(\xi) = \begin{cases} 1, & r < a - \varepsilon, \\ 0, & r > a + \varepsilon. \end{cases} \tag{14.6.2}$$

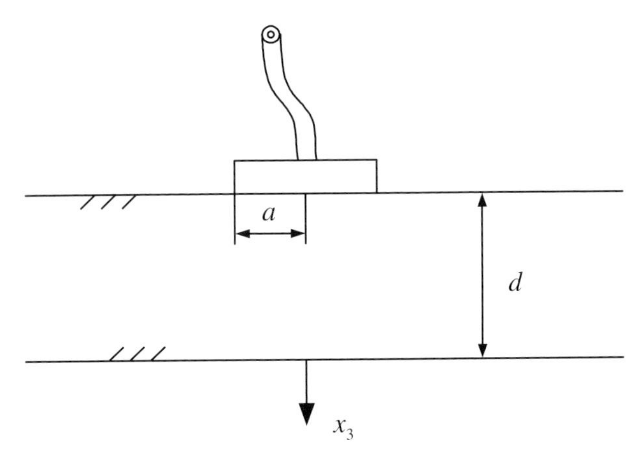

Figure 14.2 Application of electromechanical reciprocity relation to reflection of transducer beam by a free surface.

The function $\chi(\xi)$ possesses continuous derivatives of all orders for $r \in [a-\varepsilon, a+\varepsilon]$, and its derivatives of all orders vanish at $r = a \pm \varepsilon$. Here, $r^2 = (x_1^2 + x_2^2)^{1/2}$. Again ignoring diffraction, the reflected beam is

$$v_3^r = Ae^{-ik_L(x_3-d)}\chi(\xi), \tag{14.6.3}$$

where we have used the fact that the reflection coefficient for the particle velocity at a free surface is $+1$.

For the reciprocating wave fields we select state A as the incident wave field without the presence of the bottom surface, and state B as the total field of the incident and reflected waves. Only the initial reflection is considered; multiple reflections are left out of consideration. We then have for the mechanical fields at $x_3 = 0$,

$$\text{for state } A: \quad v_3^A = Ae^{ik_Ld}\chi(\xi), \tag{14.6.4}$$

$$t_3^A = -\rho c_L Ae^{ik_Ld}\chi(\xi), \tag{14.6.5}$$

$$\text{for state } B: \quad v_3^B = 2Ae^{ik_Ld}\chi(\xi), \tag{14.6.6}$$

$$t_3^B = 0. \tag{14.6.7}$$

For the electrical fields we may write

$$\text{for state } A: \quad \boldsymbol{E}^A = \left(1 + \Gamma_{aa}^A\right)\boldsymbol{E}^+, \qquad \boldsymbol{H}^A = \left(1 - \Gamma_{aa}^A\right)\boldsymbol{H}^+, \tag{14.6.8}$$

$$\text{for state } B: \quad \boldsymbol{E}^B = \left(1 + \Gamma_{aa}^B\right)\boldsymbol{E}^+, \qquad \boldsymbol{H}^B = \left(1 - \Gamma_{aa}^B\right)\boldsymbol{H}^+. \tag{14.6.9}$$

Analogously to Eq. (14.5.10) we have for the present case

$$\Delta\Gamma_{aa} = \Gamma_{aa}^{B} - \Gamma_{aa}^{A} = -\frac{1}{4P}\int_{S_T} t_3^A v_3^B \, dS. \tag{14.6.10}$$

The integration has been carried out over the plane $x_3 = 0$, and it has been taken into account that the plane $x_3 = 0$ is free of surface tractions. Here S_T is the area of the transducer, and because diffraction has been ignored it is also the area on $x_3 = 0$ that has been excited. We find

$$\Delta\Gamma_{aa} = \frac{\rho c_L}{2P} |A|^2 \pi a^2 e^{i2(k_L d + \alpha)}, \tag{14.6.11}$$

where $A = |A|e^{i\alpha}$. The argument α is assumed to be introduced by the transducer. The change in voltage, symbolized by ΔV, is proportional to $\Delta\Gamma_{aa}$, namely, $\Delta V = C(\omega)\Delta\Gamma_{aa}$, where $C(\omega)$ is a function of the transducer only and not the scatterer.

If the transducer were perfectly matched to the solid, then $\Gamma_{aa}^{A} = 0$. Moreover, if it were lossless (α real), then the time-averaged incident power is simply $P = \frac{1}{2}\rho c_L |A|^2 \pi a^2$ and, therefore, $\Gamma_{aa}^{B} = e^{i2(k_L d + \alpha)}$, which is the particle velocity reflection coefficient for a perfectly free surface with appropriate adjustments to the argument to account for the position and properties of the transducer.

14.7 Reflection and transmission by a fatigue crack

As our final example we consider the case of a fatigue crack located in an elastic layer, in a plane parallel to the faces of the layer. Whereas an open crack is a reflector of wave signals, a fatigue crack, whose rough faces have many points of contact, can also transmit wave signals.

The geometry is shown in Fig. 14.3. A transducer of radius a is placed on the surface of the layer, in such a way that the crack is in the beam of the transducer. The transducer can generate and receive longitudinal waves. As in Section 14.6, it is assumed that diffraction spreading is negligible. The objective is to determine $\Delta\Gamma_{aa}$ in terms of the geometrical and system parameters of the configuration of Fig. 14.3. A similar problem was considered by Thompson (1994).

After an appropriate choice of states A and B, the desired result follows from Eq. (14.5.10). Following the development of Section 14.5, state A is taken as the excitation of the system by an electromagnetic wave carrying time-averaged power P in coaxial line a in the absence of the flaw, and State B as corresponding to excitation by the same incident power, also in coaxial line a, in the presence of the flaw; see Fig. 14.3(a). A new feature enters here in that the traction t^B

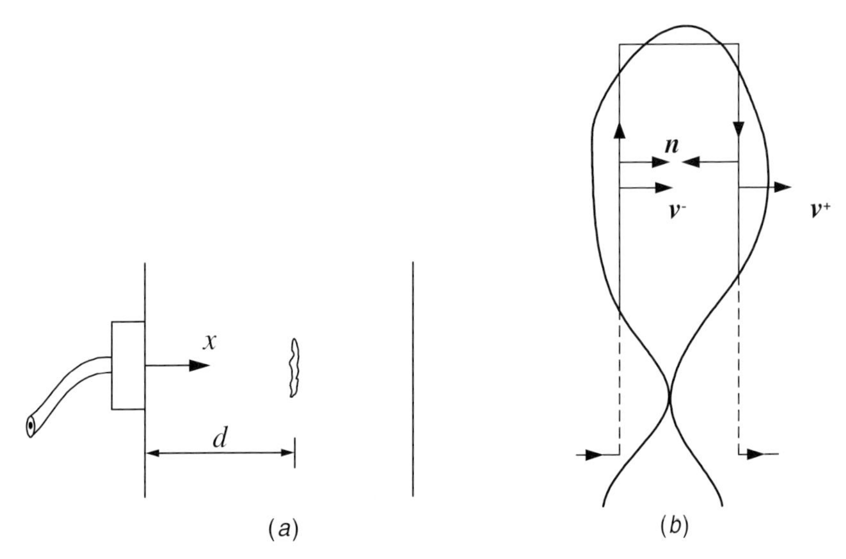

Figure 14.3 (*a*) Fatigue crack in the beam of the transducer, and (*b*) enlarged part of the rough crack faces with an integration contour for the electromechanical reciprocity relation.

is continuous across the crack faces. When the crack faces are touching, the traction is assumed to be uniformly continuous, as it is when the crack is open and the tractions on the crack faces vanish. The integration over S_F is taken along the rectangular path shown in Fig. 14.3(*b*), and the continuity of the tractions $\boldsymbol{t}^B$ across the crack faces then implies that Eq. (14.5.10) reduces to

$$\Delta\Gamma_{aa} = -\frac{1}{4P}\int_{S_F} \boldsymbol{t}^A \cdot \Delta\boldsymbol{v}^B \, dS. \tag{14.7.1}$$

In Eq. (14.7.1), $\boldsymbol{t}^A$ is simply the known traction corresponding to the incident wave in the absence of the crack. However, $\Delta\boldsymbol{v}^B$ defines the unknown time rate of opening of the crack. To find a more specific expression for $\Delta\Gamma_{aa}$, $\Delta\boldsymbol{v}^B$ must be determined either exactly or in some approximate way.

Here we will calculate an approximation for $\Delta\boldsymbol{v}^B$ by the use of the Kirchhoff approximation. In this approximation, the fields on the crack are computed as if the crack were an unbounded plane. This is generally found to be acceptable for specular reflection and when the characteristic length parameter of the crack is large compared with the wavelength of the incident wave (i.e., at "high" frequencies). As an additional approximation, the contact crack-face roughness that may occur between the faces of a fatigue crack is simulated by a layer of springs, of spring constant K, connecting the crack faces. For the Kirchhoff approximation we then have to consider normal incidence on an interface of infinite extent consisting of a layer of springs.

At $x = d$, where d is defined in Fig. 14.3(a), we have for state A

$$v^A = Ae^{ik_L d}, \tag{14.7.2}$$

$$t^A = -\rho c_L Ae^{ik_L d}. \tag{14.7.3}$$

For state B, which is now also generated at coaxial line a, we have

$$x = d - \varepsilon: \qquad v^B = Ae^{ik_L d} + RAe^{ik_L d},$$

$$x = d + \varepsilon: \qquad v^B = TAe^{ik_L d}.$$

To determine the reflection and transmission coefficients, R and T, we must also compute $\boldsymbol{t}^B$ and apply the interface conditions. We have

$$x = d - \varepsilon: \qquad t^{B-} = -\rho c_L Ae^{ik_L d} + \rho c_L RAe^{ik_L d},$$

$$x = d + \varepsilon: \qquad t^{B+} = -\rho c_L TAe^{ik_L d}.$$

The interface conditions are continuity of traction, which implies

$$-T = -1 + R,$$

and the spring condition

$$-\rho c_L T = K(T - 1 - R)\left(\frac{-1}{i\omega}\right),$$

where the term $(-1/i\omega)$ has been added because the spring condition is expressed in terms of displacements. We obtain

$$R = \frac{\rho c_L}{\rho c_L + 2iK/\omega}, \tag{14.7.2}$$

$$T = \frac{2iK/\omega}{\rho c_L + 2iK/\omega}$$

and

$$\Delta v^B = \frac{-2\rho c_L}{\rho c_L + 2iK/\omega} |A| e^{i(k_L d + \alpha)}. \tag{14.7.3}$$

Finally,

$$\Delta\Gamma_{aa} = \frac{1}{2P} \frac{(\rho c_L)^2}{\rho c_L + 2iK/\omega} |A|^2 A_F e^{i2(k_L d + \alpha)}, \tag{14.7.4}$$

where A_F is the area of the crack. Note that for $K = 0$, i.e., no contact between the crack faces, Eq. (14.7.4) reduces to the result Eq. (14.6.11) given in the previous section.

An example similar to that of this section was given by Auld (1979). Using Eq. (14.7.1), Auld worked out the details for the backscattering of a Rayleigh surface wave by a surface-breaking crack. In applying Eq. (14.7.1) he assumed a crack in the following forms: a half-circle of radius a with diameter along the surface; a part-circular crack with radius a and depth d; and a part-elliptical crack with major and minor semi-axes a and b parallel and normal to the surface and with depth d. The traction t^A is the unperturbed value at the position of the crack *in its absence*. The quasistatic approximation was used, in which the perturbed particle velocity jump Δv^B is evaluated on the basis of the elastostatic displacement produced by low-frequency applied stresses having the same distribution as t^A above, but under conditions where the crack dimensions are small compared with the elastic wavelength.

References

Abramowitz, M. and Stegun, E. A., 1964. *Handbook of Mathematical Functions*. National Bureau of Standards, US Government Printing Office, Washington DC.

Achenbach, J. D., 1973. *Wave Propagation in Elastic Solids*. Elsevier Science, Amsterdam.

1998. Lamb waves as thickness vibrations superimposed on a membrane carrier wave, *J. Acoust. Soc. Am.* **103**, 2283–2285.

2000. Calculation of surface wave motions due to a subsurface point force: an application of elastodynamic reciprocity, *J. Acoust. Soc. Am.* **107**, 1892–1897.

Achenbach, J. D. and Kitahara M., 1986. Reflection and transmission of an obliquely incident wave by an array of spherical cavities, *J. Acoust. Soc. Am.* **80**, 1209–1214.

Achenbach, J. D. and Xu, X., 1999a. Wave motion in an isotropic elastic layer generated by a time-harmonic point load of arbitrary direction, *J. Acoust. Soc. Am.* **106**, 83–90.

1999b. Use of elastodynamic reciprocity to analyze point-load generated axisymmetric waves in a plate, *Wave Motion* **30**, 57–68.

Achenbach, J. D., Gautesen, A. K. and McMaken, H., 1982. *Ray Methods for Waves in Elastic Solids*. Pitman Advanced Publishing Program, Boston, Massachusetts.

Achenbach, J. D., Kitahara, M., Mikata, Y. and Sotiropoulos, D. A., 1988. Reflection and transmission of plane waves by a layer of compact inhomogeneities, *PAGEOPH* **128**, 101–118.

Angel, Y. C., and Achenbach, J. D., 1985. Reflection and transmission of elastic waves by a periodic array of cracks, *Wave Motion* **7**, 375–397.

Auld, B. A., 1973. *Acoustic Fields and Waves in Solids*, Vols. I and II. Reprinted R. E. Krieger Publ. Co. 1990, Malabar, Florida.

1979. General electromechanical reciprocity relations applied to the calculation of elastic wave scattering coefficients, *Wave Motion* **1**, 3–10.

Banerjee, P. K. and Kobayashi, S. (eds.), 1992. *Advanced Dynamic Analysis by Boundary Element Methods*. Elsevier Applied Science, London and New York.

Belousov, Y. I. and Rimskii-Korsakov, A. V., 1975. The reciprocity principle in acoustics and its applications to the sound fields of bodies, *Sov. Physics – Acoustics* **21**, 103–109.

Beskos, D. E., 1987. Boundary element methods in dynamic analysis, *Appl. Mech. Rev.* **40**, 1–23.

Betti, E., 1872. Teori della elasticita, *Il Nuove Ciemento* (Series **2**), 7–10.

Bleistein, N., 1984. *Mathematical Methods of Wave Phenomena*. Academic Press, Orlando, Florida.

Block, G., Harris, J. G. and Hayat, T., 2000. Measurement models for ultrasonic non-destructive evaluation, *IEEE Trans. Ultrasonics, Ferroelectrics and Frequency Control* **47**, 604–611.

Blok, H. and Zeylmans, M. C. S., 1987. Reciprocity and the formulation of inverse profiling problems, *Radio Science* **22**, 1137–1147.

Bonnet, M., 1995. *Boundary Integral Equation Methods for Solids and Fluids*. John Wiley and Sons, New York.

Burridge, R. and Knopoff, L., 1964. Body force equivalents for seismic dislocations, *Bull. Seismol. Soc. Am.* **54**, 1875–1888.

Chao, C. C., 1960, Dynamical response of an elastic half-space to tangential surface loadings, *J. Appl. Mech.* **27**, 559–567.

Chimenti, D. E., 1997. Guided waves in plates and their use in materials characterization, *Appl. Mech. Rev.* **50**, 247–284.

Christensen, R. M., 1972. *Theory of Viscoelasticity – An Introduction*. Academic Press, New York.

Cohen, J. K. and Bleistein, N., 1977. An inverse method for determining small variations in propagation speed, *SIAM J. Appl. Math.* **32**, 784–799.

Collin, R. E., 1960. *Field Theory of Guided Waves*. Reprinted IEEE Press 1991, New York.

Courant, R. and Hilbert, D., 1962. *Methods of Mathematical Physics* Vol. II. Interscience Publishers, New York.

Cremer, L., Heckl, M. and Ungar, E. E., 1973. *Structure-borne Sound*. Springer-Verlag, New York.

Crighton, D. G., Dowling, A. P., Ffowcs Williams, J. E., Heckl, M. and Leppington, F. G., 1992. *Modern Methods in Analytical Acoustics*. Springer-Verlag, London.

d'Alembert, J.-le-Rond, 1747. Investigation of the curve formed by a vibrating string (transl.). In Lindsay R. B. (ed.), 1972, *Acoustics: Historical and Philosophical Development*, Dowden, Hutchinson and Ross, Stroudsbury, Pennsylvania, 119–130.

de Hoop, A.T., 1995. *Handbook of Radiation and Scattering of Waves*. Academic Press, London.

DiMaggio, F. L. and Bleich, H. H., 1959. An application of a dynamic reciprocal theorem. *J. Appl. Mech.* **26**, 678–679.

Dowling, A. P. and Ffowcs Williams, J. E., 1983. *Sound and Sources of Sound*. Ellis Horwood, Chichester, UK.

Eringen, A. C. and Suhubi, E. S., 1975. *Elastodynamics*, Vol. II, *Linear Theory*. Academic Press, New York.

Ewing, W. M., Jardetzky, W. S. and Press, F., 1957. *Elastic Waves in Layered Media*. McGraw-Hill, New York.

Fahy, F., 1985. *Sound and Structural Vibration*. Academic Press, London.

Fokkema, J. T. and van den Berg, P. M., 1993. *Seismic Applications of Acoustic Reciprocity*. Elsevier Science, Amsterdam.

Foldy, L. L. and Primakoff, H., 1945. A general theory of passive linear electroacoustic transducers and the electroacoustic reciprocity theorem I, *J. Acoust. Soc. Am.* **17**, 109–120.

Graff, K. F., 1975. *Wave Motion in Elastic Solids*. Ohio State University Press, Columbus, Ohio.

Graffi, D., 1946. Sul teorema di reciprocita nella dinamica dei corpi elastici, *Mem. Acad. Sci. Bologna* **10**, 103–111.

Gubernatis, J. E., Domany, E., Krumhansl, J. A., and Huberman, M., 1977. The Born approximation in the theory of the scattering of elastic waves by flaws, *J. Appl. Phys.* **48**, 2804–2811, 2812–2819.

Harris, J. G., 2001. *Linear Elastic Waves*. Cambridge University Press, Cambridge.

Jones, D. S., 1986. *Acoustic and Electromagnetic Waves*. Clarendon Press, Oxford.

Junger, M. C. and Feit, F., 1972. *Sound, Structures and Their Interaction*. MIT Press, Cambridge, Massachusetts.

Keller, J. B. and Karal, F. C. Jr., 1964. Geometrical theory of elastic surface-wave excitation and propagation, *J. Acoust. Soc. Am.* **36**, 32–40.

Kino, G. S., 1978. The application of reciprocity theory to scattering of acoustic waves by flaws, *J. Appl. Phys.* **49**, 3190–3199.

Knopoff, L. and Gangi, A. F., 1959. Seismic reciprocity, *Geophysics* **24**, 681–691.

Kobayashi, S., 1987. Elastodynamics. In *Computational Methods in Mechanics* (ed. D. E. Beskos), Handbooks in Mechanics and Mathematical Methods Vol. 3, North-Holland, Amsterdam.

Kupradze, V. D., 1963. Dynamical Problems in Elasticity. In *Progress in Solid Mechanics* Vol. 3 (eds. I. N. Sneddon and R. Hill). North-Holland, Amsterdam.

Lamb, H., 1888. On reciprocal theorems in dynamics, *Proc London Math. Soc.* **19**, 144–151.

1904. On the propagation of tremors over the surface of an elastic solid, *Phil. Trans. Roy. Soc. London A* **203**, 1–42.

1917. Waves in an elastic plate, *Proc. Roy. Soc. London A* **93**, 114–128.

Liang, K. K., Kino, G. S. and Khuri-Yakub, B. T., 1985. Material characterization by the inversion of *V(z)*, *IEEE Trans. Son. Ultrason.* **32**, 266–273.

Love, A. E. H., 1892. *A Treatise on the Mathematical Theory of Elasticity*. Dover, New York, 1944.

1911. *Some Problems of Geodynamics*. Dover, New York, 1967.

Lyamshev, L. M., 1959. A method for solving the problem of sound radiation by thin elastic plates and shells, *Sov. Physics – Acoustics* **5**, 122–123.

Lyon, R. H., 1955. Response of an elastic plate to localized driving force, *J. Acoust. Soc. Am.* **27**, 259–265.

Mal, A. K. and Knopoff, L., 1967. Elastic wave velocities in two component systems, *J. Inst. Math. Applic.* **3**, 376–387.

Maxwell, J. C., 1864. On the calculation of the equilibrium and stiffness of frames, *Phil. Mag.* **27**, 294.

McLachlan, N. W., 1961. *Bessel Functions for Engineers*. Clarendon Press, Oxford.

Mikata, Y. and Achenbach, J. D., 1988. Interaction of harmonic waves with a periodic array of inclined cracks, *Wave Motion* **10**, 59–72.

Miklowitz, J., 1962. Transient compressional waves in an infinite elastic plate or elastic layer overlying a rigid half-space, *J. Appl. Mech.* **29**, 53–60.

1978. *The Theory of Elastic Waves and Waveguides*. Elsevier Science, Amsterdam.

Mindlin, R. D., 1960. Waves and vibrations in isotropic elastic plates. In *Structural Mechanics*, pp. 199–232 (eds. J. N. Goodier and N. J. Hoff), Pergamon Press, New York.

Morse, P. M. and Ingard, K. U., 1968. *Theoretical Acoustics*. McGraw-Hill, New York.

Pao, Y.-H. and Mow, C.-C., 1973. *Diffraction of Elastic Waves and Dynamic Stress Concentrations*. Crane Russak, New York.

Payton, R. G., 1964. An application of the dynamic Betti–Rayleigh reciprocal theorem to moving-point loads in elastic media, *Q. J. Appl. Math.* **XXI**, 299–313.

Pekeris, C. L., 1955. The seismic surface pulse, *Proc. Nat. Acad. Sci. USA* **41**, 469–480.

Pierce, A. D., 1981. *Acoustics: An Introduction to its Physical Principles and Applications*. Acoustic Society of America, Woodbury, New York.

Primakoff, H. and Foldy, L. L., 1947. A general theory of passive linear electroacoustic transducers and the electroacoustic reciprocity theorem II, *J. Acoust. Soc. Am.* **19**, 50–120.

Rayleigh, Lord, 1873. Some general theorems relating to vibrations, *Proc. London Math. Soc.* **4**, 357–368.

1877. *The Theory of Sound*, Vol. II. Dover reprint, Dover Publications, New York, 1945.

1887. On waves propagated along the plane surface of an elastic solid, *Proc. London Math. Soc.* **17**, 4–11.

Santosa, F. and Pao, Y.-H., 1989. Transient axially asymmetric response of an elastic plate, *Wave Motion* **11**, 271–296.

Schenk, H. A., 1968. Improved integral formulations for acoustic radiation problems, *J. Acoust. Soc. Am.* **44**, 41–58.

Stokes, G. G., 1849. On the dynamical theory of diffraction, *Trans. Cambridge Phil. Soc.* **9**, 1.

Tan, T. H., 1977. Reciprocity relations for scattering of plane elastic waves, *J. Acoust. Soc. Am.* **61**, 928.

Thompson, R. B., 1994. Interpretation of Auld's electromechanical reciprocity relation via a one-dimensional example, *Res. Nondestr. Ev.* **5**, 147–156.

Vasudevan, N. and Mal, A. K., 1985. Response of an elastic plate to localized transient sources, *J. Appl. Mech.* **52**, 356–362.

von Helmholtz, H. L., 1860. Theory des Luftschalls in Rohren mit offenen Enden, *Borchardt-Crelle's J.* **57**, 1–70.

1886. Ueber die physikalische Bedeutung des Princips der kleinsten Wirkung, *Borchardt-Crelle's J.* **100**, 137–166, 213–222.

Weaver, R. L. and Pao, Y.-H., 1982. Axisymmetric elastic waves excited by a point source in a plate, *J. Appl. Mech.* **49**, 821–836.

Weston, V. H., 1984. Multifrequency inverse problem for the reduced wave equation with sparse data, *J. Math. Phys.* **25**, 1382–1390.

Zhang, Ch. and Gross, D., 1998. *On Wave Propagation in Elastic Solids with Cracks*. Computational Mechanics Publications, Southampton, UK.

Zhang, M. and Achenbach, J. D., 1999. Simulation of self-focusing by an array on a crack in an immersed specimen, *Ultrasonics* **37**, 9–18.

Index of cited names

Subject index